THE BEAST IN SHEEP'S CLOTHING

THE BEAST IN SHEEP'S CLOTHING

Exposing the Lies of Godless Human Science

BY DAVID MICHAEL LINDSEY

PELICAN PUBLISHING COMPANY
GRETNA 2005

Library of Congress Cataloging-in-Publication Data

Lindsey, David Michael, 1955-
The beast in sheep's clothing : exposing the lies of godless human science / by David Michael Lindsey.
p. cm.
Includes bibliographical references and index.
ISBN-13: 978-1-58980-288-9 (alk. paper)
ISBN-10: 1-58980-288-8 (alk. paper)
1. Apologetics–History. 2. Christianity–Philosophy. 3. Philosophy. 4. Evolution. 5. Psychology. I. Title.
BT1109.L56 2005
239'.7–dc22
2005006369

Printed in Canada

Published by Pelican Publishing Company, Inc.
1000 Burmaster Street, Gretna, Louisiana 70053

To all those who have put their faith and trust in human science instead of God

Contents

Introduction: The Second Beast of the Apocalypse

> "Human science is vanity because to increase human knowledge only is anguish and affliction of the spirit and he who multiplies [human] science multiplies such anguish."
>
> —*Poem of the Man God,* Vol. 2, p. 729

Worshipping at the Altar of Godless Human Science

I can never understand why atheists and unbelievers constantly criticize Christianity for the Church's wrongful censoring of Galileo in 1633 for his belief in a heliocentric (sun-centered) universe, yet never criticize or condemn godless human science for being wrong all the time. Human science makes many, many mistakes, much more so than the Church, and is not the harbinger of infallible truth as science worshipers like the late Dr. Carl Sagan would have you believe. But if you listen to the liberal media these days, science can do no wrong. So why is there so much scrutiny and public criticism in the liberal press against the Church and not human science? Because people today worship and trust human science much more than they worship and trust Jesus Christ.

In 1633, many of the most prominent scientists believed, like the Catholic Church, in a geocentric (earth-centered) universe, mainly because of the numerous phrases in the Bible regarding man's central place in the universe and primary importance to God.[1] That one, single misinterpretation of the Bible as literal fact rather than symbolic truth by the Catholic Church four hundred years ago is constantly being used today by skeptics and unbelievers to condemn Christianity as both untruthful and harmful. There is a double standard in today's world regarding Christianity and science.

It's high time Christians and members of other religious faiths start holding human science to the same high standards as unbelievers hold religion. Time after time the scientific elite have been proven wrong and have had to make embarrassing retractions of what they originally claimed to be infallible truth but wasn't. The world should hold these elites more accountable for their lies and mistakes, but don't expect the liberal media to be on the side of the Christian faithful and expose the errors and obvious hypocrisy and bias against religion by the scientific elite. The scientific elite invariably claim that there is an inherent process of self-correction built into human science that allows them to make mistakes without any need for scandal or apology to anyone, and the liberal press agrees. Science has the right to make mistakes, but not Christianity—not even once.

Liberals, atheists, and the scientific elite invariably use the Church's mistake about our heliocentric solar system to discredit Christianity, but if the truth be known, Galileo was just as wrong as the Catholic Church. Our sun is just one of several billion stars orbiting the Milky Way Galaxy, which itself is orbiting around trillions of other galaxies. So neither the sun nor the earth is at the center of the whole universe, which makes the dispute between Galileo and the Church unimportant. Even though they know this, worshipers of human science like Dr. Sagan continually bring up the Church's inquisition of Galileo as proof positive that Christianity is in error and is not to be believed or trusted. Atheists like him never give up beating the dead horse of Galileo because their real goal is to draw people away from Christianity into the worship of godless human science, in direct violation of God's first and second commandments: "Thou shalt not have strange gods before me," and "Worship no idols."

Of all the scientific elites within the last twenty years, by far the most celebrated popularizer and worshiper of science was Dr. Carl Sagan (1934-1996), the famed professor of astronomy and space sciences from Cornell University. He was a frequent guest on the Johnny Carson show where he became legendary for his over-glorifications of godless human science and peculiar enunciations of words like "billions and billions" (which he strangely pronounced "beeleons and beeleons").

Besides being a cult personality on late-night television, Sagan was a world-renowned Pulitzer-prize-winning author, lecturer, NASA scientist, and secular humanist whose most popular book, *Cosmos* (1980), was the best-selling science book in the English language in the twentieth century (on the best-seller list for seventy weeks). Sagan's *Cosmos* was based on a thirteen-part public television series by the same name, which became the most widely-watched limited series on any subject viewed on American public TV, and was seen by over 500 million people in sixty countries.[2] In both the book and the television series, Sagan made it absolutely clear that he was a died-in-the-wool atheist and materialist, and more Darwinist than Charles Darwin himself. "The Cosmos is all there is or ever was or ever will be," remarked Sagan in the first line of the first chapter of *Cosmos* about his fervent disbelief in God or a spiritual realm outside the material world.[3]

Although born and raised a Jew, Sagan as an adult was a hardened atheist[4] who couldn't keep from chiding and castigating faithful believers in the books he wrote, which included such popular titles as: *Intelligent Life in the Universe* (1966); *Cosmic Connection* (1973); *The Dragons of Eden: Speculations on the Evolution of Human Intelligence* (1977, Pulitzer Prize winner 1978); *Broca's Brain: Reflections on the Romance of Science* (1979); *Cosmos* (1980); *Comet* (1985); *Contact* (1985, which was later made into a movie starring Jodie Foster); *A Path Where No Man Thought* (1990); *Shadows of Forgotten Ancestors* (1992); *Pale Blue Dot: A Vision of the Human Future in Space* (1994); *The Demon-Haunted World: Science as a Candle in the Dark* (1995); and *Billions and Billions* (published a year after his death in 1996).

Replete with intelligent-sounding jargon meant to impress readers of his scientific acumen, Sagan's books on science provided clever camouflage for his true motivation, which was not promulgating the cause of science but destroying belief in God. This became all too apparent with Sagan's publication of *The Demon-Haunted World,* finished just a year before he died from a long and painful battle with cancer in 1996. *Demon-Haunted World* (*DHW*) was a blistering attack on many of the time-honored religious beliefs of traditional Christianity, such as miracles, saints, apparitions, prophecies, the Bible, God, heaven and hell, the devil, demons, exorcisms, near-death experiences, and the like. In *DHW,* Sagan delighted in attacking Christianity as illiterate and superstitious. He compared the most famous reports of miracles, apparitions, and other supernatural phenomena to phony UFO sightings, alien abductions, sleep walking, hallucinations, false memories, and other psychotic delusions of the mentally disturbed.

Regarding the most famous apparitions of the Blessed Virgin Mary in Lourdes and Fatima, which hundreds of millions of Christians and non-Christians believe in, Carl Sagan stated in *DHW* that the young visionaries who received the purported apparitions from Mary were in cahoots with the local Church and secular authorities to swindle pilgrims out of their hard-earned money after they traveled long distances to the shrines and basilicas built in Mary's honor. Other motives for the false reports of Mary, according to Sagan, were the increased social status of the visionaries' family and the validation of the Christian faith against other faiths. Said Sagan about the intentions of visionaries and their supporters:

> Possible motives for inventing and accepting such stories are not hard to find: jobs for priests, notaries, carpenters, and merchants, and other boosts to the regional economy in a time of depression: augmented social status of the witness and her family; prayers once again offered for relatives buried in graveyards later abandoned because of plague, drought, and war; rousing public spirit against enemies, especially [Islamic] Moors; improving civility and obedience to canon law; and confirming the faith of the pious.[5]

Despite what Sagan said about the supposedly evil intentions of the young visionaries and the Catholic Church, there is ample evidence proving that the Marian apparitions in Lourdes and Fatima and other famous places are genuine and worthy of belief (refer to my book *The Woman and the Dragon: Apparitions of Mary* for the evidence). Still, with all the evidence brought to us by true science, Sagan refused to consider the volumes of documented facts by the numerous scientists who examined the apparitions and found them worthy of belief, primarily because of his unfounded bias and hatred towards God and the Christian faith.

Although he made himself out to be an authority figure on science and religion and "false" reports of the supernatural, Carl Sagan warned his readers in *The Demon-Haunted World* to distrust all authorities, meaning, of course, the teaching magisterium of the Catholic Church. "One of the great commandments of science is, 'Mistrust arguments from authority,'" Sagan said in an obvious reference to the Catholic Church.[6]

As proof of the untrustworthiness of Christianity, Sagan in *DHW* repeatedly used the Church's inquisition of Galileo, all the while defending human science's right to make mistakes because of its supposed "built-in error-correcting machinery."[7] "We [referring of course to the scientific elite] are constantly prodding, challenging, seeking contradictions or small, persistent residual errors, proposing alternative explanations, encouraging heresy. We give are highest rewards to those who convincingly disprove established beliefs," Sagan said, proudly boasting of the superiority of scientists.[8]

And even though he lived in a time when the worship of human science had all but replaced the worship of Jesus Christ, Sagan feared that the old beliefs of conservative Christians would make a resurgence in the twenty-first century and topple the idols of human science that he and his liberal colleagues worked so hard to construct. "I worry that, especially as the Millennium edges nearer, pseudoscience and superstition will seem year by year more tempting, the siren song of unreason more sonorous and attractive," Sagan said.[9]

In this Sagan had reason to fear, as recent discoveries in genetics have made the case for intelligent design and supernatural creation far more believable than the random Darwinian evolution that Sagan so fervently championed.

Godless Human Science: The Second Beast of the Apocalypse

This book, *The Beast in Sheep's Clothing: Exposing the Lies of Godless Human Science,* is a sequel to my first book, *The Woman and the Dragon: Apparitions of Mary,* which was a detailed socio-political interpretation of the meaning of chapters 12, 13, and 17 of the Book of Revelation—the chapters that chronicle the fight between the Woman Clothed with the Sun, who symbolizes Mary and the Church, and the Great Red Dragon, which symbolizes Satan. In *The Woman and the Dragon,* I went to great lengths to identify the First Beast of Revelation 13 as "human power," but just touched upon the identity of the Second Beast as "human science." This book, however, will go to much greater lengths in identifying the Second Beast of the Apocalypse as the godless human sciences that deny the divinity of Jesus Christ.

The Second Beast of Revelation represents the human sciences that deny Jesus Christ is Lord and is used by the First Beast and the Great Red Dragon (Satan) to promote its interests and spread its lies. He is symbolized in the Book of Revelation chapter 13 as the one that looks like a lamb but is really a ravenous beast in disguise:

> Then I saw another wild beast come out of the earth; it had two horns like a sheep but it spoke like a dragon. It used the authority of the first beast to promote its interests by making the world and all its inhabitants worship the first beast, whose mortal wound had been healed. It performed great prodigies; it could even make fire come down from heaven to earth as men looked on. Because of the prodigies it was allowed to perform by authority of the first beast, it led astray the earth's inhabitants, telling them to make an idol in honor of the [first] beast that had been wounded by the sword and yet lived. The second beast was then permitted to give life to the [first] beast's image, so that the image had the power of speech and of putting to death anyone who refused to worship it. It forced all men, small and great, rich and poor, slave and free, to accept a stamped image on their right hand and forehead. Moreover, it did

> not allow a man to buy or sell anything unless he was first marked with the name of the beast or the number that stood for its name.
>
> A certain wisdom is needed here; with a little ingenuity anyone can calculate the number of the beast, for it is a number that stands for a certain man. The man's number is six hundred sixty-six.[10]

The Second Beast promotes man as the ultimate authority of knowledge and truth, rejects a higher power and authority other than man, deifies nature as a god, elevates man as something to be surpassed, and together with godless human power, the First Beast of Revelation 13, is in league with the devil—the Great Red Dragon of Revelation 12 who wages war against the Woman and her offspring (those who keep God's commandments and give witness to Jesus Christ).

As the eternal mocker of God, Satan holds a form of court in his kingdom of hell that mimics the Holy Trinity of God, in that he is one in purpose with the First and Second Beasts in wanting to destroy the good works of God and bring humanity to ruin like him. Ever since Adam and Eve, Satan has tempted mankind with knowledge and power to get him to rebel against God. Satan promised Eve in the Garden of Eden that if she ate the forbidden fruit of the tree of the knowledge of good and evil her eyes would be opened and she would not die like God warned but would be like God.

As you know, God expelled Adam and Eve from the metaphorical Garden of Eden for having disobeyed his commandment to remain faithful and innocent and having listened to the devil's temptation to gain knowledge and wisdom his way instead of God's. Thus, from the dawn of civilization, Satan has been telling us lies and giving us information, be it truthful or false, to get us to rebel against God.

While most people still have enough good in them not to idolize human power like they did in Nazi Germany and the Soviet Union, they do worship the godless human sciences that deny Jesus Christ is Lord. The Second Beast, therefore, is the silent killer that destroys far more souls than the First Beast because it enters and destroys without being detected, especially in the young. It is a master seducer and false prophet like the devil. It is believed to be truthful when in fact it is a liar like its master, Satan. On the outside it appears good, innocent, and gentle like a lamb, while on the inside it is a ravenous beast: a wolf in sheep's clothing that devours the souls of the naive and unwary.

How many youngsters have gone off to college only to have their faith in Jesus Christ utterly destroyed by liberal professors who promote the atheistic tenets of rationalism, Darwinism, moral relativism, and Freudian and Jungian psychology? Legions. By first destroying our belief in Jesus Christ via our high school teachers and college professors, the Second Beast then tries to get us to worship human power instead of Jesus. The godless ideologies of Nazism and communism, which gave rise to tyrants like Hitler, Lenin, and Stalin, had their beginnings in the minds of intelligent, well-educated men like Marx, Engels, and Nietzsche. They, in turn, got their inspiration from the Second Beast in Sheep's Clothing, the fallen archangel equivalent to the good archangel Gabriel who relays messages and gives information to man from God. (The First Beast, of course, is the demonic equivalent to Michael the Archangel, Power of God.)

While public schools cannot teach any religion as the one true religion or that Jesus is truly God, they can do comparative studies of the various religions and their sacred books based on historical or philosophical education. Thanks to the powerful and tremendously evil influences of Darwinism, rationalism, materialism, moral relativism, and all the other atheistic "isms" of the past few hundred years, the triune God of Christianity has been expelled from our public schools and governmental institutions, and will most likely never be allowed to return, as witnessed by the outlaw of the teaching of the Bible in public schools by the Supreme Court of the United States in 1963, prayer in public schools in 1962, the posting of the Ten Commandments in public schools in 1980, and the attempt by the Ninth Circuit Court of Appeals in 2002 to outlaw the Pledge of Allegiance in public schools.

Using the godless human philosophies that I have outlined in chapter one of this book, defenders of atheism and moral relativism like the ACLU, the People for the American Way, Americans United for Separation of Church and State, the National Education Association, and various other secular humanist organizations throughout America, have managed to persuade the government of the United States that any mention of the name of Jesus Christ or the teachings of his doctrines in public schools and other government institutions unconstitutionally violates the "wall of separation between church and state." Nowhere in our constitution does it say that there exists a "wall of separation between church and state." On the contrary, the Constitution of the United States specifically mandates the "free exercise" of religion without the government's interference, and forbids the "establishment" of any state religion or Church, such as England's Anglican Church (the Church of England), which all British citizens were once required by law to join and swear allegiance to. The never-ending recitation of the supposed constitutional "wall of separation of church and state" by liberals, which they constantly invoke to keep God out of our lives, comes not from the Constitution of the United Sates but from a letter written by Thomas Jefferson to the Danbury Baptist Association in 1801, in which Jefferson assured them of their religious liberties. In that letter, Jefferson quotes the First Amendment's non-establishment and free exercise clause, but then inserts his own opinion about how the government should stay out of Church affairs by saying that this clause "implies" a "wall of separation between and state." It is likely that what Jefferson meant in this often-quoted letter was that the government should stay out of the private affairs of individual churches, not that the worship or mention of God should be eliminated from our schools and governmental institutions. But, unfortunately, that's the way it's being interpreted by liberal activist judges today because atheism and anti-Christian bigotry has become so entrenched within modern society. Antichristianity is the last acceptable bigotry in the United States and the world. But that was prophesied to happen in the Bible, so it must take place. We are, however, obligated to fight against it.

The Beast in Sheep's Clothing: Exposing the Lies of Godless Human Science contains three chapters on the human sciences that have contributed the most to the decline of Christianity in America and throughout the world: atheistic human philosophy, Darwinian evolution, and Freudian and Jungian psychology. The fourth chapter, however, ends the book on a positive note. It gives more than enough evidence to prove that Jesus Christ is the Son of God.

If you are a non-Christian or an atheist, my greatest hope is that you will at least consider the facts that I have presented in each chapter in defense of Christianity. In no way is this book anti-science, for I have degrees in both biology and chemistry and believe that God wants us to search out the wonders of nature to appreciate him all the more. I only criticize science when it tells lies and tries to persuade us that God does not exist.

THE BEAST IN SHEEP'S CLOTHING

Chapter One

Godless Human Philosophy: Truth According to Man

"The desire of power in excess caused the angels to fall;
the desire of knowledge in excess caused man to fall."
Francis Bacon, *Advancement of Learning,* 1605

Man's Futile Search for Wisdom and Truth

The origin of the word "philosophy" comes from the combination of the Greek words *philos,* which means love, and *sophia,* which means wisdom. The meaning of the word philosophy, therefore, is a "love of wisdom." As sentient creatures made in the image and likeness of God, human beings have an inborn urge to know and understand the truth. No other creatures have this desire for wisdom and knowledge. That's what separates us from the brutes. Merely surviving from day to day like the animals do is not enough for mankind, for we cannot live in peace with ourselves if we don't ponder where we come from and where we're going when we die. God recognizes this inborn desire we have, for he was the one who gave it to us in the first place, and has revealed himself to us in order that we may know the truth, and love him and seek him out, and ultimately be reunited to him when we die.

As we have discovered from the modern sciences of archaeology, anthropology, and paleontology, the whole of human history has been a never-ending search for the truth about our creator. Primitive savages once believed that the heavenly bodies and the natural forces and elements of the earth were gods and therefore worshipped such things as the moon, the sun, the planets, earth, wind, fire, rain, crops, animals, insects, and many other things that exist in the natural realm. Savages, of course, did not yet have a fully revealed knowledge of their creator, so they were not to blame for this error.

As civilizations progressed in knowledge and understanding, stories began to originate out of man's active imagination about non-existent pagan gods and goddesses, whose characters and foibles were more human-like in nature than God's. The Greek gods were often vindictive and slaves of immoral sensuality and pleasures like man. Other than the children of Abraham, most of mankind really didn't comprehend the true nature and character of God. They didn't understand that he was holy and perfect as well as loving and merciful and generous, or that he was one in being and not many. That understanding would not be given to the world until the fullness of time, in the coming of God's son, Jesus Christ.

Gentiles (non-Jews or pagans) living in the centuries before Christ, while having an inward understanding that a higher power other than themselves exists, did not have an accurate understanding of the true nature and being of God, so they created idols of whatever they thought was divine. The ancient Egyptians worshipped Amon-Ra, the sun god; Osiris, the god of fertility; and Isis, the goddess of nature. In the lands now occupied by Syria and Iraq, Assyrians and Babylonians worshipped the higher god Ashur, the divine founder of their nations; the lesser gods Anu, Bel, and Ea; the moon god Sin and sun god Shamash; and Isthar, the goddess of the crescent moon and queen of the stars. Over in India, the Hindus worshipped Brahma, the creator god; Siva, the god of destruction and regeneration; and Vishnu the preserver god. The Buddhists worshipped nature as God. On the continent of Europe, ancient Greeks created their own mythology of gods and goddesses who were in many ways pettier, more jealous, and more evil than men. Anyone who has ever read Homer and Greek mythology knows about the jealous and capricious exploits of Zeus, Hermes, Aphrodite, Athena, Apollo, Artemis, and all the rest. It was in the Greeks, however, with all their mythological gods and goddesses, that the spark of divine wisdom appeared. It began to illuminate the minds of a few of their learned and esteemed philosophers such as Socrates and Plato and Aristotle, and through them prepare the Gentile world for the coming of God's Word five centuries later.

Socrates, Plato, and Aristotle were highly regarded and revered in their time for their wisdom. They didn't buy into the ancient pagan superstitions and mythologies of Homeric legend because they sensed that there existed somewhere out there an absolute being that transcended the human idiosyncrasies and weaknesses of Greece's mythological gods and goddesses. Of course, they had their disagreements about the true nature of the supreme being and the universe he created, but they were remarkably astute in understanding his essence and being.

The teachings and writings of these famous Greek philosophers, therefore, were very compatible with the teachings to come of Jesus Christ. When Jesus appeared four to five centuries later it was only natural for Greeks, Romans, Jews, Arabs, and Orientals familiar with Greek philosophy and language and customs to become followers of Christ.

The learned men of Zion and the surrounding Middle East were well versed in Greek language and customs following the Greek occupation of the Middle East by Alexander the Great in the fourth century B.C. The Apostle Paul, who was fluent in Greek, traveled extensively within the territories of Greece to preach the Word of God. In fact, the original extant writings of the New Testament were written in Greek, which was the universal language at that time. Thus, the world owes a great deal to the ancient Greek philosophers for paving the way for Christianity.

This chapter is a detailed history of man's search for philosophical truth, beginning in ancient Greece and ending in modern times. You will first note how astute Socrates and Plato and Aristotle were in knowledge and understanding without having the benefit of direct revelations from God that the Hebrews had, but then you will note that as the centuries progressed, and man began to reject the divine revelations and commandments given to us by Jesus Christ, human philosophers sunk deeper and deeper into philosophical errors, concocting nonsensical axioms and theories out of their own perverted imaginations, which has drawn the world farther and farther away from the divine truths revealed to us by God; so much so that man doesn't know what or who to believe anymore, or if there really is a creator. This has been especially true within the last five hundred years, when confusion and misunderstanding over religion and knowledge in general has caused the downfall of countless souls. And that's exactly what the Second Beast of Revelation (godless human science) wants: the destruction of our souls via the human sciences we worship so much, especially through the scientific discipline of philosophy. Now, here is a history of human philosophy and how we have veered off of the path of truth into apostasy and error. The factual information can be overwhelming at times, but it is important to understand what the ancient philosophers believed in, and how many of their beliefs have been reincorporated in modern society as various lies and heresies, like in the New Age movement.

Ancient Greek Philosophy

Genealogically, the ancient Greeks who gave us so many great philosophers descended from a lost tribe of people called the Minoans who lived on the Island of Crete between 2500 to 1400 B.C. These people believed in a higher being and afterlife reward or punishment, and probably got their beliefs from the ancient Egyptians with whom they traded in maritime commerce. Before the destruction of the Minoan civilization in the fifteenth century B.C., their culture spread to Greece around 1600 B.C., where they became known as the Mycenaeans, the people depicted in the legendary writings of Homer.

Following the migration of Minoans into Greece, three successive waves of people came next: first the Ionians, then the Achaeans, and finally the Dorians. The Dorians practically wiped out the Minoan-Mycenaean civilization and brought in their own Indo-European religion and customs. Vestiges of the Minoan-Mycenaean culture, however, continued to survive in Greece, especially in the lower classes, and hence the religion and philosophy of the ancient Greeks became an admixture of both Cretan and Indo-European.

The ancient Greeks learned the art of writing from the Phoenicians, but made a major improvement in writing by adding vowels to their alphabet. This, no doubt, helped contribute to the advancement in learning of the Hellenic (Greek) civilization. The first great writings of the Hellenic period were those of Homer the poet. The Homeric poems, the most famous being the adventures of Ulysses in the *Iliad* and the *Odyssey,* were written sometime between the eighth and sixth centuries B.C., and became an important part of Greek education, especially in Athens (Greece was divided into city-states, the most important being Sparta, Athens, and Corinth).

The Olympian gods of Homer were not holy like the God of Judeo-Christianity. They were immoral, capricious and whimsical, differing from mortals only in their possession of supernatural powers. The Greeks also worshipped Pan, the Greek god of fertility who had an oversized phallus and was an avaricious lover of nymphs. Dionysus (Bacchus), the son of the deities Zeus and Persephone, was also popular. He was the god of vegetation, wine, drunkenness, and orgies whom the Greeks adopted from the Thracians. The cult of Bacchanalia was founded to honor him. The Maenads, or Bacchantes, exemplify Bacchus, whose occultic mysteries inspired ecstatic, orgiastic worship. This group of female devotees left their homes to roam the wilderness in drunken idolatry to their god. They wore flimsy, revealing fawn skins and were believed to possess occult, magical powers. The name "Bacchus" came into use in ancient Greece during the fifth century B.C. It refers to the loud cries with which he was worshiped at the Bacchanalia. These events, which supposedly originated in spring nature festivals, became occasions for licentiousness and intoxication.

The celebrants danced, drank, and generally debauched themselves. The Bacchanalia became more and more extreme and were finally prohibited by the Roman Senate in 186 B.C. as dangerous to the state.

In addition to the pagan gods of Homeric legend and the cult of Bacchanalia, the ancient Greeks had their religious cult of Orphism, which was derived from the legendary writings of the Greek poet and musician Orpheus. According to Greek mythology, he was the son of the muse Calliope and Apollo, the god of music. According to Orpheus, Zeus wished to make Bacchus king of the universe, which infuriated the jealous Titans, who then dismembered and devoured the young god Dionysus. Athena, goddess of wisdom, was able to rescue Dionysus' heart, which she brought back to Zeus, who then swallowed it and gave birth to a new Dionysus. Zeus then punished the Titans by destroying them with his lightning bolts, and from their ashes he created mankind. Thus, according to the religious cult of Orphism, humans have a dual nature: the material body comes from the earth-born Titans, and the divine soul comes from the divinity of Dionysus, whose remains had been co-mingled with that of the Titans. As a result man became part divine. Believers in Orphism strove to eliminate the material part of themselves through ascetic rites and purification, which they believed was derived from the Titans, and preserve the divine or Dionysus part. Orphics also believed in a long succession of reincarnation. Once they eliminated the Titanic elements of their nature, they would be reunited with the divine.

In spite of the widespread idolatry and licentiousness of the ancient Greeks (orgies, homosexuality, and adult-child sex were common), great advances in science, astronomy, and mathematics were made. In the sixth century B.C., there was Thales of Miletus and Pythagoras of Samos, the latter a religious leader who taught the importance of studying numbers in order to understand the world, and who was the originator of the famous Pythagorean theorem. In the fifth century B.C., the great geometer Democritus of Abdera discovered the correct formula for the volume of a pyramid, and Hippocrates of Chios discovered that the areas of crescent-shaped figures bounded by arcs of circles are equal to areas of certain triangles. In the fourth century B.C., Eudoxus of Cnidus discovered a method for rigorously proving statements about areas. Euclid, a Greek mathematician who studied at Athens, founded a school of mathematics in Alexandria where he taught plane geometry, proportion in general, the properties of numbers, incommensurable magnitudes, and solid geometry. In the third century B.C., Archimedes of Syracuse, a preeminent Greek mathematician and inventor, wrote important works on plane and solid geometry, arithmetic, and mechanics. His younger contemporary, Apollonius of Perga, produced an eight-book treatise on conic sections that established the names of the sections: ellipse, parabola, and hyperbola. In the second century B.C., the Greek mathematician and astronomer Hipparchus composed tables of chords in a circle, similar to a sine table, which marked the beginning of trigonometry. Menelaus of Alexandria made great advances in astronomy. In the second century A.D., the Greek astronomer Ptolemy wrote *Almagest,* which became the universally accepted cosmological theory of the organization of the world until the time of Copernicus twelve centuries later.

In the sixth century B.C., Greek philosophers were divided into two competing camps of ideology: those who emphasized the importance of the material world and those who stressed the importance of nonmaterial forms or ideas. The first school of Greek philosophy, the Ionian or Milesian school, also known as the Pre-Socratic school, was largely materialistic. Members of this philosophical school were usually Olympians in faith rather than Dionysians or Orpheans. One of its most important members, Thales of Miletus, Asia Minor (625-546 B.C.), was the founder of Greek philosophy and considered one of the Seven Wise Men of Greece. According to Thales, the original essence of all things is water, from which everything proceeds and into which everything returns. Thales was famous for his knowledge of astronomy after he correctly predicted the eclipse of the sun on May 28, 585 B.C. He is also said to have introduced geometry in Greece. Before Thales, explanations of the origin of the universe were purely mythological, so he is credited with giving birth to scientific thought. Thales left no writings, but knowledge of his philosophical and scientific teachings comes from an account in Aristotle's *Metaphysics.*

Another important member of the Milesian school was Anaximander of Miletus (550 B.C.), who believed that the elements—earth, wind, fire, and water—were all gods. One of the first known believers in evolution, Anaximander believed that the world was not created, but that it evolved by a separation of opposites (like hot and cold and dry and wet) from the primordial material, which is eternal and indestructible. These opposite forms change and merge into one another according to the rule of justice; that is, balance and proportion. Living organisms, Anaximander said, arose from moisture as it was evaporated by sunlight. Man, therefore, evolved from fish like all other creatures. In the end, all things return to the element from which they originated. Anaximander also believed in an infinitely extended being whom he called the "Infinite," who was living and intelligent and encompassed all the worlds. Anaximander also conceived of the universe as a number of concentric cylinders, of which the outermost is the sun, the middle is the moon, and the innermost is the stars. Within these concentric cylinders is the earth, he said, unsupported and drum-shaped. Anaximander is also credited with introducing the sundial into Greece and inventing cartography.

Yet another important member of the Milesian school of philosophy was Anaximenes of Miletus (570-500 B.C.). Anaximenes believed that the universe consisted of air of different densities, and that air is the primary element to which everything else can be reduced. The soul is air, fire is rarefied air, condensed air becomes water, further condensed air becomes earth, and even further becomes stone.

Pythagoras of Samos, a.k.a. the "Samian Sage" (550-500 B.C.), was another student of the Milesian school. He went on to found a religious sect in southern Italy called the Pythagoreans. The Pythagoreans adhered to mysticism and asceticism, similar in many respects to the Orphics, and practiced silence, obedience, abstinence from beans and certain meats, simplicity in dress and possessions, and the habit of frequent self-examination. Early sources identify Pythagoras as a magician claiming to have had occult, miraculous powers. He is known to have believed in reincarnation and the transmigration of souls (metempsychosis), and an afterlife reward or punishment. Pythagoras was said to have claimed that he had been Euphorbus, a warrior in the Trojan War, and that he had been permitted to bring into his earthly life the memory of all his previous existences. He likewise thought himself semi-divine.

Pythagoreans revered numbers: they believed that the whole of the universe could be explained by mathematics, which was a form of religion to them. Among the mathematical investigations carried on by the Pythagoreans were the studies of odd and even numbers, and of prime and square numbers. Numbers became for them the ultimate principle of all proportion, order, and harmony in the universe. Numbers were tangible things, like sizes and shapes, thus when we speak of numbers being "squared" or "cubed" these designations derive from the Pythagoreans. Pythagoras also showed that concordant musical intervals such as octave, fourth, and fifth could be expressed in arithmetic ratios. He also discovered that in right triangles, the sum of the square of the sides adjoining the right angle are equal to the square on the remaining side, the hypotenuse (the Pythagorean theorem we all studied in school and which was quoted by the scarecrow in the *Wizard of Oz* when he finally got his brain).

Pythagoreans also extended their reverence for numbers to the heavens. They thought that the planets were separated from one another by numeric intervals corresponding to the harmonic lengths of strings, and believed that the movement of the spheres created a musical sound—the "harmony of the spheres." Pythagoreans were the first to consider the earth a sphere, revolving with the other planets, including the sun, around a central fire; it was a revolutionary idea for its day. Pythagoras also thought of the universe as made up of atoms, and material objects built up by molecules of atoms of different shapes and arrangements. The cult of the Pythagoreans maintained its organization until the middle of the fourth century B.C., dedicated to the reformation of political, social, and moral life, and the reverence of mathematics. Pythagoras, however, left no writings. All that is known of his doctrines comes from his disciples.

One of the first known skeptics of Greek mythology was Xenophanes of Colophon (560-470 B.C.). Xenophanes was an empiricist, believing there was no possible way to know something outside the realm of direct experience. He ridiculed the anthropomorphic theology of Greece and the immorality of their gods, saying that each race has a god resembling their race, and if horses could draw, their god would be a horse. Xenophanes left Colophon in 545 B.C. to be a wandering poet and minstrel in Greece and Sicily. Xenophanes believed that human beings should reject Greek polytheism, and recognize instead a single nonhuman deity who controls the whole universe. Xenophanes also criticized the belief in reincarnation and deplored the Greek's preoccupation with athleticism and luxurious, carnal living at the expense of philosophical wisdom. In 536 B.C., according to tradition, Xenophanes settled permanently in the Phoenician colony of Elea, in southern Italy, where he founded the Eleatic school of philosophy. The school was expanded after his death by his student, the Greek philosopher Parmenides. Eleatics were opposed to the materialistic philosophy of the Milesian school, and also opposed the theory of universal flux proposed by Heraclitus. Eleatics believed that the universe is an unchanging unity, infinite in time and space, beyond the comprehension of man's senses. Sensory observations yield a distorted, limited view of reality, they said, and only through philosophy can the ultimate truths of the universe be known. The famous Greek philosopher Plato was greatly influenced by the Eleatics.

Another vocal skeptic of the mythological gods and goddesses of Greece was Heraclitus of Ephesus (500 B.C.). Called the "dark" or "weeping philosopher" because of the loneliness of his life and misanthropic views of his philosophy. Heraclitus believed in a state of "universal flux," where everything in the universe comes out of the one, and the one out of all things. Unity in the world is formed by a combination of opposites. Fire is the primordial source of all matter, and the universe is in a constant state of change, where a union of opposites produces objects. The soul, for example, is a combination of fire and water, the fire being the noble part and water the ignoble. Heraclitus believed in "parallelism of structure," where the behavior and structure of the universe and the soul are parallel. Whenever he mentioned a higher power he spoke in terms of a singular god instead of gods. Heraclitus is considered by historians to be one of the founders of Greek metaphysics, although the materialist ideas in his theory of universal flux are closely related to those of the Milesian school of Greek philosophy. Collected editions of all his surviving writings may be found in several modern editions, but only one work, *On Nature,* is definitely attributable to Heraclitus.

One of the greatest members of the Eleatic school of philosophy was the astronomer, biologist, and philosopher Parmenides of Elea (480 B.C.). Unlike the materialists of the Milesian school, Parmenides viewed ultimate reality in terms of thought and reason rather than matter, and was no doubt a big influence on the seventeenth-century French philosopher Rene Descartes, who became famous for the slogan "I think, therefore I am." To Parmenides, reality is in thinking, and the object of one's own thought. One's thoughts can only imagine what really exists. Ultimate reality or "True Being" is not known to the senses but is to be found in reason alone. Full knowledge about the real world and True Being is impossible. The non-existence of True Being, however, is inconceivable, and understanding of its being is equally inconceivable to thought. True Being cannot arise from Non-being, and True Being neither arises nor passes away. The phenomena we observe in nature are only apparent to our senses and subject to human error; they seem to exist, but, in fact, have no real existence because our senses are easily deceived. Sensible things appear to be real but are illusionary. The only true being is the "One," which is infinite and indivisible. Parmenides visited Athens at the age of sixty-five, and on that occasion Socrates, then a young man, heard him lecture. Parmenides' idealism had a big influence on Plato, Socrates' pupil, and he is credited as the first to use logic in metaphysics. Parmenides' only surviving work are fragments of a didactic poem, *On Nature.*

Empedocles (495-435 B.C.), the Sicilian disciple of Pythagoras and Parmenides, was a famous philosopher who claimed to be a god, and is believed by his followers to have been a prophet who could work miracles, control the wind, and raise the dead. However, a fatal leap into the volcano, Mt. Etna—a leap of faith Empedocles made to prove he was a god—quickly put an end to that delusion. Empedocles, nonetheless, is noted for his thoughts on the reality of the natural world and his belief in evolution. He taught that the two opposing forces of love and hate (affinity and antipathy), act upon the four elements (earth, wind, fire, water), combining, separating, and recombining them over and over into infinitely varied forms. Love unites everything, he said, then hate separates, all in a continuous cycle. No creation of new matter is possible, only changes in the combinations of the four elements. Everything is temporary; only the elements are permanent. Empedocles, who founded an Italian school of medicine, also believed that the natural realm is governed by chance and necessity rather than by design and purpose, which earned him a reputation as one of the first known evolutionists.

Protagoras of Abdera (490-420 B.C.) is thought to be the first Greek philosopher to call himself a "sophist" and the first to receive money for teaching. Sophists were traveling teachers of rhetoric, philosophy, and the art of good living, but their reasoning and arguments were often captious and fallacious. Protagoras won great fame for his teaching of philosophy in Greece and was invited by Pericles, the Athenian statesman, to write the constitution of the Athenian colony of Thurii. Protagoras was also an early agnostic and moral relativist, claiming that we have no proof for the existence of God or gods, and that there was no ultimate authority that we should obey. Each individual is his own authority, said Protagoras. He was famous for saying "Man is the measure of all things." Nothing was absolutely true or false, good or bad, to Protagoras.

One of the best-known philosophers from Athens was Anaxagoras (500-428 B.C.) of Ionia, who moved to Athens and became friends of the Athenian statesman Pericles. Anaxagoras was famous for his teachings on matter and mind, a subject that has fascinated and intrigued philosophers for thousands of years. Matter, Anaxagoras said, existed everywhere without void and with infinitely complex micro particles. All matter is infinitely divisible, and the tiniest bit of matter contains a portion of all the elements. That element which makes up the largest portion of matter characterizes it. Anaxagoras' theory that all matter exists mixed with other matter eventually became known in science as the "field theory." Mind, on the other hand, initiated and ordered the material universe. Mind is the Prime Mover, the infinite, omnipotent, omniscient, cosmic intelligence. Mind was not God, but existed in both animals and man, who only seems more intelligent than animals because he has hands. It was unknown whether or not Anaxagoras believed in divine Providence because he had little to say about ethics and religion.

Contrary to the field theory of Anaxagoras was the "atomistic theory" of Democritus (460-370 B.C.) and Leucippus (450-370 B.C.). The atomistic theory states that mind and matter are composed of tiny, indivisible, and indestructible particles of pure matter (*atoma,* or "indivisibles"), eternally swirling about in empty space (the *kenon,* or void). There are an infinite number of atoms, which are composed of the same matter but differ in size, shape, weight, sequence, and position. It is this quantitative difference that makes up the different things we see in the universe. As an unbeliever in God, Democritus believed that the world was created not by a higher power but by the ceaseless whirlings and collisions of atoms, which combine to form larger aggregates of matter. There is no divine purpose to the universe, only the movement of atoms governed by the laws of physics. Entirely materialistic, Democritus and Leucippus' atomistic theory anticipated the modern laws of physics on the conservation of energy and the irreducibility of matter. They were, however, mistaken in their disbelief in a higher power.

One of the wisest men who ever lived was the Greek philosopher named Socrates (470-399 B.C.). Like Jesus Christ five centuries later, Socrates left no writings of his own, only oral teachings handed down to his disciples through word of mouth (it still amazes me how students of science and history are so willing to believe in Socrates yet refuse to believe in Jesus Christ, even though we have many more contemporary witnesses for Jesus than we have for Socrates). All we know about Socrates and his teachings comes from Plato, Xenophon, Aristotle, and a contemporary critic of his, Aristophanes. Born in Athens, the son of Sophroniscus and Phaenarete, Socrates received an elementary education in literature, music, and gymnastics like the rest of the children his age. As he matured, he studied the Milesian philosophers, the rhetoric of the sophists, the ethics of morality and religion, and Athenian culture. By trade, Socrates was a sculptor like his father, and carved a statue group of the Three Graces, which stood at the entrance to the Acropolis until the second century A.D. When Socrates became of age, he was enlisted as an infantry soldier in the Peloponnesian War with Sparta, and was cited for bravery at the battles of Potidaea in 432-430 B.C., Delium in 424, and Amphipolis in 422.

Although short in stature and unattractive in physical appearance (he wore shaggy clothes and walked barefoot), Socrates was extremely witty, funny, and self-controlled. He refused to be cynical and rude in his oral arguments like the sophists and refused to take any money for his teachings. Indifferent to heat and cold, thirst and hunger, Socrates had mastery over all bodily passions, and seldom drank wine. Preferring oral arguments to writing, Socrates spent the greater part of his adult life in the public squares and marketplaces in philosophical debate with anyone who would listen. The term "Socratic method" of teaching derives from Socrates' dialectical method of the teacher letting the student discover the truth by engaging the student in oral argument. Socrates purposely stayed away from discussing politics because of its danger to one's career, and devoted his life to the teaching of philosophy and the enlightenment of souls about God. He wrote no books and established no schools. What we know of Socrates today comes mainly from his pupils Plato and Xenophon. Plato's depiction of Socrates' and his teaching in his dialogues, however, is somewhat tainted with his own philosophical biases, and Xenophon probably didn't know him well enough to give an accurate account of his life.

In his dialogues, Plato depicted Socrates as a brilliant orator with an extremely sharp acumen on any subject, feigning ignorance of a subject then destroying every argument and belief you had about it. This method of intellectual counterattack became known as "Socratic Irony." Socrates preferred the subjects of ethics and morality over anything else. He believed that evil is the result of ignorance in education, and that no person is willingly bad. To Socrates, knowledge is virtue, and those who are well educated in philosophy will be ethical and moral as adults. Much of Socrates' time was spent showing-up pretenders who thought they were wise, and for that he made himself many enemies. But Socrates never bragged about his own intelligence. "God alone is wise," he often told people.

Although he pledged allegiance to the Athenian state and was a man of deeply-held religious convictions, his enemies regarded Socrates as a traitor because of his criticisms of the corruption in Athenian politics and his disbelief in the pagan polytheism of Greece Socrates publicly denied the existence of Zeus in Aristophanes' *Clouds,* which got him in trouble with Athens. In 399 B.C., Socrates was arrested and charged with neglecting the gods of the state and of introducing new divinities, to which Socrates often credited his wisdom. Socrates was also charged with corrupting the morals of the youth, leading them away from the Athenian principles of democracy, although Athens was by no means a true democracy and was rife with corruption. He was also falsely accused of being a sophist because he had been ridiculed by the comic-poet Aristophanes (423 B.C.) in his *Clouds* as the master of a "thinking-shop" where young men were brainwashed to make the worse reason appear the better. Plato's *Apology* gave an historic account of Socrates' defense at his trial, and was a tribute to the philosopher's life. In *Apology,* Plato recounted that Socrates was condemned to death by a small majority of jurors for "speculating about the heavens above, and searching into the earth beneath, and making the worse appear the better cause." Upon receiving the death sentence, Socrates made a counterproposal to the court that he pay a small fine to the state because of his value as a philosopher, which made the court so angry that the vast majority of jurors voted for his death sentence. His last remarks to the court were recorded in Plato's *Apology:* "God orders me to fulfill the philosopher's mission of searching into myself and other men. . . . Men of Athens, I honor and love you; but I shall obey God rather than you. . . . For know that this is the command of God; and I believe that no greater good has ever happened in the State than my service to the God."

Socrates' friends wanted to break him out of prison, but he preferred to obey the law and die a martyr for his beliefs. He had no fear of death and believed that his eternal soul would ascend to a better place. His last day was spent with his friends and admirers, as described in Plato's *Phaedo,* and in the evening he calmly drank a cup of hemlock. Socrates was seventy years old when he died. Through the writings of Plato's *Apology, Crito, Gorgias,* and *Phaedo,* Socrates had a profound effect on the entire course of Western philosophy, and his wisdom, insight, philosophy, and oratorical skills were looked upon with reverence and awe by his students for years to come. The Socrates depicted in Plato's dialogues became, as the influence of Greek and Roman mythology waned, the archetype of the higher life in the ancient world. When asked who was the wisest of all men, the Oracle of Delphi is purported to have said that there was none wiser than Socrates.

The Greek philosopher who no doubt had the greatest influence on early Christian theology was a star pupil of Socrates named Plato (428-347 B.C.). Plato was born to an aristocratic Athenian family in 428 B.C. His father, Ariston, was believed to have descended from the early kings of Athens. Perictione, his mother, was distantly related to the sixth-century-B.C. lawmaker Solon. When Plato was a child, his father died and his mother married Pyrilampes, who was an associate of the Athenian statesman Pericles. As a youth, Plato wanted to become a politician, but soon grew disillusioned with the corruption of the Athenian democracy. Plato went on to become a pupil of Socrates, and was trained in Socrates' dialectic method of oral questions and answers to arrive at the truth.

Unlike Socrates, Plato was a prolific writer who cast the legendary Socrates as his mouthpiece for his own philosophical epistemology (the theory on the origin of knowledge) and philosophical ontology (the theory on the origin of being). Some of Plato's "dialogues" represent authentic thoughts of Socrates, and others used Socrates to convey Plato's own beliefs. The earliest collection of Plato's work includes thirty-five dialogues and thirteen letters, but the authenticity of a few of the dialogues and most of the letters has been disputed. Plato's dialogues are divided into early, middle, and later periods of composition. The earliest dialogues are an attempt by Plato to portray the philosophy and dialectic mastery of Socrates. In them, Socrates encounters someone who professes to be a wise man. Engaging him in discussion, Socrates professes his ignorance and asks for assistance from the supposed wise man. As Socrates begins to raise questions, however, it becomes clear that the one who thinks himself wise doesn't really know what he's talking about. Socrates emerges as the wiser of the two because he at least admits his own ignorance. Knowledge of our ignorance, Socrates would tell them, is the true beginning of wisdom. The dialogues of the middle and later periods of Plato's life are a reflection of his own philosophical ideology, and most historians attribute the ideas put forth in these dialogues to Plato (Socrates continues to be the main character but is now Plato's mouthpiece).

Included in Plato's early dialogues are *Charmides* (an attempt to define temperance), *Lysis* (a discussion of friendship), *Laches* (a pursuit of the meaning of courage), *Protagoras* (a defense of the thesis that virtue is knowledge and can be taught), *Euthyphro* (a consideration of the nature of piety), and Book I of the *Republic* (a discussion of justice). The dialogues of the middle and later periods of Plato's life are a reflection of his own philosophical ideology, and most historians attribute the ideas put forth in these dialogues to Plato (Socrates continues to be the main character but is now Plato's mouthpiece). The writings of the middle period include *Gorgias* (a consideration of several ethical questions), *Meno* (a discussion of the nature of knowledge), the *Apology* (Socrates' defense of himself at his trial against the charges of atheism and corrupting Athenian youth), *Crito* (Socrates' defense of obedience to the laws of the state), *Phaedo* (the death scene of Socrates, in which he discusses the theory of forms, the nature of the soul, and the question of immortality), the *Symposium* (Plato's outstanding dramatic achievement, which contains several speeches on beauty and love), and the *Republic* (Plato's supreme philosophical achievement on the city, state, government, and rulers, which is a detailed discussion of the nature of justice). The works of the later period include the *Theaetetus* (a denial that knowledge is to be identified with sense perception), *Parmenides* (a critical evaluation of the theory of forms), *Sophist* (further consideration of the theory of ideas, or forms), *Philebus* (a discussion of the relationship between pleasure and the good), *Timaeus* (Plato's views on natural science, cosmology, the soul, harmony of the world), and the *Laws* (a more practical analysis of political and social issues).

Plato was a deeply religious man who rejected the gods of Olympus and believed instead in the Creator God of the universe. He was, however, a dualist and believed that God didn't create everything, but only what is good, and that God created the universe out of preexisting matter instead of nothing. Plato also believed that the body and soul are separate and unique, that the soul is immortal and can reincarnate or transmigrate after death, that an afterlife reward awaits the souls of philosophers who have conquered the flesh, and while the souls of the lukewarm go to purgatory, the souls of the wicked go to hell. Souls, according to Plato, are composed of three distinct parts: the rational part, the will, and the appetites. The just person's rational part, supported by his will, is capable of subduing the appetites. Freud's theory of the unconscious superego, ego, and id was no doubt influenced by Plato's theory of the tripartite soul.

Philosophically, Plato was an idealist rather than an empiricist because he believed that true knowledge comes from reason rather than sensory experience. The senses, Plato believed, are too easily deceived. True reality supercedes appearances and is only revealed in thought through our ability to reason. Reason, when used properly, results in awareness of objects that are certain, unchanging, immutable, and permanent. There are two forms of existence in Plato's philosophy: the sensible realm or world, and the intelligible world where real intelligence emanates (the more perfect realm). The objects of our reason, said Plato, are the abiding universals, the eternal "forms" or substances that constitute the real, higher, intelligible world. The senses, on the other hand, experience only changeable phenomena in the physical world, so the objects we perceive through our senses are not proper objects of knowledge. An absolute knowledge of unchanging reality can only be achieved when the soul leaves the body at the moment of death. To Plato, knowledge is virtue and can only be learned through the understanding of forms. The ultimate Form is the Good, and knowledge of the Good is our guide in morality. To know the Good is to do good. Immorality results in ignorance in not knowing the Good. The moral person is the happy person, and since we all desire happiness, we all desire to do what is moral. Oh, if that were really so!

On politics, Plato wrote in his *Republic* that the ideal form of government was a commonwealth headed by a philosopher-king who completed the highest level of education and was able to understand all the forms and, therefore, make the wisest decisions. The ideal state is composed of three classes: the merchant class, whose unique virtue should be temperance; the military class, whose virtue should be courage; and the ruling class or philosopher-kings, whose virtue should be wisdom. The fourth virtue, justice, should characterize society as a whole. In the *Republic,* Plato called the ideal commonwealth a "utopia." In it all four virtues were practiced, wealth was abolished, the Homeric gods of Olympus were outlawed, and a rigid censorship imposed. Plato's Socrates, however, viewed the ideal commonwealth as more communistic: women were equal to men in all things, marriages were to be arranged by the state, communal marriages allowed, children should be taken from their parents and raised by the state, abortion and infanticide should be compulsory for unsanctioned pregnancies, and the practice of eugenics encouraged.

Plato's prolific writings and oral teachings in philosophy gained him an international reputation. In 387 B.C., Plato founded a school of learning in Athens called the Academy, which is generally regarded as the first European university. The Academy provided comprehensive training to students in philosophy, astronomy, biology, mathematics, and political theory. It flourished for eight centuries until the Byzantine Emperor Justinian I, a Christian who wanted to purge the Roman Empire of the last vestiges of Greco-Roman paganism, demolished it in 529 A.D.

In 367 B.C. and again in 361, Plato traveled to Sicily to tutor the new ruler of Syracuse, Dionysius the Younger, in the art of becoming a philosopher-king. Plato failed miserably and spent the remaining years of his life writing teaching at the Academy, where he died at the age of eighty.

Plato's influence on Christian theology and Western philosophy is enormous. The Christian theologians Clement of Alexandria, Origen, and St. Augustine were all enthusiastic advocates of Plato because Plato's ideology and philosophies dovetailed nicely with orthodox Christian belief. As Christianity started to overtake paganism as the dominant religion of the Roman Empire, however, an attempt was made in the third century A.D. by Porphyry of Greece and Plotinus of Rome, both enemies of Christianity, to bring back the ancient pagan practices and beliefs of Plato's Greece. These two began the heretical school of Neoplatonism, whose tenets of an infinite, unknowable, perfect One, from which a world intelligence (*nous*) and a world soul emanated from, were condemned by the Church as unorthodox. Despite being censured by the Church, Neoplatonism made a reappearance during the Renaissance in the fifteenth century, when the Italian scholar and humanist, Marsilio Ficino, at the direction of the wealthy Italian nobleman Cosimo de Medici, translated and annotated the works of Plotinus, Porphyry, and Iamblichus, and thereby founded the Florentine Academy, where students studied Neoplatonism and the writings of Plato in the original Greek. The resurgence of Neoplatonism during the European Renaissance (1300-1600) was due, in part, to a backlash against the morally rigid thirteenth century scholastic Aristotelianism of St. Thomas Aquinas, in favor of the more idealistic metaphysics of Plato.

Of the three great Greek philosophers Socrates, Plato, and Aristotle, the most educated in the natural and philosophical sciences was a student of Plato's named Aristotle (382-322 B.C.), who some say is the "Father of Modern Science and Philosophy." So famous was Aristotle that throughout history he is simply referred to as "The Philosopher." Aristotle grew up in Stagira, in Macedonia, the son of a physician to the royal court. At seventeen, he left home and traveled to Athens to study at Plato's Academy, where he remained for twenty years as student and teacher. When Plato died in 347 B.C., Aristotle moved to Assos in Asia Minor, where he was an advisor to the ruler, Hermias. Aristotle later married Pythias, the niece and adopted daughter of Hermias, but after Hermias was captured and executed by the Persians, Aristotle fled to the Macedonian capital of Pella, where he tutored the future emperor Alexander the Great. In 335, when Alexander became king, Aristotle returned to Athens and established his own school of philosophy called the Lyceum. The Lyceum became known as the Peripatetic ("walking" or "strolling") school because much of the discussions between student and teacher took place while strolling the grounds of the Lyceum.

At the Lyceum, Aristotle instituted courses in logic, metaphysics, ethics, psychology, rhetoric, dialectics, art, aesthetics, politics, physics, mathematics, biology, meteorology, and astronomy. On the subject of science, Aristotle's most famous work was *On the Heavens,* a cosmological theory on the make up of the universe with earth at its center, which predominated scientific thought until the time of Copernicus and Galileo. In addition, Aristotle also wrote *On Physics,* which deals with the science of motion and change, and *On Meteorology,* in which he made observations on comets and volcanoes.

Unfortunately, all of Aristotle's original writings on science and philosophy were lost or destroyed in Europe after the fall of the Roman Empire in the fifth century A.D. and weren't rediscovered until the time of the Crusades, when commentaries on the writings of Aristotle by the Islamic scholar Averroes (1126-98) were discovered and translated into Latin. The Crusaders' discovery in Constantinople during the thirteenth century of more of Aristotle's manuscripts in the original Greek prompted further Western interest in Aristotle's philosophy. Aristotle's scientific empiricism had a great deal of influence on Catholic theologians of thirteenth century, particularly the Catholic Dominicans

Albertus Magnus and St. Thomas Aquinas (1225-74). Aquinas placated Church authorities by showing that Aristotle's philosophy of knowledge through sensory experiences was completely compatible with the prevalent Augustinian philosophy of divine knowledge through Christian revelation and reason. The Catholic Church had banned the study of Aristotle in 1210 and threatened to excommunicate anyone who studied it, but reversed its ban because of Aquinas. The subsequent birth of Thomist scholasticism in the thirteenth century and the Catholic Church's acceptance of Aristotle owes a great deal to Thomas Aquinas' brilliant achievement of reconciling Catholic theology with Aristotelian philosophy and metaphysics.

In essence, Aristotle's metaphysics argued for the existence of a supreme being whom he called the "Prime Mover." The Prime Mover controls cosmological unity and the whole purpose of nature. It is the first cause, pure thought and intellect, perfect in unity and being, immutable in nature, and is the desire of all its creatures that want to share in its beauty and perfection. The concept of the Prime Mover, however, was incompatible with Christianity's concept of a triune God because of Aristotle's belief that other "movers" existed apart from the Prime Mover (Aristotle suggested that there were either fifty-five or forty-seven other movers). Aristotle's more materialistic theory of the soul was significantly different than Plato's theory of a separate material body and spiritual soul. Aristotle defined a soul as a "kind of functioning of a body organized so that it can support vital functions." The soul does not exist separate from the body as Plato believed, but is like Plato's concept of the form, the essence or unchanging substance in an object like the body. A part of the divine lives in the soul, so we should try to emulate it by suppressing the animal within us. Aristotle did say, however, that *nous poetikos* (active mind), which is responsible for human insight in its highest form, is not reducible to any physical process of the body. This implies a kind of immaterial, separate, transcendency of spirit, which was in keeping with Christian doctrine. Aristotle, unlike Plato, did not believe in reincarnation, which was also in keeping with Catholic faith.

Aristotle's tremendous influence on science and philosophy is unparalleled in human history. The philosophical and scientific terms we use today—syllogism, premiss, metaphysics, species, genera, potentiality, categories, dialectic, and analytical—all derive from Aristotle. In the area of logic, which he was a master of, Aristotle developed hard and fast rules for reasoning that would, if applied properly, never lead from true premises to false conclusions. Aristotle's rules of reasoning made use of syllogisms: pairs of propositions that, when taken together, give a correct conclusion. For example, the propositions that "all creatures are mortal" and "man is a creature" leads to the correct conclusion that "man is mortal."

In philosophy, Aristotle broke from the Platonic Academy in his preference for using analytical reasoning over dialectic discussions as the principle method for discovering the truth. The dialectic method of Socrates and Plato, Aristotle said, uses opinions to arrive at logical truths, while his analytical method uses empirical rationale derived from direct observation and experiences to discover truth. The analytical method, therefore, is the more reliable of the two according to Aristotle. The scientific method we use today of verifying hypotheses and theories through repeatable experimental observation comes directly from the analytic method of Aristotle. Aristotle, however, did use the dialectic method early on in his career, and wrote some dialogues like Plato, but all have since been lost.

The only surviving writings of Aristotle come from a compilation of his lecture notes on a variety of scientific subjects that were collected and arranged by later editors, in addition to excerpts from a dictionary of philosophical terms and a study of Pythagoras. Among the texts accredited to Aristotle is a treatise on logic called *Organon* (instrument); a treatise on natural science called *Physics,* which gives a vast amount of information on biology, astronomy, and meteorology; a treatise on the Prime Mover and the nature of being called *Metaphysics;* a treatise on ethics he called *Nicomachean Ethics,* in honor of his son Nicomachus, in which he says that what is virtuous is good; and three other treatises called *Rhetoric, Poetics,* and *Politics,* the latter which favors democracy over communism but justifies slavery.

Aristotle's anti-evolutionary theories on biology were largely based on the concept of immutable species. The world, he said, is made up of individuals ("substances") occurring in fixed natural kinds ("species"). Each member of a species has an inborn pattern of development and grows to maturity according to the direction of its type. The environment, therefore, plays no part in species formation, as it does in Darwinian evolution. Each species reproduces true to type. The exception occurs, Aristotle said, when some "very low" worms and flies come from rotting fruit or manure by "spontaneous generation." Life cycles, according to Aristotle, are epicyclical: the same pattern repeats itself through a linear succession of generations. The animal kingdom forms a hierarchy from simple (worms and flies at the bottom) to complex (human beings at the top). Evolution of new and more sophisticated species is not possible. According to Aristotle there are four causes of speciation: the material cause, the matter out of which a thing is made;

the efficient cause, the impetus behind motion, generation, or change; the formal cause, which is the species, kind, or type; and the final cause, the mature development and purpose of the thing.

Centuries after Aristotle's death in 322 B.C., Western science and philosophy was synonymous with Aristotelianism. When one talked of logic they were usually talking about Aristotelian logic. But with the subsequent rise in popularity of Christianity in the Roman Empire in the centuries following Christ's death there was also a rise in anti-pagan sentiment. As a result, Aristotle's Lyceum, like Plato's Academy, was closed forever in 529 A.D. by Emperor Justinian I, and Aristotle's influence on academic thought in Europe dramatically declined. Most of Aristotle's important works were either lost or destroyed in the years that followed, but after the rediscovery of Aristotle's writings he was once again considered the premier Western philosopher. His cosmology, which maintained that the universe was geocentric and that everything under the moon decays while everything above is indestructible, was the dominant cosmological theory of the universe until a young Polish upstart named Copernicus developed his own heliocentric theory in the sixteenth century, which Galileo defended in 1633.

As we have seen so far, ancient Greece has given rise to some brilliant philosophers like Socrates, Plato, and Aristotle, but it also gave rise to some real misanthropes. One of these misanthropes was Diogenes of Sinope (412?-323 B.C.), a Greek philosopher generally considered to be the founder of cynicism. The word "cynic" comes from the Greek word *kynikos,* which literally means "like a dog." Cynics traveled about like a pack of dogs and lived like homeless tramps, wearing ragged clothing, eating menial food, sleeping on the ground, and disdaining everything in society.

Diogenes was born in Sinope in modern-day Turkey, and moved to Athens where he studied under Antisthenes, who taught that social conventions and worldly pleasures should be avoided, that there should be no government, no private property, no marriage, no religion, no slavery, no luxury, and no pleasures. Captivated by the antinomian teachings of Antisthenes, Diogenes plunged himself into a life of severe austerity and self-mortifications, wandering about shamelessly like a tramp, living in tubs and pitchers. (Plato described Diogenes as "Socrates run mad.") Diogenes ridiculed Greek literature and art, despised politicians and intellectuals, hated the greed and materialism of the rich, and mocked all morals and social norms. Owing no allegiance to any god or state, Diogenes and his cynical followers declared themselves citizens of the world who lived for the here and now, not needing to know why things are. Detachment from worldly comforts and pleasures was the only virtue to them; the morals and values of the civilized world were of absolutely no use. To further show his contempt for civilization, Diogenes defaced coins and wandered the streets of Athens in broad daylight carrying a lamp saying he was "looking for an honest man." On a voyage to Aegina one day, Diogenes was seized by pirates and carried to Crete and sold as a slave, but was later set free and made tutor to his master's children. According to legend, Diogenes died at Corinth on the same day as Alexander the Great in 323 B.C.

Another radical Greek philosopher during the time of Aristotle was Pyrrho the Skeptic (360-272 B.C.), who founded the philosophical school of Pyrrhonism. Pyrrho is credited with being the founder father of skepticism, a doubting philosophy that rejects the generally accepted truths of the prevailing society. The real nature of things can never be known, according to Pyrrho and his fellow skeptics, so any claims of objective knowledge should be put in doubt or totally rejected. Modern atheism and agnosticism have their roots in Pyrrhonism. Pyrrho the Skeptic was born in Elis and studied under the Greek philosopher Anaxarchus, a student of the atomist philosopher Democritus. Pyrrho followed the Greeks into Persia during the conquests of Alexander the Great, where he became acquainted with the contradictory teachings and philosophies of the Persian magi and Indian Brahmans. When he returned, Pyrrho lived the life of a recluse and produced no writings. What we know of Pyrrho comes from the writings of his disciple Timon, who maintained that equally valid arguments could be made to support or disprove beliefs. According to Timon, Pyrrho thought philosophers should have no beliefs, no dogmas or precepts, and make no judgments. Only by living a passive, non-judgmental existence could one obtain peace of mind and tranquility of soul; which, to Pyrrho, was the highest possible virtue. Believing that nothing could be proven true, Pyrrho was like the sophists of the fifth century B.C., who maintained that "nothing is; or if anything is, it cannot be known." Of course, Pyrrho's disbelief in the existence of absolute truths is fallacious and self-contradictory because to say that there are no absolute truths is to declare an absolute truth: "nothing is absolute."

Arcesilaus of Pitane (315-240 B.C.), onetime head of Plato's Academy, was another skeptic who attacked all doctrines as impossible to prove, and falsely claimed that Socrates and Plato were really skeptics like himself. In addition, the Middle Academy, the school of philosophy and learning that developed in the third century B.C. as an offshoot to Plato's Academy, and the New Academy, developed by Carneades in the second century B.C., likewise had their schools of skepticism. However, they were somewhat less radical than Pyrrho regarding absolute truths, saying that while nothing can be proven conclusively, some things can be proven to be more probable than other things. In later antiquity, Aenesidemus of Greece (first century B.C.) taught at the school of Alexandria, where he developed ten arguments for skepticism (called *tropoi*), which maintained that all judgments should be suspended because nothing is certain, and proving an assertion would require an endless process of proofs. Sextus Empiricus, a third century A.D. Greek physician and philosopher who wrote *Outlines of Pyrrhonism* and *Against the Dogmatists,* headed another school of skepticism. Centuries later during the Renaissance, Pyrrhonism reemerged in the writings of the sixteenth century French essayist and skeptic Michel de Montaigne, and the sixteenth century mathematician and philosopher Rene Descartes. Skepticism was made popular again in the eighteenth century by the Scottish philosopher and atheist David Hume, in the nineteenth century by the German philosopher Immanuel Kant, in the twentieth century by the German philosopher and linguist Friedrich Nietzsche and the American philosopher George Santayana. It is now fully incorporated in modern anti-Christian philosophy.

Another degenerate Greek philosopher who led many down the path of destruction was Epicurus of Samos (341-270 B.C.), who founded the philosophical school of Epicureanism. Epicureanism maintains that the greatest happiness and good in life is the pursuit of pleasures, particularly the pleasures of the mind and soul that derive from the pleasures of the flesh. The term "epicurean delights," which means love for gourmet food and luxurious pampering, is derived from philosophical Epicureanism. Epicurus suffered from ill health much of his life, so it is easy to see why he was so preoccupied with the pursuit of happiness through pleasures, which, unfortunately for him, also led him towards atheism. The fear of death and an afterlife punishment by the gods was the greatest evil according to Epicurus, whose conscience constantly worried him for the life he was leading. Religion, Epicurus said, was a damaging superstition that caused people unnecessary fear and worrying. Although he did not deny the possible existence of gods, Epicurus maintained that if they did exist they would have nothing to do with human affairs, except maybe when contemplating the lives of outstanding men.

Besides denying the existence of God, Epicurus also denied the existence of an eternal soul, saying that the universe consists entirely of matter and empty space composed of atoms of varying sizes and shapes, which move about randomly and are not under the control of any gods. Body and mind are likewise made up entirely of matter, which disperses after death, so there is nothing to fear in dying or having lived a life in the pursuit of carnal pleasure, because death is a state of nothingness.

The nihilistic despair of modern-day existentialism, which comes from the loss of hope in a better life, can be traced as far back as Epicurus. To Epicurus, the utilitarian pursuit of pleasure on earth is the highest good—the "beginning and end of the blessed life." While nothing is forbidden in one's pursuit of happiness, true happiness comes from friendships and the enjoyment of sensual pleasures, in moderation. Prudence is a virtue, as is justice and honesty, but marriage and family are a distraction from the higher pursuits of life, so we should forego love, politics, passionate activities, and gluttony and drunkenness for the pursuit of the more intellectually rewarding pleasures. Above all, avoid fear, which disturbs peace of mind. Don't ever worry about a vengeful god who doesn't exist or being sent to hell by the Almighty; we are the masters of our fate. "When we are, death is not; and when death is, we are not," Epicurus often said.

Born on the island of Samos to an Athenian family, Epicurus was home schooled by his father, a schoolteacher, and various philosophers. At eighteen, he was enlisted into military service at Athens and later rejoined his father in 322 B.C. In 311, Epicurus founded a school of philosophy in Mitilini on the island of Lesvos, and two years later he headed another school in Lampsacus (Lapseki, Turkey). In 306, he returned to Athens where he settled permanently and founded yet another school. Instructions to his friends and admirers often took place in the sumptuous gardens of his home, so his followers became known as "the philosophers of the gardens." Because both women and men frequented Epicurus' garden parties, rumors began circulating about activities of impropriety taking place on the premises. Students from all over Greece and Asia Minor began coming to Epicurus' school, attracted not only by his charm and wit but also by his liberal, free-thinking philosophy on life and death.

In matters regarding science, Epicurus was an atomist like Democritus and Leucippus. He regarded the universe as infinite, eternal, and uncreated, consisting only of atoms of varying sizes, shapes, weights, and space. The uncreated world results from the collision, aggregations, and dispersal of atoms. Like the philosophers of the Milesian school, Epicurus was a strict materialist, believing that material things cast off films or "idols" that are perceived by the senses. All information that comes to our senses are reliable, it is only in their interpretation that we make mistakes. The soul, likewise, is made up of fine particles that dissolute after death. It is therefore wrong to believe in an afterlife as moralists and believers teach. Natural forces give rise to the different species, and those species that are best able to propagate themselves survive.

Epicurus' influence on Western philosophy was substantial and long lasting, particularly amongst Greek and Roman statesmen. The Greek grammarian Apollodorus (fl.146-140 B.C.) was a prominent Epicurean, and the Roman statesman Pliny the Younger and the Roman poets Horace and Lucretius were all Epicureans. Other influential Roman writers from whom we obtain information about Epicurus were Cicero, Seneca, and Plutarch. The third century historian, Diogenes Laertius (200?-250 A.D.), who wrote a biography of Epicurus, claimed that Epicurus was a prolific writer who left three hundred manuscripts, including thirty-seven treatises on physics and numerous works on love, justice, the gods, and other subjects. Of these writings only three letters and a number of short fragments survive, all in Diogenes' biography.

Epicureanism as an organized school of philosophy went out of existence early in the fourth century A.D. because of the growing moralistic influence of the Roman Catholic Church, but reemerged in spirit in the seventeenth century by the French philosopher Pierre Gassendi and in the nineteenth century by the English utilitarian Jeremy Bentham, both advocates of whatever is pleasurable is what is good. Today, Epicureanism, with its hedonistic and pleasure seeking lifestyle, is as popular as it ever was in ancient Greece and Rome.

Ancient Roman Philosophy

Most of the ancient Greek philosophies were assimilated into neighboring Rome, but it was the philosophy of stoicism that was the most prevalent philosophy of the ancient Roman Empire. The stoic school of philosophy actually began in Greece, where it was founded in Athens around 300 B.C. by the Greek philosopher Zeno of Citium in Cyprus (c.335-263 B.C.). Zeno was a student of Plato's Academy and studied under the cynic philosophers Crates of Thebes and Stilpon of Megara. Zeno and his best students, Cleanthes of Assos in the Troad (area surrounding ancient Troy) and Chrysippus of Soli in Cilicia, represent the first period of stoicism (300-200 B.C.). The second period (200-50 B.C.) represents the incorporation of Greek stoicism into Rome. The period of Chrysippus was followed by Zeno of Tarsus and Diogenes of Babylonia; then Antipater of Tarsus, who taught Panaetius of Rhodes (c.185-109 B.C.). It was Panaetius who introduced stoicism into Rome. Among Panaetius's pupils was Posidonius (c.135-51 B.C.) of Apamea in Syria, who was the teacher of the great Roman orator and statesman Marcus Tullius Cicero (106-43 B.C.), commonly known as "Cicero." The third period of stoicism (50 B.C. to 200 A.D.) was entirely Roman in influence. The principal adherents of Roman stoicism were the statesman Cato the Younger (95-46 B.C.), Nero's tutor Lucius Annaeus Seneca the Younger, commonly known as "Seneca" (4 B.C.-65 A.D.), the Greek slave turned Roman philosopher known as "Epictetus" (c. 55-135 A.D.), and the infamous Roman emperor Marcus Aurelius (121-180 A.D.), one of the last great stoic philosophers of antiquity, and one of the ten worst Roman persecutors of Christians.[1]

Stoicism, the most influential school of philosophy in Rome prior to the introduction of Christianity, taught that the world was made up of passive matter, which is animated in creatures by an active, fiery-like material called the *Logos*. The soul is a manifestation of the *Logos*, which in turn emanates from the divine, who is the life soul of the world. God is not separate and transcendent from the world as Christians believe: he *is* the world. Each being, therefore, contains part of the divine fire in their souls. Thus, a healthy respect for nature and its natural laws was paramount to stoics.

Natural law played an important part in Roman philosophy, morality, ethics, and jurisprudence. The virtuous will, to Romans, is the will that lives in conformity to the divine order of nature and the universe. Natural law embodies the four cardinal virtues of wisdom, courage, justice, and temperance. The stoics disagreed with the Epicureans that the pursuit of pleasure was the cardinal virtue. Ethics was the main field of study in Roman stoicism, but stoics also made advances in logic and the natural sciences. Happiness, peace, and the good life comes not from external things such as health and wealth, but is a manifestation of the internal state of one's soul. Like Plato, stoics believed that true wisdom lies in the ability to deliver yourself from the passions and worldly desires that perturb the ordinary soul, and being able to live in harmony with others. All people, regardless of rank, position, or social status, are reflections of the divine soul, and should love and respect one another as equals. Even before Christianity, stoics believed in a brotherhood and equality of man, although in practice the Roman Empire was one of the cruelest and most bloodthirsty world powers in the history of mankind.

Stoics weren't concerned with otherworldliness; what was important to them was the here and now, not a belief in an afterlife reward or punishment. They rejected Plato's dualistic theory of the immortality of the soul that separates from the material body upon death (most stoics believed the soul perishes with the body). Everything that happens on earth is governed by divine providence, even if it is a tragedy; so one shouldn't despair or worry over misfortunes or one's lot in life. One should only be concerned with doing what is right and virtuous.

The previously mentioned Greek philosophy of atomism put forth by Democritus, Leucippus, and Epicurus was similarly championed by the famous Roman poet and philosopher Titus Lucretius Carus (c. 99-55 B.C.), commonly known as "Lucretius." Lucretius is best remembered for his great didactic poem in six books called *On the Nature of Things*. In it Lucretius presents a materialistic viewpoint on the origin of the world similar to the theories of the Greek atomists. Lucretius believed that the universe is made up of atoms moving about in a void, that the soul is material like the body and not immortal, that there is no afterlife reward or punishment, and that all events on earth are by mere chance and not directed by any god. Lucretius tried to relieve people of the fear of divine punishment, and, like Epicurus, considered this fear the main cause of human unhappiness. If the gods exist, Lucretius said, they have no concern with mortals. Lucretius, however, was unable to convince people of his atheistic materialism, as witnessed by the tremendous rise in Rome a century or so later of a popular new religious sect called Christianity.

Christianity began gaining substantial ground on pagan philosophy and practices by the end of the second century A.D., and was about to completely replace it at the beginning of the fourth century, in spite of ten terrible persecutions of Christians at the hands of the pagan Roman emperors Nero (64-68), Domitian (95-96), Trajan (106-117), Marcus Aurelius (161-180), Septimus Severus (202-211), Maximin the Thracian (235-238), Decius (249-251), Valerian (257-260), Aurelian (274-275), and Diocletian and Galerius (303-311). A last ditch effort to stem the rising tide of Christianity and bring back the glory days of ancient Greek and Roman philosophy was made in the third century by the introduction of Neoplatonism into Rome. Neoplatonism actually began in Alexandria, Egypt, in the third century A.D., and was ushered into Rome about 244 by the Roman philosopher Plotinus. Plotinus (205-270) was born and raised in Egypt and studied philosophy at Alexandria with the philosopher Ammonius Saccus before moving on to Rome and founding a school of Neoplatonism. Other notable Neoplatonists after Plotinus were the Syrian-Greek scholar and philosopher Porphyry (232-304), the Syrian-Greek philosopher Iamblichus (250-330), and the Greek philosopher and mathematician Proclus (410-485), the last important Greek philosopher.

Plotinus preached the ascetic and metaphysical philosophies of Plato and Pythagoras, and through his influence convinced many a Roman to forego their riches, set their slaves free, and take up the study of philosophy and the practice of asceticism and piety. With the permission of the emperor Gallienus, at age sixty Plotinus planned to establish a communistic-style commonwealth on the model of Plato's *Republic,* but the project failed because of strong opposition. Plotinus continued to lecture and write, however, and by the time of his death he had written fifty-four treatises in Greek called the *Enneads.*

Plotinus' Neoplatonism was based primarily on Plato's philosophy of forms or ideas, but it was modified somewhat to include his own theory of emanation. Plotinus believed that power continually emanates from the Absolute Being or the "One," to the whole of creation, through *nous* or pure intelligence. From *nous* flows the soul of the world; and from it flows the souls of all creatures and material things. Human beings, therefore, live in two worlds: a world of matter and the senses and a spiritual world of pure intelligence. Man should forego the material world, because matter is evil, and pursue the world of pure intelligence through meditation and contemplation of the infinite, unknowable, perfect One. There are three universal principles according to Plotinus: the One or Good is the highest, the Intellect is the next highest, and the Soul is the lowest.

Plotinus believed that the world was "beamed" from the intelligible "there," and that souls descended from the "there" and will someday ascend back. Souls have high and low parts. The highest part is linked to the Intellect "there." In life, the "here," we choose what part of our soul we like best, the high or low part, and that choice determines our afterlife as plants, animals, demons, or gods. A soul that has chosen its lower sensual part rather than its higher spiritual part can still obtain salvation by reversing course and seeking union with the One.

Porphyry, whose original name was Malchus, was a Greek scholar who grew up in Palestine where he studied under Cassius Longinus. Porphyry went to Rome, where he studied Neoplatonism under Plotinus. Unlike Plotinus, who didn't say much about Christianity, Porphyry was vigorously opposed to it. In his defense of paganism, Porphyry wrote *Adversus Christianos.*

The third important Neoplatonist was Iamblichus, a Syrian philosopher and a student under Porphyry in Rome. Iamblichus established a school of Neoplatonism in Syria, where he combined selected elements of Plato, Pythagoras, Plotinus, and Oriental mysticism, turning them into a mystical, pagan religious philosophy. Among his works are *On the Egyptian Mysteries* and *On the Pythagorean Life.*

Proclus was the last important Greek philosopher of antiquity and an important member of the Athenian school of Neoplatonism. Born in Constantinople, Proclus studied in Alexandria, Egypt, with the Greek philosopher Olympiodorus, and later joined the Academy in Athens, eventually becoming its director. Proclus was just as opposed to

Christianity as Iamblichus, and like Iamblichus, practiced a mystical, magical form of Neoplatonism. His surviving works include *Elements of Theology* and *Platonic Theology*.

The Christian Philosophy of St. Augustine: Doctor of the Church

The writings and teachings of the aforementioned Greek and Roman philosophers, at least those who believed in a supreme being and a God-given soul, helped pave the way for the replacement of paganism with Christianity in the Roman Empire. In 313 A.D., Roman Emperor Constantine the Great signed into law his Edict of Milan, which for the first time officially recognized Christianity as a lawful religion within the Roman Empire. With the blessing of Constantine and his successors, many apologists and early church fathers contributed to the rise in popularity of Christianity within the Roman Empire, but the greatest of them all was the famous Christian philosopher known as St. Augustine of Hippo (354-430), Doctor of the Church.

St. Augustine was born on November 13, 354, at Tagaste (now Souk-Ahras, Algeria), a small village in Numidia not far from Hippo (now Annaba, Algeria). His father, Patricius, was a pagan of violent temper, who converted to Christianity before his death, but his mother, St. Monica, was a faithful, good-hearted Christian who prayed endlessly for the conversion of her husband and son. Augustine had been instructed in the teachings of Christianity as a child by his mother, but chose to live the pagan life because it was easier and more pleasurable. Augustine recorded years later in his *Confessions* that his motto was "Give me chastity and continence, but not just now."At seventeen, Augustine left home and traveled to Carthage in the year 370. There he enrolled in a school of rhetoric. Somewhere between the ages of fifteen and thirty, Augustine moved in with a Carthaginian woman whose name is unknown. In 372, she bore him a son, whom he named Adeodatus, which is Latin for "the gift of God." But around the year 385 while living in Milan, Augustine sent his lover away, even though she had remained faithful to him all that time.

In Carthage around the year 371, Augustine read Cicero's *Hortensius,* and turned his interest from rhetoric to philosophy. For the next nine years, from 373 until 382, Augustine was a devotee of Manichaeism, a theological and philosophical heresy that maintained the existence of two gods: a good god who was the cause of all good, and an evil god who was the creator of all matter which was evil. Manichaeism was less strict in morality than Christianity and was easier for Augustine to follow. After meeting the leading Manichaean teacher, Faustus, in 383, Augustine became disillusioned with Manichaeism, and decided to secretly move to Rome without his mother's knowledge to open a school of rhetoric there. After being cheated out of his wages in Rome, Augustine moved to Milan where he received a post as master of rhetoric. While in Milan, Augustine came under the influence of the charismatic and fiery Bishop Ambrose of Milan, whose sermons on Christianity, which he attended out of curiosity and respect, began to move his spirit towards Jesus Christ. Studying the spiritual teachings of Plato and Plotinus also began to move Augustine closer to Christianity.

St. Monica followed her son to Milan, wishing to see him married, but the mother of Augustine's child moved back to North Africa without marrying Augustine, leaving Adeodatus behind with his dad. Though stirred by Christianity, Augustine, still young and handsome, continued to struggle with sexual promiscuity. Upon hearing of the conversion to Christianity of the Roman Neoplatonist professor, Victorinus, and after a visit by his Christian friend Pontitian from Africa, who read to him books about the lives of St. Paul and St. Anthony, Augustine went to his garden one day and threw himself on the ground, lamenting his past sexual iniquities: "How long, O Lord?" he said, "Wilt thou be angry for ever? Remember not my past iniquities." Immediately he heard a voice like a child, singing from a neighboring house, over and over again, "Take up and read! Take up and read!" Augustine, believing this was a message from God to read the Sacred Scriptures, returned to where his friend Alipius was sitting with the book of St. Paul's epistles, opened it, and read in silence the first words that came to his eyes from Rom.13: 13-14: "Let us live honorably as in daylight; not in revelry and drunkenness, not in debauchery and licentiousness, not in quarreling and jealousy. But put on the Lord Jesus Christ, and make no provision for the flesh, to gratify its desires."

Augustine immediately gave his life to Jesus, as did his friend Alipius. Upon telling his mother that he now accepted the Lord, she wept with joy and gave praise to God for having answering the prayers that she had been saying for so many years for his conversion. This happened in September of 386. Bishop Ambrose baptized Augustine and his son on Easter eve in 387, and his mother and son would die not long afterwards.

After his conversion, Augustine gave up his school and retired to a country house near Milan, accompanied by his mother, brother, son, Alipius, and several other friends, where they all lived a communal life in service to Jesus Christ. It was during this period in Milan that, from the discussions that took place at the communal home, Augustine wrote three dialogues: *Against the Academicians, Of the Happy Life,* and *Of Order.*

Augustine planned to return to Africa, but while in Ostia his mother died in November 387. Augustine returned to Rome for a short while, and then left for Africa in September 388. He lived a communal life in Tagaste for about three more years, fasting, meditating, teaching, and doing good works. In 391, Augustine was ordained a priest and served as an assistant to Valerius, Bishop of Hippo. While in Hippo, Augustine established a monastery where he lived with St. Alipius, St. Evodius, St. Possidius, and others. During this period of time, Augustine rigorously defended Christianity against the heresies of Manichaeism and Donatism. Donatists maintained that baptisms administered by heretics is invalid, that re-baptism was necessary for salvation, and that sanctity was a requirement for church membership.

In 395, Augustine was consecrated a bishop and succeeded Valerius as Bishop of Hippo not long afterwards. Priests and deacons of Hippo were obliged to follow the communal rules of Augustine by renouncing ownership of property. From 395 to 405, Augustine was involved in a bitter and protracted battle with the Donatists and the outrages they committed. Emperor Honorius published severe laws against the Donatists, up to the death penalty, which Augustine was against. In the course of this conflict, which was long and bitter, Augustine developed his doctrines of original sin and divine grace, divine sovereignty, and predestination. The Council of Carthage in 411 marked the beginning of the end for Manichaeism and Donatism, but another heresy called Pelagianism sprung up their place. Pelagius believed that man, by his own natural powers and free will, could be sinless and become like a son of God, that Adam's sin was purely personal and didn't pass the stain of original sin on to all humanity, that we are born free of sin like Adam and Eve, that man can do anything God can do, that we don't need a redeemer because the Mosaic Law is sufficient to save us, that there were men who were sinless before Christ, that we don't need God's grace for salvation, and that St. Peter had no authority over the other apostles. The Council of Ephesus finally condemned Pelagianism in 431.

In the year 410, Alaric the Visigoth sacked Rome, and the barbarians attributed the calamities of the Roman Empire to Christianity. To counter these blasphemous slanders against God, Augustine began his greatest work, *Of the City of God* (413-426), in which he formulated a theological and philosophical history of the world, with Christianity as the natural and foreordained successor to Greek and Roman paganism. As a prolific writer on theology, history, and philosophy, Augustine was every bit as talented as St. Paul the Apostle. One of his best-known works is his autobiographical *Confessions* (c. 400), in which he confessed his earlier life of immorality and his subsequent conversion to Christianity. In 428, Augustine wrote the *Retractions*, in which he corrected some of the mistakes in his earlier writings. His other writings include the *Epistles*, of which 270 are in the Benedictine edition, variously dated between 386 and 429; his treatises *On Free Will* (388-95), *On Christian Doctrine* (397), *On Baptism: Against the Donatists* (400), *On the Trinity* (400-16), and *On Nature and Grace* (415); as well as homilies upon several books of the Bible. The Roman Catholic Church and the major Protestant denominations both hail St. Augustine as one of the most influential doctors of orthodox Christianity. His influence on the spread of Christianity is immeasurable.

Medieval Philosophers of Europe and the Middle East

The perpetual myth that the Middle Ages were the "Dark Ages" because culture and learning precipitously declined after the triumph of Christianity has been exploded time and again by scholars and historians. In the *scriptoria* (writing rooms) of European monasteries, the famous works of Latin and Greek writers, philosophers, scientists, and physicians—such as Virgil, Ovid, Cicero, Seneca, Galen, and Celsus—were preserved and taught by well-educated monks. Modern advances in medicine, mathematics, and astronomy are indebted to medieval scholars and monks in such historic sites as the Benedictine monastery at Monte Cassino, which was founded in the sixth century and contained an extensive library on medicine and science. Other noted centers of education included the famous medical schools of Salerno, Montpellier, Paris, Oxford, Padua, and Bologna. There was no such thing as the Dark Ages.

The fifth century A.D. witnessed the sack of Rome in 410 by the Germanic tribes of Alaric I, and the reign of the last Roman emperor in the West in 476, Romulus Augustulus. The barbarian invasion of the Western Roman Empire changed forever the economic, political, and military makeup of Europe, with the coalescing of the primitive tribes of the Franks, Gauls, Goths, and Visigoths into competing kingships on the lands that were once occupied by the armies of the Roman Empire. With the coming of the barbarian invasion, the well-regulated commerce of the Roman Empire was replaced by a more primitive feudal system of peasant and landlord without a centralized political authority. The only centralized authority that maintained its influence and actually grew in importance during the Middle Ages was the Catholic Church, which established numerous monasteries throughout Europe, especially in Ireland where missionaries evangelized the pagan barbarians and converted them to Christianity. Thanks to the evangelization efforts and hard work of the Catholic Church, the barbarians of Europe became civilized and converted to Christianity.

The Eastern Roman Empire managed to survive the barbarian invasion because the Roman emperor Constantine the Great had moved his capitol to Constantinople—now Istanbul, Turkey—in 324. Like the Academy and Lyceum, all other pagan philosophical schools in Rome were closed. Both non-Christian and Christian philosophers, however, still made an impact on Western education during that period. One non-Christian philosopher from Rome named Anucius Manilus Severinus (c. 480-524), commonly known as "Boethius," wrote *The Consolation of Philosophy* in 523, which contained many elements of Christian ethics and was highly respected in Europe during the or Middle Ages Boethius was translated many times over, especially in England by King Alfred the Great and the poet Geoffrey Chaucer. His works contained numerous writings on logic, music, arithmetic, and theology, in addition to commentaries on Cicero, Porphyry, and Aristotle. What little Western philosophers knew of Aristotle during the early medieval period came largely from the translations of Boethius.

In addition to Boethius, John Philoponus, also known as "John the Grammarian," was a sixth century Greek philosopher from Alexandria who helped Christianity replace Neoplatonism. Philoponus authored two treatises on grammar, helped reconcile Aristotle with Christianity, and wrote *Diaitete,* an attempt to reconcile monophytism (the belief that God is one person rather than three) with orthodox Christianity. Another important scholar was Simplicius of Cilicia (c.530 A.D.), the last major Neoplatonist who authored commentaries on Aristotle and the *Encheiridion* of Epictetus.

In spite of the rapidly growing influence of the Church, pagan ideas and unorthodox beliefs managed to surface now and again, bringing the Church into conflict with heretical philosophers for their attacks on orthodox Church doctrines. One particular Scots-Irish philosopher named John Scotus Erigena (815?-877?), the supervisor of the court school of King Charles I, was condemned by the Church at the Council of Sens (1225) for his pantheistic work entitled *Concerning the Division of Nature,* which was written between 865-870. In his *Division of Nature,* Erigena rejected the Christian doctrine that God created the world out of nothing, and believed that space and time were only manifestations of the mind of God, and that reason supercedes Church authority. Erigena's treatise *Concerning Divine Predestination,* which denied the eternal damnation of souls and God's total control over our salvation, was condemned by the councils of Valence (855), Langres (859), and Vercelli (1050). Erigena was also believed to have rejected the doctrine of the real presence of Christ in the Eucharist. Although a heretic, Erigena was highly regarded for his knowledge and translations from Greek into Latin of a number of ancient philosophical works, particularly the Neoplatonic works of Dionysius the Areopagite.

Another famous philosopher and teacher whose unorthodox views got him into trouble with the Church was Peter Abelard (1079-1142?). Abelard was a French philosopher and early scholastic theologian who gained fame for his love affair with a young woman half his age. At age forty, Abelard fell in love with his student Heloise, age twenty, the niece of the canon of Notre-Dame in Paris. Heloise bore Abelard a son whom they named Astrolabe, and they were secretly married soon afterwards. Persuading Heloise to enter a convent and become a nun because he had not yet received his major orders, Abelard infuriated the canon of Notre-Dame, who felt that Abelard had abandoned Heloise at the abbey, so he had Abelard castrated. Abelard retired to a religious retreat at the Abbey of Saint-Denis-en-France, in Paris, where he began to write. Abelard's first published work, a treatise on the Trinity (1121), was condemned and ordered burned by a Roman Catholic council that met at Soissons in the same year. Forced to leave St. Denis-en-France, Abelard founded a chapel and oratory, called the Paraclete, at Nogent-sur-Seine. In 1125, Abelard was elected abbot of the monastery at Saint-Gildas-de-Rhuis. Abelard arranged for Heloise to become the abbess of the convent there, so that she could be near him. Their exchange of love letters became classics of romance. Abelard's philosophy was that of a skeptic and rationalist. In 1140, St. Bernard of Clairvaux, an eminent French ecclesiastic who thought Abelard's teachings were dangerous, persuaded bishops to support his charge of heresy against Abelard. Abelard appealed to Pope Innocent II (r. 1130-43), but to no avail. Although judged a heretic, Abelard's contributions to Western philosophy included important treatises on ethics, theology, and dialectics.

Islamic philosophers also contributed a great deal to Western philosophy and learning during the Middle Ages. Yaqub ibn Isahaq as-Sabah al-Kindi from Baghdad (801-873), known as "al-Kindi," was one of the first Arab students of Greek philosophy and one of the first Arabs to translate Aristotle into Arabic. Highly influenced by Aristotle and Neoplatonism, al-Kindi authored over 270 works on subjects such as philosophy, medicine, mathematics, optics, and astrology. Some of his writings were translated into Latin during the Middle Ages and greatly impacted Western scholars. Al-Kindi, however, disagreed with Aristotle that the universe was eternal, believing instead that the universe had a beginning. He also believed that revelations from God took precedence over human reason. Al-Kindi wrote *On First Philosophy,* the first Arabic work on the philosophy of metaphysics.

Muhammad ibn Muhammad ibn Tarkhan ibn Uzalagh al-Farabi (c. 873-950), commonly known as "al-Farabi" or "Alfarabius" in Latin, was an Islamic philosopher from Uzbekistan who studied in Khorasan (now in Iran) and then in Baghdad, and eventually lived at the court of Sayf ad-Dawlah (916-967), the ruler of Aleppo (now in Syria). Educated by Christian Syrians, al-Farabi was one of the earliest Islamic philosophers to teach the Arab world the doctrines of Plato, Aristotle, and Plotinus, thereby greatly influencing later Islamic philosophers such as Avicenna and Averroes, who in turn reintroduced the teachings of Aristotle to Europe during the time of the Crusades. Al-Farabi was the first Islamic philosopher to believe that philosophy was superior to revelation, and that despite the differences of the world's religions, philosophical truths are universal. Al-Farabi dreamed of a universal religion, which used the truths of philosophy as its foundation. Authoring over one hundred works, many of al-Farabi's works were lost, but many others have been preserved in Latin translations.

Abu Ali al-Husayn ibn Abd Allah ibn Sina (980-1037), commonly known as "Avicenna," was an important Iranian Islamic philosopher and physician. He was born near Bukhara (now in Uzbekistan) and became court physician to the Samanid ruler of Bukhara, where he held that post until 999, when the Turks defeated Persia. The last years of his life were spent as the scientific advisor and physician to the ruler of Isfahan. Avicenna authored over two hundred books on science, medicine, philosophy, and religion. His encyclopedic book *The Canon of Medicine* was highly regarded in the Middle East and Europe as the premiere textbook on science and medicine. Avicenna's most popular philosophical work was *Book of Healing,* a collection of treatises on Aristotelian logic, metaphysics, psychology, the natural sciences, and other subjects. Combining elements of Aristotelianism and Neoplatonism, Avicenna denied the immortality of the soul, the creation of the world in time, and a God involved with the lives of human beings, which brought heavy criticism by the orthodox Muslim al-Ghazali. Much of Aristotle's philosophical works were lost in Europe until the time of the Crusades, when Avicenna's writings on Aristotle were found. Aristotle's philosophical logic presented a major challenge to the Catholic Church, which saw his materialistic viewpoints in contradiction to the accepted Augustinian theology. At one point the Church even banned the study of Aristotle, but Thomas Aquinas later reconciled Aristotle with Christian theology.

Abu Hamid Muhammad ibn Muhammad at-Tusi al-Ghazali (1058-1111), also referred to as "al-Ghazali" or "Algazel" in Latin, was an Islamic philosopher and theologian from Persia. He was appointed in 1091 by Nizam al-Mulk (1018-1092), vizier to the Seljuk sultan, to teach at Nizamiya University in Baghdad. In 1095, after a personal crisis of faith, al-Ghazali left his job and family to become a wandering ascetic. Ten years later he found the answers he was searching for and accepted another teaching position in Nishapur. His work *The Deliverance from Error,* like St. Augustine's *Confessions,* told the story of his personal crisis of faith. Al-Ghazali's most important work, *The Revival of the Religious Sciences,* was considered to be the second greatest book in the Muslim world, second only to the Koran. Al-Ghazali refuted Avicenna's Neoplatonism and Aristotelianism, which denied the creation of the world, the immortality of the soul, and a personal God who was involved with the lives of men. His *Destruction of the Philosophers* was an attack on rationalistic philosophy, as were his subsequent works *Intentions of the Philosophers* and *Incoherence of the Philosophers.*

Abu al-Walid Muhammad ibn Ahmad ibn Muhammad ibn Rushd (1126-98), also called "Averroes," was a Spanish-Arab Islamic philosopher, theologian, jurist, scientist, and physician, born in Cordoba, Spain. Averroes' monumental commentaries on the works of Aristotle and Plato were translated into Latin and Hebrew and, like Avicenna before him, greatly influenced medieval European and Jewish philosophy. Like Aristotle, Averroes denied the creation of the world, believed in a Prime Mover, and held that the individual soul emanates from a universal soul. Averroes was a rationalist like the Greek philosopher Parmenides of Elea, believing that reason is superior to religion, which led to his exile in 1195 by Abu Yusuf Yaqub al-Mansur. Averroes also believed that metaphysical truths could be explained by both philosophy and religion, but that theologians were inferior to philosophers in interpreting divine truths. His greatest work was *Incoherence of the Incoherence,* a rebuttal of the attacks on Neoplatonism and Aristotelianism by the Islamic theologian al-Ghazali.

Also influential to Western philosophy were the writings of Jewish scholars and philosophers. One important Jewish philosopher was Solomon ben Yehuda (1021?-58? B.C.), also known as "Ibn Gabirol" or "Avicebron" in Latin. Gabirol was a Spanish Jewish philosopher and poet, born in Malaga and educated in Saragossa. Gabirol wrote *Fountain of Life,* a Neoplatonic dialogue on the universality of matter, which was translated into Latin and called *Fons Vitae. Fons Vitae* was thought to have been written by a Christian philosopher, and was praised by the Scottish philosopher John Duns Scotus. Thomas Aquinas, however, was critical of its Neoplatonism.

The most important Jewish philosopher of the Middle Ages was Rabbi Moses ben Maimon (1135-1204), also called "Moses Maimonides" or, from the initials of his name, "Rambam." Maimonides was a Jewish philosopher and physician from Cordoba, Spain. After the capture of Cordoba in 1148 by the Muslim Almohads, who forced Jews and Christians in Spain to convert to Islam, Maimonides and his family emigrated to Cairo, Egypt. There he became the chief rabbi of Cairo and physician to Saladin, the sultan of Egypt and Syria who captured Jerusalem from the Crusaders. Maimonides' greatest works on Jewish law and faith were the *Mishneh Torah,* which was written in Hebrew and combined into fourteen books from 1170-1180, and his *Thirteen Articles of Faith.* Maimonides' works were so well regarded by Jewish scholars that they called him the "Second Moses." Philosophically, Maimonides greatly influenced Western philosophers by trying to synthesize rabbinic Judaism with Arabic Aristotelianism in his work *Guide for the Perplexed,* which also contained elements of Neoplatonism. Maimonides believed that pagan faiths were primitive, superstitious, and perverse. Christianity and Islam, Maimonides said, were derived from Judaism and have spread the belief in monotheism and the coming messianic age. Maimonides, however, did not believe that Jesus Christ was the Messiah. Maimonides had a profound influence on the thirteenth century scholars Thomas Aquinas and Albertus Magnus in their development of scholastic philosophy, which became the most important school of philosophy until the beginning of the Renaissance in the late 1300s.

The Thomist Philosophy of Scholasticism: Christianized Aristotelianism

Scholasticism was a major philosophical and theological movement that dominated Western thought in medieval schools and universities from the eleventh to the fifteenth century. It was named after the scholars of the major monastic and cathedral schools of learning (the first true universities in Europe) who attempted to synthesize Christian theology and Augustinian philosophy with the materialistic and rationalistic philosophy of Aristotle, whose major works had just recently been rediscovered in the Latin translations of the Arab philosophers

Avicenna and Averroes. Scholastics believed that a harmony could exist between human reason and divine revelation, that the God of Aristotle and other important Greek philosophers such as Plato and Socrates was the same as the God of Christianity, and that any disagreements between reason and revelation were because of an improper use of reason or an incorrect interpretation of revelation. Because it is the direct communication of God to man, revelation is superior to human reason. Thus, in any conflicts over truth between the philosopher and the theologian, the theological truths would always prevail over philosophical truths.

Scholastic theologians such as Thomas Aquinas valued philosophy as a way to better understand revelation. They considered Aristotle the chief authority in philosophy and science, calling him simply "The Philosopher." St. Augustine, on the other hand, was the scholastic's chief authority in Christian theology, subordinate only to the Bible, the teaching magisterium of the Church, and the *Sententiarum Libri Quatuor (Four Books of Sentences).* The latter, written by the twelfth century Italian theologian and prelate Peter Lombard, was a collection of the opinions of the early church fathers on problems in theology. The first scholastics such as Abelard were nothing more than compilers of the words of the great philosophers, while the later scholastics such as Albertus Magnus and Thomas Aquinas were skilled commentators and debaters on the works of the great ancient philosophers and theologians. Scholastics such as Magnus and Aquinas used the disputational method of appearing before faculty and peers to debate the complexities of Christian theology and Aristotelian philosophy. After the middle of the fourteenth century, the verve of public disputation as an educational tool waned, and scholasticism became more of a rigid and boring formalism. Scholastic debaters became concerned less with real content and more with trivial points of logic and unimportant minutia. Trivial disputations gave scholasticism a bad reputation amongst Renaissance intellectuals and scholars. This, along with scholasticism's rigid defense of Aristotle's empiricism, eventually caused it to be discredited in the fifteenth and sixteenth centuries. Some of Aristotle's scientific theories, such as his mistaken belief that the universe evolved around the earth, were proved to be erroneous.

The "Founding Father of Scholasticism" was St. Anselm (1033-1109), archbishop of Canterbury, who is credited with being the most important theologian since St. Augustine. Anselm was born to a wealthy family in Aosta, northern Italy, and applied at age fifteen to the local monastery, but was turned down. In 1060, at age twenty-seven, he joined the Benedictine monastery at Bec, Normandy, where the English prelate Lanfranc was prior. Some time later, after Lanfranc was called to England to become archbishop of Canterbury, Anselm was elected abbot of Bec. In this role, Anselm gained a reputation amongst his fellow monks as a person of deep piety and wisdom. The monks of Bec insisted that Anselm write out the meditations he taught them, so he composed *Monologion* (1077), in which he provided metaphysical proofs for the existence and nature of God. He also wrote *Proslogium* (1078), a contemplation of God's attributes as well as a treatise on truth, freewill, the origin of evil, and the art of reason. Anselm provided such strong arguments for the existence of God that they are still being argued over today. Anselm maintained that even those who deny the existence of God would have to have some understanding of what they are denying. Those who say there is no God know exactly in their mind who he is: the Most High living being who is above all things and which nothing greater can be thought of. According to St. Anselm, any nonbeliever who denied the existence of God would contradict himself because he is implying that something greater exists above which nothing greater can be thought of in the mind; so by necessity, God has to exist.

In 1093, with the death of Lanfranc, Anselm reluctantly became the archbishop of Canterbury. His election to the post soon brought him into direct confrontation with King William II. They debated who was in control of the abbeys and monasteries in Normandy: the king or the archbishop (the Investiture Controversy). Anselm was forced to go to Rome to plead his case before the pope, who assured him of his support. While in exile, Anselm completed one of the greatest works on the incarnation of Jesus Christ, the *Cur Deus Homo (Why God Became Man)*. Upon the death of King William II and the succession to the English throne of Henry I in 1100, Anselm returned to Canterbury amid rejoicing of the people. Anselm, however, continued to struggle with the new king over control of the Church and the payment of investitures, resulting in another period of exile for Anselm. Anselm finally returned to Canterbury in 1106, where he died on April 21, 1109. He was canonized in 1163 and declared a Doctor of the Church in 1720. His feast day is April 21.

Another important scholastic philosopher was St. Albert the Great (1206-80), also known as "Albertus Magnus." St. Albert is recognized for his voluminous works in the natural sciences and for his introduction of Greek and Arabic science and philosophy into Europe. A Swabian by descent, Albert was born into the noble family of Bollstädt at the castle of Lauingen, Bavaria, on the Danube in 1206. Albert joined the Dominican Order of Preachers of Padua in 1222, without the approval of his family. Upon hearing that the count of Bollstädt was planning to forcibly remove him from the order, Albert moved to another friary in Cologne, Germany, where he was then ordained. Albert taught for several years in various locations throughout Germany before going on to the University of Paris, where he became a master of theology in 1245, and subsequently held one of the Dominican chairs of theology. Albert was an influential teacher, church administrator, and pastor. He traveled through Western Europe and served as a provincial and, briefly, as bishop of Regensburg (1260-62) before returning to teaching and scientific research.

A preeminent authority on physics, geography, astronomy, mineralogy, botany, physiology, chemistry, and biology, St. Albert's writings fill thirty-eight quarto volumes in print. In studying the writings of the ancient Greek philosophers, Albert saw the value of Aristotle's analytical method of scientific knowledge as useful to Church theology. The full body of scientific and philosophical writings of Aristotle had been unknown to Western philosophers until they were reintroduced into Europe via Latin translations of Arabic scholars at the time of the Crusades, and had to be dealt with by the Church, which was only familiar with the logic of Aristotle and saw Aristotle's scientific theories on matter and his emphasis on natural reason over revelation a threat to the accepted doctrines of Christianity. St. Albert, however, was able to rewrite Aristotle in such a way as to make him more acceptable to his Christian critics, but it would be Albert's most famous pupil, St. Thomas Aquinas, who would completely reconcile Aristotelianism with Christianity. Albert died at Cologne on November 15, 1280. He was beatified in 1622 and declared a saint by Pope Pius XI in 1931, at which time he was acclaimed an official Doctor of the Church. In 1941, Pope Pius XII made him the patron of all who study the natural sciences. His feast day is November 15.

St. Thomas Aquinas (1225-74) was the most important figure in scholastic philosophy and the most important Roman Catholic theologian who ever lived. Because of his tremendous scholarly contributions to scholasticism, he is referred to by the Church as the "Angelic Doctor" and the "Prince of the Scholastics." Thomas Aquinas was born of a noble family in Rocca Secca, near Aquino, Italy. His father Landulf was a knight, and his mother Theodora was of Norman descent. Thomas, being big in stature, looked more northern in appearance than Italian. Thomas was educated at the Benedictine monastery of Monte Cassino and at the University of Naples. In 1243, the year of his father's death, he joined the Dominicans while still an undergraduate. His mother was totally against her son joining a mendicant order, so she had him captured and imprisoned in the family castle at Monte San Giovanni for two years. While in confinement, Thomas read the *Sentences* of Peter Lombard, memorized much of the Bible, and is said to have written a treatise on the fallacies of Aristotle. In 1245, his mother finally realized that her attempts to keep her son away from the Dominicans were in vain, so she released him; whereupon Thomas immediately journeyed to Paris to rejoin his fellow Dominicans and continue his studies. At the University of Paris, Thomas studied under the Dominican Albert the Great, and then relocated to Cologne with Albert in 1248. Because Aquinas was rather quiet at disputations at first and heavyset, his fellow students nicknamed him "The Dumb Ox." St. Albert, however, knew this kid's potential, and said in reply to the other novices that "this ox will one day fill the world with his bellowing."

Aquinas was ordained a priest in 1250, and he began to teach as a bachelor at the University of Paris in 1252. His first writings were compilations of his lecture notes, and his first major theological work was a treatise on the writings of Peter Lombard's *Sentences* called *Scripta Super Libros Sententiarum (Writings on the Books of the Sentences).* In 1256, Aquinas received the doctor's chair in theology and was appointed professor of philosophy at the University of Paris. It was about this time that St. Thomas Aquinas began to write his brilliant work, *Summa Contra Gentiles.*

Summa Contra Gentiles was originally written to persuade Muslim intellectuals of the truth of Christianity and as a rebuttal to the followers of the Islamic philosopher Averroes, who believed that philosophy was independent of and superior to revelation. Averroes was a proponent of Aristotle and wrote a great deal on Aristotelian philosophy, the works of which were translated into Latin during the thirteenth century. From the writings of Averroes, it was evident that Aristotle was an empiricist and believed in the superiority of human understanding through sensory experiences rather than spiritual revelations. The clearness, energy, and authority of Aristotle's materialism and empiricism, as presented by Averroes, posed major problems for the Catholic Church: it couldn't just brush Aristotle aside or ignore him altogether because too many Europeans believed in the superiority of the senses in gaining wisdom, particularly the Frenchman Siger de Brabant and his fellow Averroists. The heresy of Averroism, therefore, which maintained that philosophy was independent of revelation, was a threat to the unity and supremacy of Church doctrine. It had to be dealt with if Christianity was to survive. Albert the Great and other scholastics had attempted to deal with Averroism, but with little success. Aquinas, however, succeeded brilliantly. Reconciling St. Augustine's emphasis upon the human spirit as the major source of wisdom and truth with Averroism's claim of the primacy of the senses over spirit for obtaining knowledge, Aquinas insisted that the truths of Christian faith, as presented by Augustine, and those of sensory experience, as presented by Aristotle, are fully compatible and complementary. Some truths, such as the mystery of God's incarnation and his resurrection from the dead, can only be known through divine revelation, while other truths, such as the scientific understanding of the material world, can only be understood through sensory experiences; still others truths, such as God's existence, can be known by both equally. All knowledge of the truth, Aquinas said, originates in sensory experiences, but sensory data can only make sense by the actions of the intellect, which elevates thought toward the contemplation of spiritual things. To reach understanding of the highest truths, those with which religion is concerned, the aid of divine revelation is absolutely necessary. Aquinas' brilliant synthesis of Aristotelianism and Christianity in *Summa Contra Gentiles* and his later work *De Unitate Intellectus Contra Averroistas* turned the tide against the Averroists, who were subsequently condemned by the Church as heretics.

From 1259 to 1268, St. Thomas returned to Italy where he was made preacher general, and acted as adviser and lecturer to the papal court. In 1266, he began composing the most famous of his works, the *Summa Theologiae (Summary Treatise of Theology),* which was divided into three parts (on God, the moral life of man, and Christ), of which the last was left unfinished. In *Summa Theologiae,* Aquinas said that there are two ways to know God: through reason and through revelation. By our observance of motion in the universe we can reason that God exists. That "God Exists" is self-evident in itself; from causes we can infer effects, and vice versa. God is the cause of all being, goodness, and perfection. God orders things in nature for the sake of an end, a purpose that may not be evident to us. God knows all things and has knowledge of things before they existed. God knows every act we will ever perform even before we do it. And even though God knows our destiny, we make our own destiny through our freewill.

In 1269, Thomas Aquinas returned to Paris, then left Paris in 1272 and proceeded to Naples, where he organized a new Dominican school. On December 6, 1273, the feast of St. Nicholas, while celebrating Mass, St. Thomas received a revelation from God, which affected him so much that he wrote and dictated no more. "The end of my labors is come," he said. "All that I have written appears to be as so much straw after the things that have been revealed to me." In March 1274, while traveling to the Council of Lyon, to which he had been commissioned by Pope Gregory X (1210-76), Aquinas fell ill. He died on March 7,

1274, at the Cistercian monastery of Fossa Nuova near Terracina. Thomas Aquinas was canonized by Pope John XXII in 1323 and proclaimed a Doctor of the Church by Pope Pius V in 1567. In 1880, Pope Leo XIII declared him the patron of all universities, colleges, and schools. The greatest Catholic theologian of the medieval period, St. Thomas Aquinas wrote more than eight million words in eighty works on metaphysics, ethics, morality, epistemology, the mind, and religion. It was he, more than any other theologian of his time, who was able to reconcile the truths of Greek, Roman, Islamic, and Jewish philosophy with Christianity.

The Philosophy of Nominalism: What's in a Name?

The golden age of scholasticism gave rise to such famous scholastics as Albertus Magnus, Thomas Aquinas, the English monk and philosopher Roger Bacon, the Italian prelate and theologian St. Bonaventure, the Scottish theologian and philosopher John Duns Scotus,[2] and the Belgian secular priest Henry of Ghent. Another philosophical school called nominalism, however, challenged scholasticism in number of followers. The most famous names in nominalism were Roscelin de Compiegne (1050-1125), the French theologian and scholastic philosopher regarded as the founder of nominalism, the scholastics Bernard of Chartres (1090-1130) and William of Champeaux (1070-1121), the outspoken and controversial English theologian and scholastic philosopher William of Ockham (1285-1349), and the French philosopher and theologian Nicholas of Autrecourt (c. 1300-1350?).

Nominalism, which gets its name from the Latin word *nomialis* meaning "pertaining to names," is a philosophical doctrine that says that abstractions, known as universals (Plato's eternal "forms" or substances that he said constitute the real world and are the objects of our reason), are without essential reality, and that the mind can grasp no single idea or image corresponding to any universal because only individual objects have real existence. For instance, the universal "circle" has no real existence in itself; we can only ascribe the name circle to things that are round. Nominalism is contrary to the extreme realism philosophy of Plato, who said that universals have a real and independent existence prior to and apart from particular objects.

Nominalism first evolved from Aristotle who disagreed with Plato's extreme realism philosophy, believing that reality consists in individual objects, not universal archetypes. The founding father of true nominalism, however, was the French theologian and philosopher Jean Roscelin de Compiegne, also called "Roscellinus." Roscelin taught that only individual objects are real, whereas universals or abstract concepts are merely words. Misapplying nominalism to theology, however, got Roscelin in trouble, when he said that the Christian concept of the Trinity, one God in three persons (Father, Son, and Holy Spirit), was not real either. There must exist, Roscelin said, three separate gods who are not one in being and who are not consubstantial. Needless to say, the Church was very angry with Roscelin and condemned his polytheism theology as heresy at the Synod of Soissons (1092). After his condemnation, Roscelin was forced to recant or face punishment. Roscelin fled to England, where the popular archbishop of Canterbury, St. Anselm, challenged him on his heresies. As a result of Anselm's rebuttal, Roscelin left England and went to Rome, where he recanted in the end and was reconciled with the Church.

The most notable supporter of Roscelin's theological nominalism was William of Ockham (1285-1349), nicknamed by his followers as the "Unconquerable Doctor" and "Worthy Initiator." Ockham was a maverick English philosopher and Franciscan scholastic who vigorously opposed Thomist and Scotist scholasticism. Born in Surrey, England, Ockham joined the Franciscan Order and studied and taught at the University of Oxford from 1309 to 1319. Ockham got himself in trouble early on with Pope John XXII for his unorthodox teachings, and was held under house arrest for four years (1324-28) at the papal palace in Avignon, France. After being excommunicated in 1328 by Pope John XXII over a dispute involving evangelical Franciscan poverty, Ockham fled to Italy, and then to Munich, where he spent the rest of his life until he died of the plague.[3] With Michael of Cesena, the Franciscan general, Ockham joined sides with Louis IV, the Holy Roman Emperor from Bavaria, in contesting the temporal powers of the pope. Ockham's *Dialogus* (1343) argued for sovereign independence from Rome of temporal (home) rule. In his philosophical defense of nominalism, Ockham maintained that such abstract entities as universals, essences, or forms are merely second intentions of words; that is, references of words to other words rather than to actual things. Ockham argued that the real is always individual, not universal, and that universals have no real existence in themselves, but are only abstract terms. Ockham, the man who more than anyone else helped put an end to Thomist scholasticism, however, is mostly remembered today for his fundamental principle of modern science and philosophy called "Ockham's razor," which says that one should not assume the existence of more things than are logically necessary (i.e., the most likely explanation to a problem is usually the best).

Humanism: Man Displaces Christ in Importance

The first inklings of the decline of Christianity in Western society began during the European Renaissance (late 1300s-1600), which introduced a new man-centered philosophical ideology called humanism. The humanism of the Renaissance must not be confused with the secular humanism of today; for at that time there were not yet any publicly professed atheists. Humanists of the Renaissance wanted to focus in on the major achievements of man rather than Christ, by reviving the aesthetic, opulent, artistic era of the pagan Greeks and Romans. The recent rediscovery of the ancient works of Aristotle and other important Greek philosophers from the translations of the Arab philosophers Avicenna and Averroes in the twelfth and thirteenth centuries sparked a resurgent interest within Europe for classical art, philosophy, and culture. Statues and ancient architecture of Greece and Rome were idolized and worshipped for their beauty and magnificence. Great schools of art and architecture opened up in Italy, producing such incredibly talented artists as Giotto (1267-1337), Botticelli (1445-1510), and Leonardo da Vinci (1452-1519). Italy was the birthplace of the Renaissance (which means "rebirth") in Europe because that was where the ancients lived and there was plenty of architectural ruins, statues, and works of art left for Renaissance artists to study and emulate.

A renewed vitality and spirit of human accomplishment emerged in Europe during the Renaissance. Great works of literature were written by Dante Alighieri (1265-1321), Giovanni Boccaccio (1313-75), and Francesco Petrarch (1304-74), which glorified humanism and the ancient classics. Magnificent achievements in architecture, especially the construction of huge, ornate basilicas and palaces, were also made. Churches, basilicas, palaces, and the residences of the nobility and Church hierarchy were adorned with exquisite paintings of the human body and sculptures from Italian masters. Explorers from Italy, Spain, Portugal, and England circumnavigated the world in search of new lands and wealth to finance the Renaissance. Urban cites and centralized political institutions began replacing the rural fiefdoms of the Middle Ages to accommodate the growing populations and increased wealth of the merchant class. Secular scholars began replacing monks and priests as doctors of higher education and learning at the major universities. Scientists made important new discoveries, especially in the disciplines of medicine, archeology, mathematics, and astronomy. Historians began to record time in terms of antiquity, the Middle Ages, and the Renaissance instead of the creation, incarnation, and Second Coming of Jesus Christ. Man started to replace Christ in importance.

The study of the works of the ancient philosophers coming in from the scholars of the Christian Byzantine Empire after its fall to the Muslim Ottoman Empire in 1453 produced a slavish imitation of the classics by humanists. Plato and Aristotle, who were once looked upon with suspicion by the Church during the Middle Ages, were adored and revered. The ethics of ancient Greek and Roman philosophers became the common standard of morality for humanists, and a feverish search to locate the manuscripts of ancient Greek and Roman philosophers was undertaken by humanist researchers and their supporters.

Schools of humanistic studies began to appear throughout Italy during the fifteenth century, encouraged and financed by the wealth of the Medicis of Florence, the Estes of Ferrara, the Sforzas of Milan, the dukes of Urbino, the Gonzagas of Mantua, the doges of Venice, and even papal Rome. The most notable supporter of Renaissance humanism was Cosimo de Medici (1389-1464), also known as Cosimo the Elder. Cosimo was an Italian banker and statesman who gathered at his court the leading artists, architects, and intellectuals of his day. Among those who enjoyed his patronage was the famous humanist philosopher and theologian Marsilio Ficino (1433-99), who reintroduced Neoplatonism to Europe.

During the earlier time of the scholastics, Aristotle was the most notable and respected philosopher of antiquity. Plato didn't have a strong following until the Renaissance when Marsilio Ficino, secretary to Cosimo de Medici, was commissioned by Cosimo to translate Plato's complete philosophical works into Latin (1462-69). Ficino, an Italian physician and philosopher who later became a priest, established a new Platonic Academy in Florence in 1462, which taught Plato, Plotinus, and the works of other Neoplatonists, as well as classical art and architecture. Ficino's Florentine Academy, therefore, was the greatest contributor to the Neoplatonic revival during the Renaissance.[4]

Students of the Florentine Academy, although they studied Neoplatonism, were nonetheless Christians. As they studied the works of the early Church fathers and ancient philosophers, they also studied the newly discovered writings of a legendary, fictional character named Hermes Trismegistus (whose name means "three times greater than Hermes," the fabled messenger of Zeus). Hermes Trismegistus was believed by Cosimo de Medici and other members of the Florentine Academy to have been an ancient Egyptian sage and contemporary of Moses who wrote a body of writings called the *Corpus Hermeticum,* or the "Hermetic Corpus." The Hermetic Corpus, which was actually written around the third century A.D. and rediscovered about 1460, contains Gnostic writings on magic, astrology, alchemy, the occult, and supposed correspondences from Hermes to his students (*Pimander, Asclepius*).[5] In fact, Cosimo de Medici, knowing that he only had a short time to live, ordered Ficino to stop translating Plato and concentrate solely on translating the Hermetic Corpus. The fictional Hermetic Corpus had a profound effect on Renaissance writers like Pico della Mirandola and Giordano Bruno, who revered these phony writings more than they did the Bible. Hermeticism, the mistaken belief in the fabled writings of Hermes Trismegistus, was one of the major causes for the decline in popularity of Christianity during the later part of the Renaissance, a fact that few people are aware of.

Another influential humanist and philosopher of the Renaissance was a student of Marsilio Ficino's named Giovanni Pico della Mirandola (1463-94). Pico della Mirandola was a gifted student of the Florentine Academy who studied the Kabbala, Arabic, Hebrew, philosophy, science, mathematics, and theology. At the age of twenty-three in 1486, he nailed a list of nine hundred theses to the cathedral door in Florence, publicly challenging anyone to an open debate, but was prevented to do so by the Church. For his unorthodox cabalistic beliefs, he was accused of heresy by Pope Innocent VIII in 1487, but cleared in 1492 by Pope Alexander VI and was reconverted to orthodoxy by the controversial Dominican monk Savonarola. In 1486, Pico della Mirandola got himself in trouble with the Church when he published *De hominis dignitate oratio (Oration on the Dignity of Man),* which said that man can do anything by himself: a typical creed of modern secular humanists.

Another Italian humanist who is worthy to note is Lorenzo Valla (1407-1457). Valla was secretary to the king and protégé of Pope Nicholas V and Pope Calixtus VI. He was fluent in Latin and Greek, and translated Homer, Herodotus, and Thucydides. He was often called upon to authenticate ancient texts and manuscripts, including the New Testament. Valla discovered that the famed Donation of Constantine, which gave all temporal authority of the Western Roman Empire over to the papacy, was actually a forged document. It was supposedly written in the fourth century A.D. by Constantine, but Valla discovered that it had been written around the eighth century A.D. because it was written in later Latin grammar rather than earlier. This important new discovery gave substantial ammunition to secular rulers who believed in state's rights over Church rights, and helped fuel the Protestant Reformation in the late Renaissance era and the subsequent decline in influence of the Roman papacy.

One of today's heroes of freethinking and secular humanism during the Renaissance is Giordano Bruno (1548-1600). Bruno was an Italian philosopher and poet from Naples who joined the Dominican Order in 1565, where he studied Aristotle and Thomist scholasticism. Egotistical, rebellious, and argumentative by nature, Bruno fled the Dominicans in 1576 to avoid prosecution for his unorthodox beliefs. Influenced by the phony writings in the Hermetic Corpus, Bruno believed in magic and said that Christ was a magician and the philosophical heir to Hermes Trismegistus. Wandering around Europe teaching, preaching, and writing anonymously about such topics as the Hermetic Corpus, pantheism, Neoplatonism, atomism, a plurality of worlds, an infinite universe, and the Copernican cosmology which said that the sun is the center of the universe instead of the earth as the Church maintained, Bruno raised the ire of the Catholic hierarchy. In fact, if Bruno had never championed Copernicus, the Catholic Church probably would have never censured Galileo in 1633 for his support of the Copernican theory. It was Bruno, not Galileo, who gave the Copernican theory a bad reputation within the Church. In 1584, Giordano Bruno gave a lecture at Canterbury and was very insulting and demeaning to the scholastics, but he had many friends in high places in London and Paris who disliked the Catholic Church and papacy as much as he did, and offered him protection against his enemies. Bruno's career, however, was about to come to an end. Invited to Venice to be the private tutor to the nobleman Giovanni Moncenigo, Bruno was turned in to the Inquisition in 1592 (a possible setup) and imprisoned while undergoing questioning on charges of blasphemy, immoral conduct, and heresy. First recanting, then affirming, his heretical ideas, Bruno was condemned and burned at the stake in Campo dei Fiori on February 17, 1600, for being an "impenitent heretic." Giordano Bruno has since become an icon and martyr for freethinkers, rebels, secular humanists, and atheists ever since. Late in the nineteenth century, a statue was erected on the site of his execution in his honor, which has become a memorial magnet for humanists worldwide.

Renaissance humanism in Italy was concerned primarily with the revitalization of European art and architecture, and the translations and study of ancient philosophers such as Plato and the Neoplatonists. Humanism in central and northern Europe, however, extended much broader into the fields of education and theology, and had a profound impact on the birth of the Protestant Reformation in Germany at the start of the sixteenth century. The invention of the Gutenberg printing press with movable metal type in 1455 helped spread humanism through the dissemination of the Greek classics (a critical study of the Bible in the original Greek and Hebrew language was paramount to the Reformation cause of *sola scriptura,* or the Bible only). In Germany,

Johann Reuchlin (1455-1522) was a leading humanist and scholar of Greek and Hebrew who wrote *De Rudimentis Hebraicis* in 1506, the first text on Hebrew grammar written by a Christian. Although personally opposed to the Reformation, Reuchlin's nephew, Philipp Melanchthon (1497-1560), was a leading German Reformer. Elected to the chair of Greek at the University of Wittenberg in 1518, Melanchthon befriended Martin Luther, the founder of Protestantism, and became a well-respected theologian and major voice for the Protestant cause.

One of the key figures in the enlargement of humanism in France and England was the Dutch cleric and humanist Desiderius Erasmus (1466-1536). Erasmus was born in Rotterdam, the illegitimate son of a priest, Roger Gerard, and a physician's daughter. Erasmus went to strict monastic schools in Deventer and s'Hertogenbosch and, after his father's death, became an Augustinian canon at Steyn. Erasmus was ordained a priest in 1492, and while employed by the bishop of Cambrai, studied scholastic philosophy and Greek at the University of Paris. Unhappy with living the rigorous religious life of an Augustinian, Erasmus petitioned the pope for a papal dispensation to work as a secular scholar without wearing religious apparel, which he was granted. Beginning in 1499, Erasmus traveled from city to city throughout Europe, working as a tutor and lecturer and constantly writing correspondences to the higher ups in Europe, while at the same time searching out ancient Greek and Latin manuscripts. Erasmus had a profound affect on English education. During four trips to England, Erasmus befriended such humanists as John Colet, founder of Saint Paul's School in London; Thomas Linacre, founder of the Royal College of Physicians; St. Thomas More, Catholic author and lord chancellor of England who was executed by King Henry VIII in 1535 for his allegiance to the pope; Bishop John Fisher (1459-1535) who was also executed by the king for his papal allegiance; and William Grocyn (1446-1519), lecturer in Greek at Oxford University (Erasmus also taught Greek at Cambridge). Erasmus' most important writings include his *Adagia,* a collection of Latin proverbs, *The Manuell of the Cristen Knyght,* and *The Praise of Folie,* both of which criticize the Church and advocate practical Christian piety without the need for strict religious observances. Erasmus believed that religion was far too important to be left to the theologians, and that men of letters like himself could purify Sacred Scripture by researching the ancient sources. Erasmus' Greek New Testament (1516), which was based on newly discovered manuscripts, contained critical notes and a new Latin translation, and was a more accurate version of the New Testament than the Latin Vulgate written by St. Jerome in the fourth century. Erasmus' New Testament greatly influenced Protestant reformers of the time who sought to use the Bible as their sole source of authority (*sola scriptura*). While still maintaining that he was a Catholic, Erasmus befriended many Protestants and criticized the Catholic Church in his *Colloquies* (1518) for what he saw as her errors and superstitions. This led many within the Church hierarchy to suspect that he was secretly a Lutheran, which he vehemently denied. Erasmus refused to take sides publicly and purposely stayed out of theological debates. Accused of concealing his true beliefs for fear of condemnation by the Church, Erasmus responded to his critics by writing a complete declaration of his theological position, *De Libero Arbitrio (On the Freedom of the Will,* 1524), which contained a witty and erudite attack on Martin Luther's belief in the satanically enslaved will. A vicious counterattack by Luther elicited a final polemic by Erasmus, *Hyperaspistes* (1526). Despite Erasmus' claim that he was a Catholic, the Catholic Church has always regarded Erasmus as a Protestant reformer. His books were listed in the Index of Forbidden Books by the Council of Trent (1545-63) during the Catholic Counter Reformation.[6]

We have seen how the rise of humanism during the Renaissance began to draw man's interests away from Jesus Christ. Now let's see how another insidious human philosophy known as rationalism, the most destructive philosophy in human history, put mankind on a collision course with God himself.

Rationalism: Human Reason Supercedes Divine Revelation

Rationalism is a human philosophy that glorifies man's reasoning ability to understand truths and obtain knowledge over and above sensory experiences and divine revelation. Rationalism is derived form the Latin word *ratio,* which means "reason." In ethics, morality, and religion, rationalists believe that divine revelation is unnecessary because human reason alone can define what is good and bad, right and wrong. Man doesn't need commandments or direct intervention from any god. In fact, any claims of divine revelation or any moral mandates that can't be rationally proven by science should be looked upon with suspicion, or dismissed altogether out of hand as spurious, superstitious, and evil.

The seventeenth century birth of the irreligious spirit of rationalism caused a skeptical crisis in faith throughout Europe, particularly in France and England where it gave birth to anti-christian deism, the Age of Enlightenment, the Age of Reason, and the French Revolution of the eighteenth century. Since its inception, rationalism has been linked to many, if not all, antireligious movements in the West, and is primarily responsible for the modern expulsion of Christianity from our schools, governments, and public institutions. History has recorded the devastating consequences of rationalism: the French Revolution (1789-1794), the rise of communism and the European Revolution of 1848, the birth of Darwinism in 1859, the French Communard uprising of 1871, the rise of Bolshevism in 1917, and the rise of Nazism in 1919. Hundreds of millions have died because of rationalism, and yet it is as strong as ever.

The never-ending argument over how we obtain knowledge and understanding of the world, whether by reason or sensory experiences, and whether or not there is such a thing as absolute truths revealed by God, is as ancient as philosophy itself. Diogenes the Cynic (412?-323 B.C.), Pyrrho the Skeptic (360-272 B.C.), and the Greek physician and philosophical skeptic Sextus Empiricus of the third century A.D. were three of the first known philosophers who could be considered rationalists. The founding father of the modern philosophical school we now call rationalism is the seventeenth-century French mathematician, scientist, and philosopher Rene Descartes (1596-1650). Descartes was born in La Haye, Touraine, to a noble family of learned scholars. At the age of eight, he was enrolled in the Jesuit school of La Fleche in Anjou, where he remained for eight years. In addition to the usual studies of classical Greek and Latin scholars, Descartes received instruction in mathematics and philosophy. The education he received from the Jesuits exerted a strong influence on Descartes throughout his entire life. His later philosophical discourses were attacks on the prevalent scholasticism of his day. Upon graduation from Jesuit school, Descartes studied law at the University of Poitiers, graduating in 1616, but he would never practice law. Following a brief stint in the military, Descartes traveled around Europe lecturing and teaching. In 1619, while in Germany, Descartes had a vision and three dreams that told him that every truth up until now should be doubted and what was needed was an entirely "new philosophy" which he himself would find out.

Back in France, inspired by his prophetic vision and dreams, Descartes plunged himself into the study of philosophy. In 1628, Descartes moved to the Netherlands, where he lived for the remainder of his life. It was here that he composed his major philosophical works: *Le Monde* (1633), *Discourse on Method* (1637), *Meditations on First Philosophy* (1641; rev. 1642), and *The Principles of Philosophy* (1644).

Descartes' first work, *Le Monde (The World),* was a treatise on physics and cosmology that thoroughly attacked the Christian philosophies he learned in Jesuit school. In favor of a more mechanistic, materialistic explanation of the world, Descartes, in *Le Monde,* contended that only two things exist in the world—mind and matter. On the complex subject of matter, Descartes maintained that matter is inert, has extension with length, breath, and width (there are no empty spaces or voids as Aristotle believed), that there exists a plenum rather than a vacuum in space, that celestial matter and earthly matter are of the same substance, that matter has motion and is continually moving, that this motion must necessarily come from the Prime Mover, that there is a conservation of motion in the universe (contrary to Aristotle who believed that things tend to rest), that objects in motion move rectilinearly unless acting upon by forces, and, finally, that there exists three kinds of matter in the world, which differ only in size and shape. He called these first, second, and third matter. Third matter is visible and large; second matter consists of small, invisible "spheres" that fill in the spaces between the third matter; first matter is tiny, irregular, and invisible, and fills in the spaces between the spheres.

There are no empty spaces in the universe, according to Descartes. Matter, like the planets, moves in circular, enclosed paths called vortices. This was Descartes' vortex theory of planetary motion. Vortices revolve counterclockwise in the same plane (this was the first time anyone addressed the problem of why the planets move in the same plane). At the center of each vortex is a star. Eventually all stars die out. Sunspots are impure matter of decay, which will eventually overwhelm the star. Dead stars are cast out of their paths and appear as comets. Comets will eventually be caught in a vortex and form a planet. Descartes was one of the first to say that the earth was once a ball of fire that encrusted.

Descartes' mechanistic theory of the universality of matter and his support of Copernicus' heliocentric theory of the universe were, in 1633, considered to be heretical by the Church. The Church believed like Aristotle that celestial matter is permanent and indestructible, that earthly matter corrupts and decays, and that the earth, according to the literal interpretation of the Bible and a majority of scientists of that time, is the center of the universe. In light of Galileo's recent prosecution by the Roman Inquisition in 1633 for his espousal of Copernicus' heliocentric cosmology, Descartes wisely elected to cautiously withdraw *Le Monde* from publication, at least for the time being.

On the controversial subject of mind, Descartes attempted to show that mind is superior to the senses in reliability regarding the confirmation of truths. Senses can often be deceived, he said, like when a stick appears to be broken when immersed halfway in water. In his "Discourse on the Method of Rightly Conducting Reason and Reaching the Truth of the Sciences" (1637), the fourth part of his *Essais Philosophiques,* which also contains essays on geometry, optics, and meteors, Descartes attempted to build a new philosophy of truth by doubting everything at first, than trying to find things that can't be doubted, such as exists in mathematics and geometry, in which field he was a genius.[7] On the most fundamental question in philosophy, regarding whether or not we really exist and have being, Descartes stated that he can be certain he exists by using the reasoning power of his own mind. Said Descartes: *"Cogito, ergo sum."* ("I think, therefore I am.") This has since become the single most famous philosophical quote in all of history.[8]

The problem with Descartes' rationalistic philosophy of doubting everything at first, then trying to prove what is true, is that there exist many truths that cannot be proven by either reason or the senses, and must be accepted by faith as given to us by God through divine revelation. This includes the Christian doctrines of the Trinity, the virgin birth, the incarnation, and the real presence of Jesus Christ in the Eucharist, which, by the way, Descartes didn't believe in. Descartes, in fact, got himself in trouble with the University of Paris and the Catholic Church for his denial of the Catholic doctrine of transubstantiation in his *Meditations on First Philosophy.* The real presence is a traditional Catholic doctrine, which says that at the moment of consecration the bread and wine are turned into the actual body and blood and soul and divinity of Jesus Christ. As you will see in the final chapter of this book, there are many miracles of the bread and wine given to us by God to strengthen our faith in the real presence.

The contemporary critics of Descartes argued that his rationalistic oversimplification of everything in the physical universe to just matter and mechanical causes, and his advocacy of the philosophical skepticism of Catholic faith, took the mystery out of the universe and the faith away from the people. As a result of Descartes' rationalism, atheism, pantheism, agnosticism, and hostile antichristianity, all of which were just beginning to gain acceptance in the universities of Europe at the end of the seventeenth century, took off like a rocket, ending forever the Christ-centered era of Thomist scholasticism.

Rationalism had many champions beside Descartes, but he it is considered the founder. Baruch Spinoza (1632-77) was another important philosophical rationalist whose godless philosophical writings didn't become popular until a century after his death, when the irreligious spirit of rationalism was at its peak. Spinoza, whose first name Baruch or Benedictus means "Blessed," which he definitely wasn't, was the first major philosopher to publicly advocate abandoning the traditional beliefs of Judeo-Christianity.

Spinoza was born to Spanish-Portuguese Marranos (forced converts to Christianity living secretly as Jews) and resided in Amsterdam, Holland. He was given a traditional Jewish education, but Spinoza repudiated the Jewish religion of his ancestors following his introduction to the rationalistic philosophy of Rene Descartes, and the mechanistic and naturalistic philosophies of the Englishman Thomas Hobbes. After quitting the local synagogue, Spinoza was cursed and excommunicated by the local rabbis in 1656, and banished from Amsterdam. He then lived on the outskirts of the city and eked out a meager living grinding lenses.

During the next five years as a lens grinder, Spinoza wrote three major philosophical works espousing Cartesian rationalism: *Treatise on God and Man and His Happiness, Theologicopolitical Treatise* (published anonymously out of fear of reprisals), and *On the Improvement of Understanding.* In 1674, Spinoza wrote his fourth and most famous philosophical work on pantheism, entitled *Ethics Demonstrated with Geometrical Order.* According to this treatise, the universe is identical with God, who is the uncaused "substance" of all things. Substance, according to Spinoza, was not a material reality but a metaphysical entity that we think of as God. Substance has infinite attributes, but only two are perceptible to human minds: "extension" (the world of material things) and conscious "thought." Thought and extension, in turn, are dependent on and exist in an ultimate reality, which we think is God. All material objects are the modes of God in the attribute "extension"; all ideas are the modes of God in the attribute "thought." Modes are *natura naturata,* "nature begotten," or nature in the multiplicity of its expressions. Substance or God is *natura naturans,* "nature begetting," or nature in its formative unity, acting as the causal factor of its own modes. Modes are short-lived, and their existence assumes transitory form. God, however, is eternal, transcending all modal modifications. He is both an infinite physical thing and an infinite thinking thing. Material objects and conscious ideas are finite and unstable, but God is indestructible and infinite.

Spinoza rejected the orthodox Judeo-Christian idea of a personal, spiritual God who providentially watches over and judges man, whom he made out of nothing to share in his love and happiness in heaven for all eternity, or hell if he is wicked. No, nature and God were one and the same to Spinoza. God did not create nature but *is* nature. The natural laws of the universe are God, the material objects of the universe are God, we are God, and everything is God. Orthodox Judeo-Christianity, which teaches that God is separate from and transcendent over the universe he created, was intolerant and hateful to Spinoza, but it does provide a moral foundation for the ignorant masses who were incapable of forming a "philosopher's" intellectual love of God, which Spinoza claimed is the true salvation. Organized religion, said Spinoza, is life-denying in its promise of eternal happiness for righteous living on earth, which he didn't believe in either. Man's primary aim should be enjoying life in the here and now without fear or worry about condemnation. We should be one with God (nature), which is true salvation. Responding to his critics, who were many, Spinoza urged tolerance from them and freedom of thought in matters of faith and religion. Spinoza believed that the mythical stories and miracles in the Bible were meant to teach ethics and morality and weren't meant to be taken literally, a popular antichristian belief that has now become known as "modernism" or "post-modernism."

Spinoza's pantheism and disbelief in orthodox Judeo-Christianity was looked upon with horror and outrage by the religious leaders of his day, but have since been adopted and espoused by the majority of history's scientific elites, godless philosophers, and liberal intellectuals. Among the most famous philosophers and intellectuals who were proud followers of Spinoza were Goethe, Lessing, Heine, William Wordsworth, Percy Shelley, Nietzsche, Bertrand Russell, George Eliot, Sigmund Freud, Albert Einstein, and George Santayana, to name a few.

Another important rationalist was Baron Gottfried Wilhelm von Leibniz (1646-1716), a famous German scientist, philosopher, mathematician, and statesman of the seventeenth century. Leibniz is best remembered as a brilliant mathematician who, along with Isaac Newton, invented integral and differential calculus. He also wrote several important works on philosophy and theology. Both Leibniz and Spinoza were proponents of an ancient philosophical theory called "monism," which says that everything in the universe is of one substance. There is no Platonic dualism of material body and immortal spiritual soul in monism; mind and matter are the same stuff.

Leibniz' philosophy of monism said that the world is composed of innumerable conscious centers of energy, known as "monads," which lack spatial parts and whose properties are functions of their perceptions and appetites. Each monad represents an individual mini-world, mirroring the universe in varying degrees of flawlessness, and developing independently of all other monads. The universe that the totality of these monads constitute is the harmonious result of God's divine plan. To answer his critics on why God, if he is omniscient, omnipotent, omnipresent, and perfect in every way, created this perfect monadal world filled with so much evil, Leibniz replied in his *Theodicy* (1710) that if a better world could have been made God would have created it. Therefore, God must have had "sufficient reason" to create a world such as ours. Voltaire, the infamous cynic and antichristian deist of the

French Enlightenment, brutally satirized Leibniz's "best of all possible worlds" in his infamous novel *Candide* (1759), a wicked satire of Judeo-Christianity in particular and all religions in general.

Yet another champion of seventeenth and eighteenth century rationalism was a German mathematician and philosopher with an ironic name, Christian Wolff (1679-1754). Inspired by the rationalistic doctrines of Descartes and Leibniz, Wolff became professor of mathematics and natural philosophy at the University of Halle in 1706. In 1721, Wolff delivered a controversial lecture denying the necessity of a Christian education for learning ethics and right moral behavior, citing the moral tenets of the Chinese sage Confucius (551?-479 B.C.) as proof that human reason alone could attain moral truths. Wolff's antichristian lecture got him into serious trouble with faculty heads at Halle, and he was subsequently banned from Prussia for teaching atheism and fatalism" (all events are fixed in advance by some mysterious power, and human beings are powerless to change them). Despite Wolff's banishment from lecturing, the dominant philosophy of the German universities for the next century was Leibniz-Wolffian rationalism. Wolff came up with no new ideas of his own, but he was important as an organizing and systematizing philosopher of antichristian rationalism. His volumes of writings include the infamous work *Rational Thoughts on God, the World, and the Souls of Men* (1719).

On the heels of rationalism, a plethora of antichristian philosophies emerged in Europe and North America that attempted to explain the truth about reality without reference to Christianity, which drew mankind further and further into error and sin. The remainder of this chapter will attempt to define and explain each one of these philosophical errors, and how they've contributed to the downfall of this world we so lament. If you don't believe in God or Christianity, what you do believe will be contained within one or more of these atheistic philosophies, so pay attention.

Empiricism: I Must See, Smell, Touch, Taste, or Hear It to Believe It

Empiricism is a philosophical doctrine that says that all knowledge comes from sensory experience rather than any spontaneous ideas or *a priori* thought. In essence, empiricism is a systematic process of observation and experimentation, which entails observing the material world neutrally and dispassionately, without prejudice or bias, and without uncorroborated testimonies, especially in regards to reports of supernatural miracles. Empiricism is different than rationalism, but is equally skeptical towards religion and revelation. Empiricists say that all we can know about the truth is that which the natural world wishes to reveal to us through our five senses. If you can't see it, smell it, touch it, taste it or hear it, you can't say it exists. Any theological statements of fact regarding metaphysics and the supernatural realm are senseless to empiricists because they profess agnosticism.

Philosophical empiricists believe that our minds are like a blank tablet, a *tabula rasa,* on which information is imprinted via sensory inputs from the moment we are born. The problem, however, with saying that all knowledge and information comes from sensory input that is recorded on our minds like a blank CD is that it denies the mind's uncanny ability to foreknow things unknown, mysterious, or not yet experienced, such as the existence of a higher power outside ourselves, the existence of another world outside our space-time continuum, and the undeniable reality of ESP precognitions, premonitions, intuition, mental telepathy, and other extra-sensory gifts such as prophetic dreams and visions. Even pets who have no eternal soul have an uncanny ability to understand things they've never experienced before, such as finding their way home from far-away distances, when an earthquake or natural disaster is about to strike, or sensing when a family member is in trouble or a stranger has evil intentions. Empiricism, therefore, denies the spiritual nature of God's creatures.

In its heyday during the seventeenth, eighteenth, and nineteenth centuries, Great Britain was home to most of the world-famous empiricists, including Francis Bacon, John Locke, Thomas Hobbes, and David Hume. Francis Bacon (1561-1626), the first Baron Verulam and Viscount Saint Albans, was a preeminent English lawyer, philosopher, and statesman, and is considered to be the founder of modern scientific thought, although he was not a scientist himself. Bacon was born on January 22, 1561, at York House in the Strand, London, and educated at Trinity College, University of Cambridge. Elected to the House of Commons in 1584, he served until 1614. In 1613, Bacon was appointed attorney general. In 1616, he became a privy councilor, and in 1618, he was appointed lord chancellor and raised to the peerage as Baron Verulam in 1621. But that same year he was charged with bribery and confessed that he was "heartily and penitently sorry." He submitted himself to the mercy of his peers who ordered him fined, imprisoned, and banished from Parliament. In September 1621, the king pardoned him but prohibited his return to Parliament or the court. Bacon spent the rest of his life writing at his family residence at Gorhambury and died in London on April 9, 1626, while trying to determine whether cold could preserve meat.

Bacon's most important philosophical works were *The Advancement of Learning* (1605) and *Novum Organum* (1620). In *Advancement of Learning,* Bacon said that we should stop looking back to the ancient classics and authority figures for knowledge and turn to the moderns. At the time, the ancients were the accepted authorities of the scientific community, but by the end of the seventeenth-century modern scientists, like Newton and Boyle, had replaced them as the new authorities. In *Novum Organum,* Bacon said that what we need is a "new method" of discovering things to progress in knowledge; a new, empirical method of scientific inquiry not beholden to any authority. Bacon believed that Aristotle's deductive method of scientific inquiry, reasoning from preconceived prejudices and attitudes (which he called idols) to account for observable experimental results, is a hindrance to making new scientific discoveries. What we need, Bacon said, is to first begin by objectively collecting facts through scientific observation and experimentation, without any preconceived conclusions or theories. Then we can draw our own conclusions and theories from what we've learned. According to Bacon, we should never have to form a hypothesis about anything. In fact, he argued against them.

Bacon's method of scientific observation and experimentation preceding the formulation of any theory or conclusion is known as the scientific method,[9] and is credited with stimulating the scientific revolution of the seventeenth and eighteenth centuries. Scientists in that period wholeheartedly embraced Bacon's scientific method, and governments and universities rushed to organize new societies for the advancement of scientific learning. For instance, in 1620, the year Bacon wrote his *Novum Organum,* there were no state-sponsored scientific societies. By 1660, the Royal Society of London was founded, and France's Royal Society in 1666. As a result of these new societies, tremendous scientific discoveries were made in England, France, and the rest of Europe, simply by doing what Bacon recommended—observing what happens in carefully controlled experiments.

The English philosopher, physician, politician, and statesman John Locke (1632-1704) is actually considered to be the founder of the philosophical school of empiricism, because he was the first to systematize the empiricist principles laid down by Francis Bacon. Locke was born in the village of Wrington, Somerset, on August 29, 1632, and was educated at Westminster School and Christ Church at the University of Oxford, where he lectured on Greek, rhetoric, and moral philosophy from 1661 to 1664. Liberal in politics and Protestant in faith, Locke held a couple of minor government appointments and served on the Board of Trade until he resigned in 1700 because of ill health. Locke's most important philosophical work was *An Essay Concerning Human Understanding* (1690), which took him twenty years to complete. In his *Essay,* Locke was the first to say that the mind is like "white paper" at birth, on which experience imprints knowledge. Locke did not believe in intuition, innate conceptions, instinctual understanding, or any type of inborn knowledge. All knowledge and all ideas, he said, come from experience. Locke did, however, try to include some of Descartes' rationalism in his philosophy. Senses alone, he said, do not impart knowledge; it is reason that works out the connections between ideas obtained from sensory experience that give us knowledge. Without reason, Locke said, all we have is belief, not knowledge. Locke, therefore, was empirical about obtaining information yet rational and liberal about interpreting it. Regarding Christianity, Locke's philosophy in his *Essay* elicited many criticisms for his skepticism towards religion. His polemics against innate ideas were taken as dangerous for religion and morality, and the role he gave to reason in religion was taken as to imply an impersonal deism. His weird theory that non-living matter might "think" was also seen as scandalous and heretical.

Another infamous empiricist, rationalist, and materialist from England was Thomas Hobbes (1588-1679). As the secretary to Francis Bacon, Hobbes was a controversial philosopher and political theorist whose mechanistic and naturalistic theories got him in trouble with both church and state. Born in Malmesbury, Hobbes was educated at Magdalen Hall, University of Oxford. In 1608, he became the tutor of English explorer William Cavendish, the third person to circumnavigate the world. In the following years, Hobbes traveled to continental Europe, where he visited and conversed with Galileo, Descartes, Gassendi, Mersenne, and others. Back in England, Hobbes became involved in politics and the civil war between Parliament and the king. In support of the monarchy, in 1637 Hobbes wrote *The Elements of Law, Natural and Politic,* which was circulated and read by members of Parliament. Hobbes feared that Parliament would arrest him for his *Elements,* so he fled to France in 1641 and lived in voluntary exile for the next decade. From 1646 to 1648, he was a mathematics tutor to the prince of Wales, later King Charles II, who was living in exile in Paris. In 1651, Hobbes completed his most important political work called the *Leviathan; or, The Matter, Form, and Power of a Commonwealth Ecclesiastical and Civil.* Hobbes' *Leviathan* was seen by the French authorities and British royalists living in exile in France as being too supportive of Oliver Cromwell's commonwealth and too critical of the papacy, so

Hobbes returned to England for fear of arrest in France. Hobbes had enemies on both sides of the English Civil War. Royalists saw Hobbes' denial of the divine right of kings as anti-king, and Parliament saw his support of the king over Parliament as anti-commonwealth. When Cromwell finally died in 1658, and his former student, Charles II, returned to England and became king, Hobbes regained favor with the English royalists. In 1666, however, the House of Commons passed a bill that included Hobbes' *Leviathan* on a list of books to be investigated on charges of teaching atheism. Similarly, the Catholic Church included *Leviathan* on their Index of Forbidden Books.

Hobbes' *Leviathan* and his philosophical trilogy of *De Cive* (1642), De Corpore (1655), and *De Homine* (1658), which covered topics on logic, language, and optics, as well as moral and political theory, clearly showed that Hobbes held a materialistic viewpoint about the world. For instance, Hobbes believed that all human psychology, such as sense, behavior, and tastes, are due to matter in motion inside the body. In matters pertaining to ethics, Hobbes was utilitarian and based his morality on psychological egoism, in that he believed that all things desired are good, and that every man seeks what is good for him. Although he said he believed in God and tried to justify his philosophical and political viewpoints by quoting Scripture, Hobbes felt that God was completely nonessential as the source of morality and ethics. Atheists and deists, he said, are subject to natural law and civil law just like theists, but not necessarily the commandments of God. To Hobbes, reason by itself provides a sufficient guide to right moral conduct, just as the rationalist Rene Descartes had said.

Skepticism: Doubt Everything, Especially Christianity

Skepticism (from the Greek word *skeptesthai,* which means "to examine") is an ancient philosophical doctrine that denies the possibility of knowing any reality apart from human perception, and usually means doubting what is generally regarded to be the truth, especially divine truths. The birth of skepticism as a major philosophical movement in the West is attributed to the Greek sophists in the fifth century B.C., who, for the most part, doubted everything. The most familiar sophist, Protagoras of Abdera, who was famous for saying "Man is the measure of all things," taught that human beings can know only their perceptions of things, not the things in themselves. The Greek sophist Gorgias is similarly known for his saying "Nothing is; or if anything is, it cannot be known." Gorgias said that all statements concerning reality are false, and that even if true, their truth can never be demonstrated. Pyrrho of Elis, who founded the philosophical school of Pyrrhonism in the fourth century B.C., maintained that human beings can know nothing of the real nature of things and, consequently, the wise person should suspend all judgment. Timon of Philius, Pyrrho's pupil, said that equally good reasons can be given both for and against any philosophical proposition. Carneades, a more moderate skeptic, maintained that no beliefs could be proved conclusively, but that some can be shown to be more probable than others. Skeptics of later antiquity included the Greek philosopher Aenesidemus, who classified ten arguments in support of the skeptical position, and the Greek physician Sextus Empiricus, who emphasized observation and common sense as opposed to hypothesis. During the Renaissance, skepticism was advocated by the French writer Michel de Montaigne (1533-92), and a few centuries later during the Enlightenment it was staunchly defended by the English philosopher David Hume.

Inevitably, the scientific revolution of the seventeenth century led to a revival of philosophical skepticism, which, in turn, initiated a tidal wave of antichristian sentiment. Influenced by the empirical and rationalistic philosophies of the seventeenth century, orthodox religion came under increasingly vicious attacks in the eighteenth century by skeptics, atheists, agnostics, and deists who saw the scientific revolution as proof positive that organized religions, especially Judeo-Christianity, were mere myths invented by the authorities to retain power and keep the ignorant masses in subjugation. Bolstered by their pride and arrogance in man's achievements in mathematics and the sciences, especially the vindication of Galileo, skeptical philosophers glorified the period they were living in as the "Age of Reason" and the "Age of Enlightenment." Not afraid anymore to publicly condemn the Church, many skeptics who still wanted to believe in some higher power opted to believe in a new, nature-based religion called deism. Deists believed that God or some other higher power created the universe but left it alone to run by itself, much like a watch. Deists didn't believe in the Bible or its accounts of supernatural miracles and divine intervention and revelations, but believed instead that each person, aided by reason alone, had within themselves an inherent understanding of the difference between right and wrong.

The skeptical, antichristian philosophy of deism first emerged in England around the beginning to middle of the seventeenth century (around the time of the English Revolution and the beheading of King

Charles I at Whitehall Palace in 1649), where it quickly gained a following. Lord Edward Herbert of Cherbury (1582-1648), English philosopher and ambassador to France from 1619 to 1624, is considered by many to be the "Father of Deism." Lord Herbert wrote in his most important metaphysical work, *Of Truth* (1624), that there exists a universality of belief in some higher power, which is indicative of man's capability to understand right from wrong through the power of his reason alone. Lord Herbert and other English deists such as John Toland (1670-1722), who coined the word "pantheism," and Charles Blount (1654-93), criticized the supernatural claims in the Bible and advocated a new, rationalistic religion based on the existence of nature instead of Jesus Christ as evidence of a higher power.

Deism was taken up in England in theeighteenth century by Anthony Collins (1676-1729), Thomas Chubb (1679-1746), and Matthew Tindal (1655-1733), and in France was championed by the likes of Denis Diderot (1713-1784) and Francois Marie Arouet (1694-1778), a.k.a. "Voltaire." In America, deism had its champions in Benjamin Franklin (1706-1790), Thomas Jefferson (1743-1826), Ethan Allen (1738-1789), and Thomas Paine (1737-1809). Paine wrote the controversial book *The Age of Reason*,[10] a philosophical digest of antichristian beliefs in which Paine boasted of a glorious and peaceful future after deism overturned, as his friend and fellow philosopher Joseph Priestley put it, "the old errors of superstition," meaning, of course, Christianity.[11]

The Age of Reason, however, which was so glorified by Thomas Paine as an enlightened period of intellectual awakening and rationalistic antichristianity, should be remembered as the "Age of Pain" because it witnessed more violence, more wars, and more destruction then any other century. There were four French and English Wars waged in North America (1689-1763); there was the War of Spanish Succession in Europe (1701-14); the Swedish-Russian War of 1709; the Austrian-Turkish War in the Balkans in 1717; the War of the Polish Succession (1733-35); the Russian-Austrian War in the Balkans in 1739; the War of Austrian Succession (1740-48); the Scottish Jacobite Rebellion of 1746; the four Carnatic Wars fought between England and France in India (1740-63); the Seven Years War (1756-73) fought over the control of Germany by Prussia, Great Britain, and Hannover on one side and Austria, Saxony, France, Russia, Sweden, and Spain on the other; the American Revolutionary War (1755-81); the Russian defeat of Polish uprisings in 1768 and 1793-94; the French Revolution (1789-94); the French Revolutionary Wars in Europe and the subsequent Napoleonic Wars (1789-1815). Millions of lives were lost in wars and revolutions during the eighteenth century, especially in France, which, ironically, was lauded centuries earlier by the early church fathers for being the "Eldest Daughter of the Catholic Church."

France, one of the first European nations to convert to Christianity, was in the eighteenth century home to more deists and antichristians than any other European nation. Pompous French intellectuals and philosophers known as *philosophes* gathered together at extravagant evening parties in the homes (salons) of prominent persons, and mercilessly attacked Christianity and the French monarchy with their hateful propaganda and liberal ideas. Antichristian and anti-monarchial rhetoric coming from the salons lit the fire of the French Revolution of 1789-94, and the horrible and bloody persecution of the Catholic Church, the overthrow of the French monarchy, and the guillotining of King Louis XVI and his wife Marie Antoinette.

Among the leading French *philosophes* of eighteenth century were Charles Montesquieu, Voltaire, Jean-Jacques Rousseau, and Denis Diderot. Baron Charles Montesquieu (1689-1755) was a French writer and jurist who was born in the Chateau of La Brede, and educated at the Oratorian school at Juilly and later at Bordeaux. He became counselor of the Bordeaux Parliament in 1714 and was its president from 1716 to 1728. Montesquieu was a landed aristocrat from a noble family, part of the resurgent nobility after the reign of King Louis XIV (1638-1715), and became famous for his satirical work *Persian Letters* (1721), which was based on the fictional letters of two Persian patricians traveling about Europe. In this cynical work, Montesquieu poked fun at French politics, social conditions, ecclesiastical affairs, literature, education, and other French institutions, which won Montesquieu instant fame in France among all those who hated the French monarchy and the Catholic Church. Montesquieu's second most important work was *The Spirit of Laws* (1748), a political treatise that said that the French monarchy had overstepped its bounds, and must be brought back in line. Power, he wrote, must be shared by the monarchy and other institutions, such as the Estates General (the French Parliament which hadn't convened since 1614), the local estates, organized nobility, charter towns, and the Church. Governmental powers, he said, should be separate and balanced to guarantee individual rights and freedom. Both *Persian Letters* and *The Spirit of the Laws* were widely read in France, and helped precipitate a climate of rebellion, which climaxed in the French Revolution.[12]

The one person who most typified the cynical spirit of skepticism and antichristian bigotry during the Enlightenment in France was Francois Marie Arouet (1694-1778), who called himself Voltaire. Born to a middle-class family in Paris, November 21, 1694, Voltaire was educated by the Jesuits at the College Louis-le-Grand and went on to become an avant-garde philosopher and compulsive letter writer. One of the leading figures of the Enlightenment, Voltaire was a prolific writer whose collected works on various topics numbered more than seventy volumes. Voltaire had a biting tongue and a quick, sarcastic wit, which he seemed to relish but which got him in trouble with the authorities on many occasions. His lampooning of the French regent Philippe II, duc d'Orleans, accusing him of heinous crimes he didn't commit, caused him to be imprisoned in the Bastille from 1717 to 1718. A quarrel with a member of an illustrious French family, the Chevalier de Rohan, resulted in Voltaire's second incarceration in the Bastille in 1726, from which he was released within two weeks on promise to leave France and go to England, where he quickly mastered the English language and infected England with his pernicious, irreverent philosophies. In 1728, Voltaire returned to France, where he immersed himself in writing during the day and at night frequented the salons of Paris and Versailles where he entertained the ladies with his charm and wit and the *philosophes* with his political and theological liberalism. Voltaire got himself in trouble again with the French authorities with his publication of *Letters Concerning the English Nation* (1734), which covertly attacked the Church and political institutions of France, and once again he was forced to leave Paris. This time Voltaire took refuge at the Chateau de Cirey in the independent duchy of Lorraine. There he formed a close relationship with the aristocratic and learned Gabrielle Emilie Le Tonnelier de Breteuil, marquise du Chatelet (1706-49), who exerted a strong intellectual influence upon him. Voltaire's exile in Cirey was not without interludes of travel, though. He often traveled to Paris and to Versailles, where, through the influence of the Marquise de Pompadour, the famous mistress of Louis XV, he became a court favorite. At the court of Versailles, Voltaire was appointed historiographer of France, and then a gentleman of the king's bedchamber. Finally, in 1746, he was elected to the French Academy. Following the death of Madame du Chatelet in 1749, Voltaire finally accepted a long-standing invitation from Frederick II of Prussia to become a permanent resident of the Prussian court. He journeyed to Berlin in 1750, but stayed there only two years because his smart-aleck mouth clashed with the king's laconic German temperament, which led to frequent disputes between the two. For a while, Voltaire wandered about Europe, but he ended up in Ferney, near Geneva, Switzerland, where he spent the remaining twenty years of his life.

In his most memorable novel, *Candide* (1759), Voltaire satirized the Church as cruel, vindictive, and evil, and made fun of their belief that the pain and suffering in the world is propitiatory and all part of God's redemptive plan. Voltaire's main characters in *Candide* are unrealistically struck with terrible tragedies time after time, but continually repeat Leibniz's optimistic and sanguine phrase that this is "the best of all possible worlds." Notable throughout Voltaire's letters and writings was his obsessive hatred of Christianity, saying in his letters that it was necessary to "crush the infamous thing," meaning, of course, the Catholic Church. But Voltaire didn't just hate Catholics; he was equally hateful of Protestants, believing that all sects of Christianity stifle freedom of thought and social progress. Politically, Voltaire insisted that the state should be supreme over the Catholic Church and demanded universal religious toleration (Voltaire was a deist), abolition of censorship and the death penalty, lenient punishment of criminals, and a strong government acting only under general rules of law to safeguard social progress and individual liberty. Voltaire believed in strong government and wasn't against monarchies *per se;* he was only against what he called "unenlightened monarchies."

Another very influential social and political skeptic of eighteenth century France, whose writings likewise inspired the overthrow of the French monarchy by the commoners, was Jean-Jacques Rousseau (1712-78). Rousseau was actually born in Geneva, Switzerland, but lived in France most of his life. Coming from a working-class Protestant family, Rousseau never really felt at ease in France. He was paranoid that his French friends were really out to get him. Rousseau had no money, no social status, and fit the classic profile of the deluded outsider, pathetically maladjusted yet convinced of his virtue and righteousness in the face of heavy criticism. Rousseau's mother died a few days after his birth and he was subsequently raised by an aunt and uncle. Apprenticed at the age of thirteen to an engraver, Rousseau ran away at sixteen and became secretary and companion to an older woman, Madame Louise de Warens, who had a profound influence on his life and writings. In 1742, Rousseau went to Paris, where he earned a living as a music teacher, music copyist, and political secretary. He became close friends with the French philosopher Denis Diderot, who commissioned him to write articles on music for the multi-volumes French *Encyclopedie, ou dictionnaire raissone des sciences, des arts et des métiers (Encyclopedia, or Systematic Dictionary of Sciences,* published between 1751-1772), a biased and often inaccurate encyclopedia of science, philosophy, politics, and religion written by the leading *philosophes* and scholars of France, such as Montesquieu, Voltaire, and Baron von Grimm.

Rousseau's scathing criticism of French society, the aristocracy, and the Catholic Church instilled a deep sense of distrust and hatred amongst the French lower classes for the upper classes and the Church. In his *Discourse on the Origin of Inequality Among Mankind* (1755), Rousseau expounded the view that science, art, and social institutions have corrupted mankind, and said that the savage, natural state of primitive man was superior to the civilized state of modern man. Real virtues are a product of nature, not society, Rousseau said, which is too artificial and pompous. Voltaire, who disliked Rousseau, said that reading Rousseau's *Discourse* made him feel like "going on all fours." Rousseau, in turn, attacked Voltaire's "God of Reason." Reason, he said, is a false god when followed alone, and needs to be balanced by "feelings." Rousseau's emphasis on feelings, intuition, and consciousness over reason greatly influenced French society and helped popularize the Romantic movement of the nineteenth century and the later development of the mind sciences, like Freud's psychoanalysis. Thanks to Rousseau, the common man and the ordinary life were now the ideal. Sophisticated French ladies began breast feeding their own children again, Marie Antoinette started gardening and milking cows, men began crying in public, and the worship of nature replaced Christianity. Regarding Christianity, Rousseau believed that organized religions were corrupt and hypocritical. Virtue, he said, came not from organized religions and social institutions but from an appreciation of nature. Nature is to be loved. Nature is to be adored. Nature is God.

Rousseau's most famous work, *Emile* (1762), a dialogue on experimental education between a priest of Savoy and a student, expounded a new liberal theory of education, emphasizing the importance of self-expression rather than academic discipline to produce a well-balanced, freethinking child. Sensitivity, Rousseau said, comes before intelligence, feelings before ideas. "To exist is to feel," he said. Rousseau's permissive theory of education was later adopted by many liberal educators in France, Germany, and Switzerland, in spite of the well-known fact that Rousseau had five illegitimate children from a young, illiterate barmaid named Therese Levasseur, and put them all in orphanages. So much for Rousseau's feeling and caring about children.

Rousseau's other famous work, *Social Contract* (1762), championed the will of the people over the divine right of the king, and, therefore, both *Emile* and *Social Contract* were condemned by the Parliament of Paris in 1762. As a result, Rousseau fled to Switzerland in 1762, then to London in 1766 with his friend David Hume, the famous English skeptic. They soon became enemies, though, and Rousseau returned to France under an assumed name in 1768. Rousseau spent the remainder of his life in France, going insane for the last ten to fifteen years before his death. Rousseau's last major work was *Confessions*, completed in

1770, an autobiography of the many mistakes he made in his life. There is no doubt that Rousseau's widely-read attacks on the Church, the aristocracy, the monarchy, and the like were all driving forces behind the anarchy of the French Revolution.

The principal editor of the controversial and rationalistic French *Encyclopedie* was a Frenchman by the name of Denis Diderot. Diderot (1713-84) was a French philosopher, teacher, novelist, essayist, playwright, and art and literary critic from Paris who was considered one of the leading skeptics and antichristian materialists of pre-Revolutionary France. Born in Langres on October 5, 1713, and educated by Jesuits, Diderot's first major work, published anonymously, was *Pensees philosophiques* (1746), which was a philosophical apologia of his rationalistic deist beliefs. Diderot's second and third works, *Les Bijoux indiscrets (The Indiscreet Jewels,* 1748) and *Lettre sur les aveugles (Letter on the Blind Men,* 1749), particularly the latter, landed him in prison for its atheistic materialism. After a brief incarceration, Diderot was released to work with Jean le Rond d'Alembert, French mathematician and philosopher, on the *Encyclopedie,* a powerful propaganda tool of the French *philosophes* who tried to use science and philosophy to argue against Christian doctrines, monarchial conservatism, and peasant-landlord society. In turn, Diderot and his associates became the objects of clerical and royal attacks. In 1759, the Conseil du Roi formally suppressed the first ten volumes (published from 1751 onward) and forbade further publication. Nevertheless, Diderot continued work on the remaining volumes and had them secretly printed. The *Encyclopedie* was completed in 1765, with plates and supplements added until 1780. As history records, Denis Diderot,[13] like Jean le Rond d'Alembert[14] (1717-83), Paul Henri d'Holbach (1723-89), Julien Offroy de La Mettrie[15] (1709-51), and other atheistic materialists, was one of the leading provocateurs of the French Revolution.

Over in England, David Hume (1711-76), the one-time friend of Jean-Jacques Rousseau, was eighteenth-century England's most controversial philosopher and religious skeptic, nicknamed by his critics as "The Great Infidel." Born in Edinburgh, Scotland, on May 7, 1711, Hume was educated at home in a strict Presbyterian household. A child prodigy, Hume at age twelve matriculated to the University of Edinburgh where he excelled in his studies. Somewhere along the line, however, Hume renounced his belief in Christianity. When asked by a friend just before he died when he first became an unbeliever, Hume said that he was religious when he was young, and went away to Edinburgh as a believer, but lost his faith in college.

In 1729, Hume gave up the study of law to become a philosophical skeptic, saying that a tremendous awakening transported him beyond measure, as he engrossed himself in study rather than the pursuit of the customary pleasures of young men his age. Long hours of study and deprivation, however, affected him mentally. Often plagued by depression, anxiety, and the pain of poverty, Hume suffered from what he called "the disease of the learned." On one night, during a fever, he ranted and raved about the devil, hell, and damnation, and tried to commit suicide by drowning himself in a well.[16]

After four years of intense study, during which he had a nervous breakdown, Hume left Scotland in 1734 and settled in France in La Fleche, a small town in Anjou where Rene Descartes and Marin Mersenne had attended Jesuit school. From 1734 to 1737, Hume totally immersed himself in the study of philosophy, and during this period he wrote his most important philosophical work, *A Treatise of Human Nature* (1739-40), which characterized the core of his skeptical ideas. Because of its difficult-to-read style and poor editing, Hume's *Treatise* never gained popularity. Hume himself admitted that it was "dead-born" from the beginning. Careful not to repeat his mistake, Hume's later works were written in the easier-to-read essay or dialogue forms that were the vogue of his day.

After the publication of *Treatise,* Hume returned to his family estate in Berwickshire, Scotland and wrote a work on ethics and politics entitled *Essays: Moral and Political* (1741-42), which attained him some degree of success. Hume then applied for an appointment to the chair of ethics and philosophy at the University of Edinburgh, but was turned down because of his atheistic views. Hume went on to become tutor to the insane marquis of Annandale and then judge advocate to a British military expedition to France. In 1748, Hume published his *Philosophical Essays Concerning Human Understanding,* which after 1758 was called *An Enquiry Concerning Human Understanding,* an edited version of the *Treatise* with a section on miracles that was left out of the original. Hume's *Essays* proved to be a better seller than the *Treatise.* Like Timon of Philius, Pyrrho the Skeptic's pupil, Hume believed reason has limits because there exists equally valid arguments for and against any belief. In *Essays,* Hume said that he was "ready to reject all belief and reasoning, and can look upon no opinion even as more probable or likely than another." Most knowledge of matters of fact depends upon cause and effect, Hume said, and since no logical connection exists between any given cause and its effect, one cannot hope to know any future matter of fact with certainty. Even the most trustworthy laws of science might not remain true forever, which was a revolutionary and unheard of thought in Hume's time.

All ideas about truth and knowledge, Hume said, are derived from sensory "impressions" (sensations, passions, and emotions), not reason. The ideas we form in our minds that we think come from reason are, in fact, "the faint images of these [impressions] in thought." Therefore, because all ideas are derived from sensory impressions, no opinion or belief is more valid than any other. Reason, therefore, does not give us our beliefs. Beliefs come from habits, customs, principles, or imaginations that originate in sensory impressions.

> Reason can never show us the connexion [sic] of one object with another, tho' aided by experience, and the observation of their conjunction in all past instances. When the mind, therefore, passes from the idea or impression of one object to the idea or belief of another, it is not determined by reason, but by certain principles, which associate together the ideas of these objects and unite them in the imagination.

Thus, according to Hume, knowledge of matters of fact about anything is impossible. Hume, however, acknowledged that people had to think in terms of cause and effect and accept the validity of their beliefs, else they would "go mad." What Hume really meant was that it is all right to believe in God, just know that he's all in your mind.

Hume wrote in his *Enquiry,* Part II, in the section "of miracles," that he doubts the existence of God can be proven, and definitely denies the existence of miracles, that "there is not to be found, in all history, any miracle attested by a sufficient number of men, of such unquestioned good sense, education, and learning, as to secure us against all delusion in themselves."

Showing his true prejudice and bias against people of faith, Hume sarcastically remarked in the same section of *Enquiry* that believers in God and miracles were nothing but ignorant heathens. "It forms a strong presumption against all supernatural and miraculous relations, that they are observed chiefly to abound among ignorant and barbarous nations," he wrote.

David Hume returned to Edinburgh in 1751 and published *Political Discourses* in 1752. That same year, having again failed to obtain a university professorship, Hume received an appointment as librarian of the Advocates Library in Edinburgh. During his twelve-year stay in Edinburgh, Hume worked chiefly on his six-volume history of England, which appeared in parts from 1754 to 1762. In the years 1762 to 1765, Hume served as secretary to the British ambassador in Paris. There he gained fame as a literary genius with French intellectuals and *philosophes,* forming a strong friendship with Jean Jacques Rousseau.

Hume brought Rousseau back with him to England in 1766. Rousseau, however, was paranoid that Hume was trying to defame him, so their friendship ended and Rousseau returned to France, where he eventually went insane. After serving as undersecretary of state in London (1767-68), Hume retired to Edinburgh and there spent the rest of his life. Hume contracted a disease of the bowels that, like his mother, would end in his death. On his deathbed, while attended by his physician, Dr. Black, Hume admitted to an interviewer, David Boswell, that he lost his faith in college after reading Locke and Clarke, and henceforth believed that all religion was bad and immoral for mankind: "He [Hume] said he never had never entertained any belief in Religion since he began to read [John] Locke and [Samuel] Clarke. . . . [17] He then said flatly that the Morality of every Religion was bad . . . that when he heard a man was religious, he concluded he was a rascal.'"[18]

David Hume died an unbeliever on August 25, 1776, and on August 29 his coffin was carried out on St. David Street and buried amidst a violent rainstorm. Having once said that "The Church is my aversion,"[19] Hume's attacks on Christianity continued well after his death, with the posthumous publication of his autobiography in 1777, and his *Dialogues Concerning Natural Religion* in 1779. Hume, some say, cowardly waited until he died to publish his *Dialogues* because of its blatant skepticism of religion. Atheists ever since have been in love with Hume and have considered him a hero of antichristianity.

Idealism: Mind Is What Matters

"What is mind? never matter. What is matter? never mind."

—Unknown

Idealism is a human philosophy that maintains that the primary nature of reality lies in man's consciousness and reason. Only that which is perceptible to the human mind through internal ideas and external sensations are real. The mind, therefore, plays the key role in understanding and experiencing the realities of the world. It is both rational and empirical in theory, holding that the understanding of sensory experiences is necessarily dependent on *a priori* reason. The extreme form of idealism is a philosophy called solipsism, which says that nothing exists outside of the mind. To solipsists, the entire world is a figment of one's imagination, like a dream. The absurdity of solipsism is easy to prove by asking solipsists to step out on the freeway and see whether or not traffic is real, which they will never do because they know, despite what they pretend, the reality of external objects. The vast majority of idealists are not solipsists, and they give full credence to the existence of external things. They do say, however, that the mind is paramount in importance in producing and sustaining modes of beings or realities that otherwise would not be possible; such as faith, ethics, morality, justice, art, music, color, sounds, smells, language, and mathematics. Idealists believe that the mind is crucial in observing how external objects truly exist in nature, and is indispensable to scientific observations and experimentations.

The philosophy of idealism can be traced all the way back to Plato, who postulated the existence of forms or ideas, genuine substances that constitute the real, higher, intelligible world. Ideas, Plato said, are more reliable than sensations because senses can be deceived. External objects imperfectly reflect the more clearly intelligible and more real ideal forms that exist within our minds. Plato's forms, therefore, are, to him, more real and more perfect than external things.

In the eighteenth century, idealism was breathed new life by the English philosopher and Anglican clergyman George Berkeley (1685-1753), the founder of the modern school of idealism, as an argument against the growing movement towards skepticism and atheism. Born in county Kilkenny, Ireland, March 12, 1685, Berkeley studied at Trinity College, Dublin, where he became a fellow in 1707. In 1709, he published his *New Theory of Vision* and the following year *A Treatise Concerning the Principles of Human Knowledge.* When his *Treatise* failed to win widespread support for his defense of idealism, he published a more popular version in 1713 written in the old dialogue style, *The Three Dialogues Between Hylas and Philonous,* which gained him more notice.

Berkeley was ordained a deacon in the Anglican Church of Ireland in 1724 and became increasingly prominent within his church. In 1728, he went to Bermuda to found a missionary college. Although he abandoned his plan in 1732, Berkeley had a great effect on higher education while in America, assisting in the development of Yale and Columbia and a number of other schools. It is important to note the highly secular Ivy League schools of today were founded as Christian universities. In 1734, Berkeley became bishop of Cloyne. He died on January 14, 1753.

Berkeley was a devout Anglican who objected to the skeptical, scientific worldview of his day, and the empirical philosopher John Locke was the primary target of his criticisms. Berkeley believed that Locke's empirical beliefs, which maintained that all ideas are formed in our mind as a result of sensory experiences of matter, necessarily leads to materialism, skepticism, and atheism. In an effort to downplay empiricism, Berkeley tried refuting the reality of matter, which is nonsensical philosophically. Berkeley theorized that material things have no reality in themselves except as perceptions in our minds (a radical philosophy known as immaterialism).[20] "To be is to be perceived," Berkeley would often say. To Berkeley, only things that are perceived exist in reality. To answer the question on whether or not things exist if we don't perceive them, Berkeley said that they do because they're always perceived by God, who is all knowing and perceives everything. Our minds, of course, don't conjure external objects into existence, but the ideas of them are caused in the human mind by God.

Berkeley's espousal of immaterialism never gained much of a following. In fact, most of Berkeley's contemporaries in England thought his theory on ideas rather foolish. Berkeley would, however, greatly influence the development of German idealism, a major philosophical movement in eighteenth- and nineteenth-century Germany, in the persons of Immanuel Kant, Johann Fichte, George Hegel, and Friedrich Schelling, four very important philosophers who significantly altered the course of human history.

The rise of Germany as the dominant world military power in 1871 after the defeat of Napoleon III's army at the Battle of Sedan, followed its rise as the dominant intellectual center of Europe. Nineteenth-century Germany was home to some of the most famous, or, rather infamous, philosophers in history. Among them were the idealists (Kant, Hegel, Fichte, and Schelling), the pessimist Arthur Schopenhauer, the materialist and antichristian zealot Ludwig Feuerbach, the communists Marx and Engels, the nihilist and self-proclaimed antichrist Friedrich Nietzsche, and the existentialist Martin Heidegger.

The most celebrated German idealist, Immanuel Kant (1724-1804), is often called the greatest philosopher of modern times because the majority of nineteenth- and twentieth-century philosophers owe at least part of their philosophical theories to him. Kant's idealistic and antichristian philosophies influenced the development of two opposing philosophical movements in Germany: on the right was German romanticism, Aryan nationalism, and jack-booted totalitarianism, which can be traced back to Kant's disciples Johann Fichte and Friedrich Schelling; on the left was Marxist communism, which can be traced back to Kant's disciple George Hegel, whose dialectic theory was taken from Kant and later adopted by Karl Marx and Friedrich Engels in their formulation of historical materialism (communism). Both of these radical philosophical movements (Nazism and communism) originated from Germany and owe their existence to one man, Immanuel Kant.

It is important to note that Immanuel Kant had an extremely

symbolic name, which fits his role in human history as the primary philosophical impetus behind godless Nazism and communism. Immanuel, as everyone knows, is the Hebrew word for "messiah," which means "God is with us." Kant, on the other hand, is the German variant of the English word "cant."[21] According to Webster's English dictionary, cant, as a noun, is the private language of the underworld, and as a verb means to speak hypocritically and use pious words insincerely. Combining the two names, Immanuel and Kant, transliterates to "the messiah who tells demonic lies."

I truly believe that Satan was mocking Jesus Christ, the Word of God Made Flesh, through Immanuel Kant and antichristian philosophers like him. Let's examine the evidence. Kant renounced his Lutheran faith not long after entering college and began teaching a totally false conception of God and Christianity. For instance, Kant believed that the Bible was only meant to teach morality and not to be taken literally, and he rejected the dogmatic prayers and devotional exercises of his pietist Lutheran upbringing, calling them "hypocritical" and "delusional." Religion, Kant said, was "ingratiating" and "fawning" to the Highest Being. However, Kant was careful to make a clear distinction between believing in the moral teachings of Christ, which he claimed he did, and believing that Jesus was God, which he didn't: "Now it is very clearly seen that the apostles regarded the side-doctrine of the Gospel as its fundamental teaching, and . . . instead of extolling the practical religious doctrine of the holy teacher [Jesus] as essential, they preached reverence for the teacher himself, and a kind of ingratiation by flattering and by eulogizing him."[22]

To Kant, Jesus was an exemplary "archetype lying in our reason . . . of a course of life well-pleasing to God,"[23] but he was not God. This declaration soon brought Kant into trouble with the German authorities. Kant's writings and lectures on religion were seen by the Germany authorities as too antichristian, so Kant was forbidden by Frederick William II, king of Prussia, to teach or write on religious subjects. Near the end of his life, the king sent a letter to Kant on October 1, 1794, when Kant was seventy years old, that reproached him for having "misused" his philosophy over a long period of time "for the distortions and debasing . . . of Holy Scripture and Christianity." Kant disagreed with the king that he debased Christianity, but agreed to stop teaching about religion for five years until the death of the king, and then felt released from his liability and continued publishing his unorthodox views about God, which caused a lot of Germans to renounce their belief in Christianity.

Kant said he believed in God, but one has to be very cautious when people like him say that they believe in God because the word "god" is very fuzzy and generic, and can be used to designate any number of things other than the God of the Bible. To some people, god can mean nature, ethics and morality, a universal world spirit, demons, or even man himself.[24] While "god" can mean anything, there is no mistaking Jesus Christ, who claimed to be God. (See chapter 4 for the evidence of Jesus' divinity.)

Many intellectuals and philosophers say that they believe in God but refuse to believe in Jesus Christ. They hate Jesus because he sets limits on their freedom to do as they please. Kant was one of these people. He loved his freedom and felt that Christianity was far too constraining. So he, like many philosophers before and after him, renounced orthodox Christianity and remade God in their own preference. These people are the false messiahs and false prophets of the latter days whom Jesus Christ warned us of,[25] those who multiply words upon words, play games with language, invent obtuse and hard to comprehend theories and philosophies in order to impress and convert others to rebellion like them. But they, in fact, are liars. So too was Kant.

Immanuel Kant was born in Königsberg (now Kaliningrad, Russia) on April 22, 1724, and received his education at the Collegium Fredericianum and the University of Königsberg. At Fredericianum, Kant studied Latin and the ancient classics, and at Königsberg he studied the hard sciences, physics and mathematics. After his father died, Immanuel was compelled to halt his university career and earn his living as a private tutor. In 1755, Kant resumed his studies and obtained his doctorate. From that time on, he taught at the university, lecturing first on science and mathematics, and then philosophy. Kant received a chair at the University of Königsberg in 1770, when he was made professor of logic and metaphysics, and for the next twenty-seven years he was a philosophy instructor who attracted a wide following.

Early on in his career, Kant was a student and follower of the German rationalist Christian Wolff, but later on became a follower of Locke, Rousseau, Berkeley, and Hume. Kant was the first philosopher to attempt to combine the antichristian philosophies of the rationalists, empiricists, idealists, and skeptics into one philosophy. He came up with a brand new epistemology he called "transcendental idealism," which he proudly thought to be as important as Copernicus' revolutionary theory that the sun was at the center of the universe. According to Kant, the world is comprised of two different principles: tangible "phenomena," such as rocks, which are touchable and sensible and therefore knowable and transcendent "principles," such as God, the soul, freedom, and immortality, which are thinkable but not knowable.

Transcendent principles, Kant said, are "things in themselves" or *noumena.* They cause sensations in our minds that we truly experience as ideas, but we can only know them as they appear to be, not as they truly are, because they do not exist in space and time. It is the mind that furnishes the forms and categories (space, time, causality, substance, and relation) to these sensations and experiences of things in themselves. Transcendental idealism, by Kant's own definition, is agnostic and skeptical because it denies the possibility of knowing ultimate realities like God; it is empirical in that it asserts that all knowledge arises from experience; it is rationalistic in that it maintains the *a priori* (intuitive) character of this empirical knowledge; and it is idealistic in that it believes that the mind gives meaning to sensations and experiences. Thus, Kant's transcendental idealism was an amalgam of all the popular secular philosophies of the last four hundred years.

Kant believed that there was a problem with using reason alone (rationalism) when defining ultimate realities such as God. Reason, Kant said, inevitably falls into contradiction when it contemplates these kinds of extrasensory *noumena* (Kant's problem of "antimonies" or opposing viewpoints). For example, Kant pointed out that there are equally plausible arguments for something (the "thesis") and against something (the "antithesis"), such as an infinite universe or an eternal being (a concept first proposed by Timon of Phlius, a student of Pyrrho the Skeptic). Kant underscored the negative implications of these supposed contradictions in his theory of the transcendental "dialectic," which, he thought, provided a systematic account of metaphysical "illusions." In an obvious jab at Christianity, Kant said that metaphysics must solve the problem of antimonies if it is to survive. To demonstrate the problem of antimonies and the supposed illusions of orthodox religion, Kant set out to destroy every traditional argument for the existence of the biblical God in his *Critique of Pure Reason* (1781). Then, in his *Critique of Practical Reason* (1788), Kant gave reasons for believing in his type of god as a moral ideal rather than a personal being revealed to us through Jesus Christ. This tearing down of the Biblical God, of course, won Kant many friends on the liberal side, but many enemies on the conservative. One of these liberals was a man by the name of Karl Marx, the co-founder of communism.

Kant's theory of antimonies and their resolution by way of the dialectic method of thesis and antithesis was adopted in a revised form by the German idealist George Hegel, who in turn greatly influenced Karl Marx's theory of dialectical materialism (communism). Marx's theory maintained that the material (economic) basis of a reality are constantly changing in a dialectical process, and that matter (and its unequal distribution) takes precedence in determining the course of human history (Marx's class struggle for material goods). In this way Kant's transcendental idealism greatly contributed to the evolution of Marxist communism by way of his theory of antimonies and their resolution by means of the transcendental dialectic.

On the subject of ethics and morality, Immanuel Kant insisted in his work *Metaphysics of Ethics* (1797) that the free will of human beings was in itself sufficient for determining morality. Kant was totally against a morality of ethics given by any theologian or natural scientist. He wanted, instead, an "autonomous" morality of reason that has its own *a priori* knowledge of right and wrong. To Kant there was only one "categorical imperative" of morality. "Act only on that maxim whereby thou canst at the same time will that it should become a universal law," he wrote.[26]

Two other important idealists, Johann Gottlieb Fichte and Friedrich Schelling, were students of Kantian idealism. Fichte (1762-1814), the first of the post-Kantian idealists, was born on May 19, 1762, at Rammenau in Saxony, and was educated at Pforta, Jena, and Leipzig. Fichte anonymously published an essay entitled *Attempt at a Critique of All Revelation* (1792), which at first was thought to be written by Immanuel Kant, and which later gained him the philosophy chair at Jena in 1793. In 1799, however, Fichte, like Kant, was charged with espousing atheism and forced to resign his post. In 1805, Fichte was appointed to the philosophy chair at Erlangen, and in 1810 he became the first rector of the new University of Berlin before he died in Berlin on January 27, 1814.

Fichte was greatly influenced by Kant's transcendental idealism, but he substituted Kant's "things in themselves" (*noumena*) with his "unconscious self-limitation of the I." The "I" is not a thing or a substance, Fichte said, it is activity that posits itself (the thesis). To be aware of itself, though, the I must posit non-I (the antithesis). A contradiction arises when I both posits itself and negates itself. To resolve this contradiction, Non-I, in part, negates the I, and I, in part, negates the non-I (the synthesis). Fichte's version of thesis, antithesis, and synthesis reappeared in Hegel's works, although Hegel didn't like to use Fichte's terminology of thesis, antithesis, and synthesis.

A summary of Fichte's doctrines appeared in his work *Vocation of Man* (1800), which transforms Kant's critical idealism into absolute idealism by making the human will the ultimate reality. Fichte maintained that the world is created by an absolute ego, of which the human will is a partial manifestation and which tends toward God as an unrealized moral ideal. God is the moral order of the world and we exist "only in

God and through God." But Fichte's belief that God was the moral order of the world and not a personal being as maintained by Judeo-Christianity caused him to be accused of atheism and lose his chair at Jena.

There were two important consequences of Fichte's philosophies. Fichte's absolute idealism, with its emphasis on man's ego and will as the sole basis of reality, was adopted by German ultra-nationalists in their espousal of German racial purity and superiority and the triumph of the will. On the other hand, Fichte's system of resolving philosophical contradictions by using thesis, antithesis, and synthesis was adopted by George Hegel, who, in turn, influenced the birth of Marxist communism or dialectic materialism.

Friedrich Schelling (1775-1854) was one of the leading proponents of romanticism, a literary, artistic, and philosophical movement that started around 1750 and became very popular in Europe during the first half of the nineteenth century. Romanticism, which was idealized in the writings of Rousseau and the German poet, novelist, and dramatist Johann Wolfgang von Goethe (1749-1832), sought to break away from the philosophical and scholarly regimentation of the classics and neo-classics and move on to a more open society of emotions, imaginations, freedom of expression, the idolization of nature, and the elevation of the common man to the ideal. Schelling, a leading European idealist, romanticist, and pantheist, believed that the evolution of nature (God) culminated in the human organism, which is an emanation of the self-realization of the Absolute. Schelling's unorthodox ideas, of course, were influenced by Spinoza, Kant, and Fichte, and would later influence the Nazi's romantic idealization of the tall, blond-haired, blue-eyed Aryan man.

One of the leading Western philosophers and thinkers of the nineteenth century was the German idealist George Wilhelm Friedrich Hegel (1770-1831). Hegel was born in Stuttgart on August 27, 1770, the son of a revenue officer with the civil service. Hegel was brought up in a strict Lutheran household and attended the Stuttgart gymnasium preparatory school where he obtained the usual education in Greek, Latin, and the classics. Encouraged by his father to become a clergyman, Hegel entered the seminary at the University of Tuebingen in 1788. There he developed a close friendship with Friedrich Schelling. Having completed his courses in philosophy and theology at seminary, Hegel decided not to become a pastor. Instead, he became a private tutor in Berne, Switzerland, in 1793. In 1797, Hegel assumed a similar position in Frankfurt, and in 1801 he went to the University of Jena, paid for by an inheritance from his father, where he became a lecturer on philosophy. At Jena, Hegel completed his most important philosophical work, *The Phenomenology of Mind* (1807). Hegel remained at Jena until October 1806, when the city was taken by Napoleon and he was forced to flee. Having spent his father's inheritance, Hegel became editor of the *Bamberger Zeitung* in Bavaria. After that, he moved to Nuremberg, where he served for eight years as headmaster of a gymnasium. While at Nuremberg, Hegel published an important work over a period of several years, *The Science of Logic* (1812, 1813, 1816). In 1816, Hegel became a professor of philosophy at the University of Heidelberg. Soon after, he published in summary form a systematic statement of his entire philosophy, entitled *Encyclopedia of the Philosophical Sciences in Outline* (1817). In 1818, Hegel transferred to the University of Berlin, where he stayed for the remainder of his career. He died there on November 14, 1831, during a cholera epidemic.

Hegel is one of the hardest philosophers to read and understand, and his ideas and theories can be deciphered in many different ways. Nevertheless, Hegel had a tremendous following in the first half of the nineteenth century. Hegelians were divided into two camps: Left (Young) Hegelians and Right (Old) Hegelians. Right Hegelians were conservatives who thought that Hegel's theories lent support to their belief in orthodox Christianity and the status quo in government. Left Hegelians, on the other hand, were radicals, antichristians, and anarchists who thought Hegel supported their revolutionary and anticlerical beliefs.[27] Among the most infamous of the Left Hegelian radicals were Karl Marx, Friedrich Engels, Bruno Bauer,[28] Ludwig Feuerbach, and Max Stirner.[29]

Influenced by the antichristian idealism and dialectics of Kant and Fichte, Hegel in his *Lectures on the History of Philosophy* (1833-36)[30] stated that there was a philosophical conflict in the course of human history, which began with a thesis, which was then contradicted by an antithesis, which in turn would finally be resolved by a synthesis.[31] History's thesis started during the time of Socrates when the science of ethics and morality begin to emerge in pagan Greece, and culminated in the triumph of Christianity during the Middle Ages.[32] The historical conflict or antithesis began during the time of the Protestant Reformation when individual consciousness began to challenge the authority of Church dogmas and doctrines. The synthesis or resolution of philosophical and ideological conflicts will come, Hegel said, only when a rationalistic, harmonious community emerges out of the conflict that fosters both freedom of thought and respect for individual rights.[33] Both Marx and Engels borrowed from Hegel's idea of the historical dialectic, but they said that the course of human history has been shaped not by ideology but by the struggle for material goods (dialectical or historical materialism). Therefore, class struggle, Marx said, has determined the events of world history, and the struggle would only end in the overthrow of bourgeoisie capitalists and the governments that support them by the violent uprising of the proletariats. The proletariat revolutionaries would then establish a community where material goods are divided equally, "each according to his abilities and each according to his needs" (the basic theory of Marxist communism). Then, and only then, would world conflicts and class struggles end forever.

Karl Marx espoused Hegel's theory of the historical dialectic, but he criticized what he called Hegel's "mysticism" regarding the existence of what Hegel termed the "Absolute Spirit." In his two works *The Phenomenology of Mind* and *The Science of Logic,* Hegel said that the Absolute Spirit "is" the world. The world is whole, said Hegel, and the whole is Absolute Spirit. Dialectically speaking, Absolute Spirit is Pure Being, the thesis; but pure being without any qualities is Non-Being, the antithesis. The union of Being and Non-being is Becoming, the synthesis—the Absolute evolving from an abstract, undifferentiated being into a more concrete reality by the dialectical process. The Absolute, therefore, is in the process of Becoming; developing over the course of history towards an end.

Hegel's idea of Absolute Spirit is somewhat confusing. Is it the transcendent God of the Bible who created the universe and remains separate from it, or is it some sort of cosmologic, universal world force? Hegel said the Absolute is Thought thinking about itself. This Thought or Reason is the true substance of the universe, eternal and all-powerful in essence. It is infinite spirit or Mind, what German philosophers call the *Geist.* All philosophical and intellectual achievements up until now, Hegel said, are the result of Mind coming to know itself as ultimate reality. Mind alone is all that is real. Every human mind is part of this universal Mind, which guides the course of man's intellectual accomplishments, moral ethics, and formation of national

Hegel's philosophy of Absolute Spirit or Mind is very similar to Spinoza's pantheism, the heretical belief that all of nature and the universe is God. We know that Hegel studied in great detail Asian history and Eastern mysticism, as well as Spinoza, and was no doubt greatly influenced by both of them. And although he held up Christianity as the ideal religion in terms of teaching ethics and morals to children, Hegel was by no means a Christian. Hegel's unorthodox theories about the Absolute Spirit being equal to the world helped popularize pantheism and Eastern mysticism in Europe, much to the detriment of Christianity, and likewise influenced the rise of godless atheistic communism, two contradictorily yet equally antichristian philosophies attributable to both Kant and Hegel.

Materialism: Matter Is All There Is and All That Matters

The aforementioned philosophy of idealism maintains that ultimate reality exists only in man's consciousness and reason. The material world exists, but is only perceived through the formation of objects in the mind via sensory inputs. Contrary to idealism is the human philosophy known as materialism, which says that matter is the only thing that truly exists. There are no eternal souls or spirits in human beings, and nothing exists outside of the material universe; no gods, no angels, no heaven, nothing. Feelings, morals, thoughts, and reason are purely physical processes that take place in the brain from biochemical actions and reactions in neurons. Materialism is atheistic and unbelieving, yet it is one of the most widely accepted philosophies in our educational system today, especially amongst liberal scientists and philosophers.

Materialism as a human philosophy has several categories. Ancient philosophers such as Democritus, Leucippus, Epicurus, and Lucretius subscribed to a type of materialism known as "hylozoism," which maintains that matter and life are one and the same. Cosmological materialism, the kind most modern philosophers and scientists adhere too, is a term often used to characterize the usual materialistic interpretation of the universe: the universe consists of nothing but matter. The most radical form of materialism, the concern of the remainder of this section, is antireligious materialism. Antireligious materialism is motivated by an extreme hatred and hostility toward the theological dogmas and doctrines of organized religion, especially Christianity.

As we have already seen, the growing antireligious sentiment of the Age of Enlightenment was in large part responsible for the French Revolution and the Napoleonic Wars. Antireligious materialists such as the eighteenth century French *philosophes* Denis Diderot, Paul Henri d'Holbach, Jean le Rond d'Alembert, and Julien Offroy de La Mettrie hated Christianity and advocated the overthrow of the Catholic Church. But the coming of the nineteenth century saw an even greater hostility towards Christianity by a new generation of antireligious materialists in Germany, in the persons of Ludwig Feuerbach, Karl Marx, and Friedrich Engels. These three atheists were not content to just criticize and control religion, they wanted to forcibly dethrone God and put man in his place: the original crime of Lucifer.

Ludwig Andreas Feuerbach (1804-72), whose last name in German means "fire brook," was a hero to Karl Marx, Friedrich Engels, Friedrich Nietzsche, and Sigmund Freud. He is considered by many to be the first major philosopher to try to disprove God's existence and the credibility of Christianity through a comparative study of science and anthropology. Feuerbach maintained that religion and belief in God are really projections of the highest ideals of man, from the dawn of civilization to modern times. The pagan polytheism of ancient cultures, Feuerbach said, was a glorification of the wonder and beauty of nature, and modern Christianity is really the glorification of man himself. Thus, God is really the glorification of man's highest principles in himself—the mirror image of man. Man makes God in his own image, rather than God making man in God's image.

Feuerbach was one of the most important philosophers in history, in that his writings were widely read and espoused by communists, Nazis, atheistic psychologists, secular humanists and many other groups and organizations hostile to Christianity. An understanding of Feuerbach's life and writings are vital to the understanding of the scourges and catastrophes that befell the twentieth century.

Ludwig Feuerbach was born in Landshut, Germany, and studied Protestant theology at Heidelberg as a youth. Feuerbach later moved to Berlin, where he studied philosophy for two years under Hegel, and then obtained a teaching post at Erlangen, where he was a leading Left Hegelian. In 1830, Feuerbach published his first atheistic work, *Thoughts on Death and Immortality,* which argued against immortality and the existence of God. This book caused him to fall out of favor with the faculty and lose his post at Erlangen, and ended his career as a teacher. Henceforth, Feuerbach withdrew into private life to write. In 1839, Feuerbach published *Towards a Critique of Hegel's Philosophy,* and two years later he published his famous diatribe against religion entitled *The Essence of Christianity,* which detailed his atheistic belief that Christianity was false but psychologically reassuring to the naive and weak-minded.

Religion and the belief in God, Feuerbach said, is only useful in that it satisfies a psychological need, our preoccupation with ourselves. The worship of God, therefore, is actually worship of our idealized self. Only when we finally realize this will our mental health and overall lot in life improve. "The turning point of history will be the moment when man becomes aware that the only God of man is man himself," said Feuerbach.

In his controversial *Essence of Christianity,* Feuerbach said that his main goal in life was to replace the love of God with love of man, and faith in God with faith in man (a popular atheistic philosophy known as secular humanism). To accomplish this goal, Feuerbach felt it necessary to exorcize the "ghosts" of theology and idealism past by a comparative study of science and anthropology. Then and only then would man be totally free to realize his full potential as a god. Said Feuerbach, "The goal of my work is to make men no longer theologians but anthropologists, to lead them from the love of God to the love of men, from hopes for the beyond to the study of things here below; to make them no longer the base religious or political servants of a monarchy and an aristocracy of heaven and earth, but free and independent citizens of the universe."[34]

Feuerbach felt that the material needs of people should be the primary concern of society, and providing more food to the hungry would be a good starting point. "Man is what he eats," Feuerbach would quip.

Feuerbach's materialistic and atheistic beliefs had a tremendous influence on two of his most notorious disciples, Karl Marx and Friedrich Engels, the evil co-founders of communism. The history of communism is extremely important to know and understand because of the tremendous destruction it has caused the world. In the nineteenth century, tens of thousands were massacred in the communist-inspired European Revolution of 1848 and the Paris Commune uprising of 1871, when Marxist communists tried to overthrow Europe's monarchial governments. In the twentieth century well over 100 million people died in Europe, the Soviet Union, Asia, Africa, and elsewhere, in the wars, purges, and famines of Lenin, Stalin, Mao Tse-Tung, Pol Pot, Ho Chi Minh, and others. In fact, communism has caused more death and destruction than any other ideology in history, and it hopes to make a comeback once Christianity is overthrown. For those of you who berate Christianity for causing more death and destruction than any other ideology, the truth is that the Crusades and Inquisitions caused the deaths of hundreds of thousands, while communism and Nazism caused the deaths of hundreds of millions. There is no comparison.

Marx and Engels were firmly convinced that communism was supported by history and science.[35] Independently, these two researched science, sociology, politics, economics, world history, philosophy, anthropology, and other subjects in formulating their theory of historical materialism; which they arrived at separate of each other at the same time in different locations—Engels in Germany and Marx in Paris. Together they joined forces and changed the course of world history forever.

Karl Marx (1818-83), whose first name means "man," was born in Trier, Germany, on May 5, 1818, the second of seven children of Jewish parents. Although his parents were Jewish, Karl was baptized a Lutheran at the age of six. Karl's father was a lawyer and a middle-class bourgeoisie. After attending a Protestant high school in Trier, Karl's family sent him to three universities, Bonn, Berlin, and Jena, where he majored in philosophy and history. In college, Marx was a boisterous rowdy and a heavy drinker. He once fought a duel and once was jailed for drinking. Despite his hooligan ways, Karl Marx obtained his doctorate in philosophy in 1841.[36] In 1842, shortly after contributing his first article to the Cologne newspaper *Rheinische Zeitung*, Marx became its editor. Although a political radical, Marx was not yet a communist. Marx's political editorials got him in trouble with the authorities, and in 1843 he was compelled to resign his editorial post. Marx left Germany for Paris, where he continued his studies in world history, philosophy, economics, and political science. It was in Paris that Marx formulated his communist ideology. In 1844, Friedrich Engels left Germany for Paris to meet Marx. The two had previously become friends in Berlin in 1842. In Paris, Marx and Engels found that they had independently arrived at identical theories on socialism (communism), so they agreed to collaborate in the formulation of the theoretical principles of communism and in the organization of an international working-class movement dedicated to communist principles. In the movement, Marx would research political thought, political economy, and economic history, while Engels would concentrate on that plus the natural sciences, which he knew more about.

Engels (1820-95), whose first name means "peaceful ruler" and whose last name means "angels," was born in Barmen, Germany (now Wuppertal). The son of a cotton-cloth manufacturer, Engels came from a wealthy Protestant family. At an early age, Engels wanted to have a career in literature but his father insisted that he work in the family business. As a young man, Engels was influenced by the writings of the radical German poet Heinrich Heine, the early socialists Charles Fourier,[37] the duc de Saint-Simone,[38] Moses Hess, the British economic reformer Robert Owen,[39] the German philosophers Hegel and Feuerbach, and the modernist theologian David Strauss, whose skeptical reinterpretation of the Gospels called *The Life of Jesus* (1835) was said to have turned Engels into an atheist. In 1839, Engels began writing on literary and philosophical topics for a number of publications, and in 1842, the same year he met Marx, he became a communist. Between 1842 and 1844, Engels studied a workers' rights movement in Britain called Chartism. Engels' research of Chartism and the poor working conditions in Britain and his vast literary research and experiences led him to formulate his doctrinal theory of communism. He first introduced his theories in a work entitled *Condition of the Working Class in England* (1844), a work that established Engels' reputation in Germany as a political revolutionary and anarchist.

In 1845, Marx was kicked out of Paris because his revolutionary activities were viewed as a threat to the French government. Marx then moved to Brussels, Belgium, and began the work of organizing and directing a network of revolutionary groups, called Communist Correspondence Committees, in a number of European cities. In connection with the consolidation of these committees in 1847 to form the Communist League, both Marx and Engels were commissioned by its charter to formulate and publish a statement of communist principles. Thus, in 1848 Marx published the *Communist Manifesto* in London, which was written from a rough draft prepared by Engels. This notorious gospel tract of anarchy and rebellion has since sold hundreds of millions of copies worldwide, second only to the Bible, and has inspired untold number of revolutionaries.

The *Communist Manifesto* is a declaration of godless principles and objectives of the Communist League, a secret European organization of emigrant German artisans and workers, intellectuals, and political revolutionaries, which Marx and Engels headed. The *Communist Manifesto* begins with a warning to the world: "A specter is haunting Europe—the specter of communism." It then proceeds to outline the theory of historical materialism, which begins in dialectic fashion with a thesis, which in turn is contradicted by an antithesis, which is finally resolved by a synthesis: worldwide communist domination and the overthrow of all religion.

History, according to Marx and Engels, is a chronicle of class struggle over the material goods of the world, a struggle between the exploiters and the exploited, between the haves and the have-nots. Said Marx in the *Communist Manifesto,* "the history of all hitherto existing society is the history of class struggles."

In every historical period the prevailing economic system by which material goods are produced determines the form of societal organization and the political, religious, ethical, intellectual, and artistic history of that period. History, Marx said, is characterized by a succession of five socio-economic periods. The first were the primitive communists, Eskimos and Indians. The second were the large Asiatic empires, the third were the slave-owning societies of ancient Greece and Rome, the fourth were the feudal societies of the Middle Ages. The last, the period Marx lived in, is the bourgeoisie socio-economic formation of modern capitalism. According to Marx, an aristocracy dominated the feudal society of the Middle Ages, where the most important thing was one's birth. In feudal society, there was a paternal, deferential relationship between landowner and serf, the aristocrat being the father and the peasant being the child. Feudal societies were characterized by a conspicuous consumption of material goods by the aristocrats, with much of the rest of society going without.

After the Middle Ages, as industry, commerce, mercantilism, and trade increased in simultaneity with population growth, a new socio-economic class emerged—the bourgeoisie, or middle, class. In the bourgeoisie socio-economic formation of modern times, the emphasis on bloodlines is replaced by an emphasis on merit, deference replaced by competition, greed and conspicuous consumption replaced by thrift. As the Industrial Revolution of the nineteenth century progressed, Marx said, the ruling bourgeoisie capitalists compete more and more for the forces of production (labor, tools, raw materials), which negatively affects the relations of production (how people engage and relate to each other). The working-class proletariats migrating into industrialized centers in search of work become more and more exploited by the competing bourgeoisie capitalists who seek larger and larger profits, which, Marx predicted, would lead to a revolution by the disenfranchised proletariats, who would unite and overthrow capitalism and form a classless society of workers where everything is distributed equally, "each according to one's abilities and needs." Agricultural and industrial communes would be set up where competition for the forces of production is replaced by cooperation. Once communism forcefully takes over, Marx said, conflict and struggles will cease forever because there will be no one left to exploit economically, and the communist state, which was formed during the revolution by a socialist, scientific elite, will simply "wither away."

To bring about this so-called workers' paradise on earth, Marx emphasized the necessity of forcibly abolishing all private property in land and application of all rents of land to public purposes, establishing a heavy progressive or graduated income tax, abolishing all rights of inheritance, confiscating the private property of all emigrants and rebels, centralizing credit in the hands of the communist state by means of a national bank with state capital and an exclusive monopoly, centralizing the means of communication and transportation in the hands of the state, extending factories and instruments of production owned by the state, ensuring equal liability of all to labor, establishing industrial and agricultural armies, combining agriculture with manufacturing industries, gradually abolishing the distinction between town and country by a more equitable distribution of the population over the countryside, establishing free education for all children in public schools, abolishing the family unit, and most important of all, abolishing all religion. Children, furthermore, were to be raised by the state, not by their parents.

Marx and Engels hated all religions, especially Christianity, which they accused of conspiring with the bourgeoisie governments to economically suppress the ignorant masses. Fellow communists, therefore, were called upon to overthrow all world religions. This fight, Marx declared, would be a "fight against the world," against the perverted world whose "spiritual aroma is religion."[40] Only when all religion is abolished, Marx said in his *Toward the Critique of the Hegelian Philosophy of the Right,* would man be truly happy. "The first requisite for the happiness of the people is the abolition of religion," he said.

Marx characterized religion as an addiction of the oppressed, similar to heroin addiction. "Religion is the sigh of the oppressed creature, the feelings of a heartless world, just as it is the spirit of unspiritual conditions. It is the opium of the people," he said.

Much of Marx's hatred of religion came from reading Feuerbach. In *The Holy Family,* co-written with Engels in 1845, Marx praised Feuerbach for having "dispelled the old quibbles" about religion and having rightly placed man on God's throne. "The religion of the workers has no God," Marx said, "because it seeks to restore the divinity of man."[41]

Immediately after the *Communist Manifesto* appeared in early 1848, with its rallying call for the "workers of the world [to] unite," revolutions broke out in France and Germany. The Belgian government, fearful that revolution would spread to their country, kicked Marx out, and he went back to Paris and then to the Rhineland. While in Cologne, Germany, Marx established and edited a communist periodical, the *Neue Rheinische Zeitung,* and organized revolutionary activities. In 1849, Marx was arrested again and tried in Cologne on a charge of inciting armed insurrection. He was acquitted of all charges but was expelled from Germany, and the *Neue Rheinische Zeitung* was put out of business. For a second time in 1849, Marx was expelled from France, whereupon he moved to London and spent the remainder of his life in Soho on Dean Street.

In England, Marx devoted himself to study and writing and to efforts to build an international communist movement. Engels became an employee again in the textile mill in Manchester, England, and throughout the years became the chief financial supporter of Marx and his immediate family. The First International Working Men's Association, later called the First International, was founded by Marx in 1864. Headquartered in London, the First International was an organization of radicals, communists, socialists, anarchists, and syndicalists. When this organization met at Geneva in 1866, it was the first international forum for the promulgation of worldwide communism. Marx made the inaugural address, wrote its statutes, and subsequently directed the work of its governing body. Retired from the textile industry, Engels moved to London in 1870, becoming a member of the General Council of the First International, and began to relieve Marx of the work of directing the council's affairs. But the First International declined after the Communard defeat in Paris in 1871, whereupon it moved to the U.S. and finally disbanded in 1876.

During this period, Marx wrote the first volume of his *Das Kapital* (1867), in which he developed the theory of the exploitation of the "surplus value" of the working class by the capitalist class. Marx also contributed articles on contemporary political and social events to newspapers in Europe and the U. S. He was also a correspondent of the *New York Tribune* whose editor was Horace Greeley, from 1852 to 1861. In 1857 and 1858 he wrote a number of articles for the *New American Cyclopedia,* edited jointly by the American writer and editor Charles Anderson Dana (1819-97) and George Ripley (1802-80).

Karl Marx died on March 14, 1883, and was buried in London's Highgate Cemetery. In his funeral oration for Marx on March 17, 1883, Engels proudly boasted that Marxist communism, which was based on "scientific" principles, was just as important as Darwin's theory of evolution in revolutionizing the world. "Just as Darwin discovered the law of evolution in organic nature, so Marx discovered the law of evolution in human history," he said.

Friedrich Engels would go on to publish volume two and three of Marx's *Das Kapital* in 1885 and 1894, as well as works of his own on communist theory. Although he did not take part in the formation of the Second International in 1889 on the centennial anniversary of the French Revolution, Engels remained active in supporting revolutionary activities until the day he died. For the rest of the story on the rise of communism in Russia and its ultimate downfall, refer to my first book, *The Woman and the Dragon.*

Utilitarianism: Good Is What's Useful, Pleasure Is Paramount

Utilitarianism is a philosophical doctrine that defines what is good as being that which is "useful." It can be traced as far back as the Greek hedonist Epicurus. Utilitarianists believe that morality and ethics are based not on human conscience, reason, and divine revelation, but on their human "utility": whatever brings the greatest happiness to the greatest number of people is the most good. Pleasure and pain, therefore, are the primary determining factors of what is right and wrong. Utilitarianism, by its very own definition, is self-centered and hedonistic, placing the utmost importance on the gratification of one's selfish desires and the seeking out of pleasures. Followers favor liberal governments that don't impose any religious system of ethics or morals, and believe that individual freedom of conscience is the highest human virtue. They oppose any religious dogma or doctrine that mandates self-denial or mortifications of the flesh as being the will of God. The pain of self-denial and self-sacrifice, utilitarians believe, is bad. Pleasure alone is good.

The obvious problem with utilitarianism is equating pleasure with goodness, because what pleases me may not necessarily please you. For example, if a majority of a society would be happy if members of a minority were blown up by an atomic bomb or pushed into the sea or burned in ovens, it wouldn't be a good thing to do so, even if it made the majority happy. But according to the tenets of utilitarianism, Hitler's holocaust of the Jews would be morally justified because it pleased a greater number of German people, the Nazis. Another obvious problem with utilitarianism is that pleasurable acts can often times have unpleasurable consequences. Having sex outside of marriage may feel good at the moment but you can contract AIDS and die, or have an unwanted pregnancy, both of which are not good. The philosophical tenets of utilitarianism, therefore, are ludicrous, but they did have tremendous influence on ethics and morality in the eighteenth and nineteenth centuries, and still do today. The popular slogan of the 1960s, "If it feels good do it," is just one example of utilitarian morality.

Morality, according to the utilitarians, should be based on the democratic concept of majority rule. While it is true that majority rule is the best form of governance, majority rule does not apply to ethics and morality. God is sovereign and he alone decides right from wrong. He has given us the knowledge of right from wrong through our consciences, through Moses and the Ten Commandments, and through his son, Jesus Christ. Up until the Age of Reason, people believed in keeping the Ten Commandments, but with the coming of the Age of Enlightenment, philosophers and theologians began to doubt Judeo-Christianity and, therefore, doubted the existence of moral absolutes. No longer having a supreme being determining right from wrong, people of the Age of Enlightenment were free to choose for themselves what they believed to be right and wrong. Utilitarianism is a predictable outgrowth of the Enlightenment and its moral relativism.

The founder of philosophical utilitarianism was a man named Jeremy Bentham (1748-1832). Bentham was a British philosopher, economist, and lawyer born in London on February 15, 1748. An enormously gifted child, Bentham was reading literature by the age of three, playing violin at five, and studying Latin and French at six. At the age of twelve, Bentham entered the University of Oxford, where he studied law and was admitted to the bar, but did not become a jurist. Bentham was disgusted with the current state of English law, and instead of practicing law he set about to reform it. As a youth, Bentham dreamed of founding a sect of followers called utilitarians, who were followers of his radical new ideas. He also wanted to have his body preserved as an "auto-icon," so that every time his followers met he would be near them. Upon his death on June 6, 1832, Bentham got his wish: his body was stuffed and put in a glass box in the University College in London, where it was openly displayed at every meeting of the Benthamites.

In 1789, the year the French Revolution started, Bentham published his *Introduction to the Principles of Morals and Legislation,* which expounded his philosophical ideals of utilitarianism, and which made him world famous. Bentham began chapter one in his *Introduction* with a seditious complaint against Nature:

> Nature has placed mankind under the governance of two sovereign masters, pain and pleasure. . . . the principle of *utility* recognizes this subjection, and assumes it for the foundation of that system, the object of which is to rear the fabric of felicity [happiness] by the hands of reason and of law.[42]

The right laws in governance or morality will produce happiness, Bentham went on to say, and will be in accordance with man's reason. Therefore, moral and legislative laws are dependent upon their utility. By using the word *utility,* Bentham meant "that property in any object, whereby it tends to produce benefit, advantage, pleasure, good, or happiness . . . or . . . to prevent the happening of mischief, pain, evil, or unhappiness." So the good, for Bentham, is the maximization of pleasure and the minimization of pain. Bentham proposed "four sanctions or sources of pain and pleasure," namely, the physical, the moral, the religious, and the political. The physical sanction, according to Bentham, is the basis of all the others, so the pleasure of the flesh is paramount. Bentham sought further to devise a scale of pleasures and pains, rating them in terms of their intensity, purity, duration, propinquity or remoteness, certainty, fruitfulness, and the extent to which pleasure and pain are shared among the greatest number of people.

As an outlet and voice for his utilitarian ideas, Bentham helped establish the British periodical *Westminster Review* in 1823. He was also the leader of the Philosophical Radicals, whose most important members included James Mill and John Stuart Mill, two other utilitarians who contributed greatly to the popularity of utilitarianism. James Mill (1773-1836) was a Scottish philosopher, historian, economist, and the leading disciple of Jeremy Bentham. Mill met Bentham in 1808, whence he adopted his utilitarian principles. Afterwards, he submitted utilitarian articles to the *Encyclopaedia Britannica* (1816-23) and to the *Westminster Review,* and applied utilitarian ideas to the education of his son, John Stuart Mill (1806-1873). John was another leading proponent of utilitarianism and is considered one of the most important British philosophers of the nineteenth century.[43] At an early age, John was subjected to a rigorous liberal education by his father with the intent of making him the torchbearer of utilitarianism. The Utilitarian Society (1823-26), which met at Bentham's house for readings and discussion on all sorts of topics, was organized by John. He was also a chief contributor to the *Westminster Review.*

John Stuart Mill's most important works on moral philosophy and utilitarianism were his *On Liberty* (1859), *Utilitarianism* (1861), and *System of Logic* (1843). John's *System* was a philosophical treatise which attempted to synthesized the irreligious empiricism of the Enlightenment with the romantic-idealist movement of his time. Regarding religion, Mill was an ardent supporter of the atheistic philosophy called naturalism, which says that nothing exists outside of nature. Naturalism denies the validity of teleology (the study of the design or purpose in nature), such as the existence of God, the supernatural realm, or anything metaphysical. Moral values in naturalism, therefore, are relative. Whatever is best for the majority is moral; whatever is useful is good. The atheistic tenets of naturalism have their roots in British empiricism, the philosophical doctrine that all knowledge is derived from sensory experience and scientific investigations, Bentham's utilitarianism, and in French positivism, another atheistic doctrine that denies any validity whatsoever to metaphysical speculation.

Positivism: Religious Beliefs Are Not Provable and Therefore Meaningless

Positivism is a philosophical doctrine originating from France that maintains that "positive" knowledge is only derived from natural phenomena and their properties and relations as verified by the empirical sciences. It claims to be neither atheistic nor theistic in belief, insisting that rational questions and answers should be restricted to observable facts alone. Since theology and metaphysics deal with the unobservable, it is meaningless, therefore, to affirm or deny the existence of God.

Positivism cites an evolutionary timeline in the achievement of "positive" or factual knowledge, characterized by three historical stages of scientific, political, and sociological development. According to positivists, the first historical stage was the "theological" or fictitious stage, which occurred during the Middle Ages and was characterized by an immature belief in a supreme being and his personal interactions in the lives and affairs of men. The second historical stage was the "metaphysical" or abstract stage, which occurred during the Scientific Revolution of the seventeenth and eighteenth centuries. During the metaphysical stage, suppositional theories and abstract ideas were employed to explain the existence of naturally occurring phenomena. The final evolutionary stage of human development is the "positive" or scientific stage. This stage foregoes any metaphysical attempt to explain cause and effect or ultimate reality. Rather, attention is focused on how naturally-occurring phenomena are related to one another, with the aim of arriving at acceptable, verifiable conclusions about reality based on scientifically observable facts. The positive stage is more concerned with the how than the why and, therefore, is the most productive and utilitarian.

The founder of philosophical positivism was a Frenchman by the name of August Comte (1798-1857), whose name means the "Grand Count." Comte was born in Montpellier on January 19, 1798. At an early age, Comte renounced his Catholic upbringing and began sympathizing with radical socialists, which got him expelled from the Ecole Polytechnique in Paris in 1816 for participating in a student rebellion. For several years thereafter Comte was secretary to Claude Henri de Rouvroy, the duc de Saint Simone, a noted socialist ideologue who influenced the development of Marxist communism. The remainder of Comte's life was devoted to the writing and teaching of his positive philosophy and "sociology," a term he invented. Relying on monetary support from his utilitarian friend John Stuart Mill and his pupil Maximilian Littre, Comte endured poverty and misery much of his life, which was marked by periods of severe mental illness. He died in Paris on September 5, 1857.

In his most famous work, *Course of Positive Philosophy* (1830-42), Comte wrote that human thought evolves through different stages, which he termed the religious stage, the metaphysical stage, and the scientific stage. The final scientific stage is the factual, "positive" stage, which, as has already been discussed, is the most incontrovertible stage in the course of human thinking. Having an aversion for Christianity from his youth, Comte's goal in his *Course of Positive Philosophy* was to extract all supernatural elements from science. He categorized the history of science as developing and maturing in different stages, forming a hierarchy of subjects, which evolved much like Darwinian evolution. The first scientific discipline to evolve was mathematics, which was proceeded by astronomy, then physics, then chemistry, then biology, and finally Comte's science of sociology, with the purpose of the intellectual, moral, and political reorganization of society. Each of the three stages of intellectual development, Comte said, is correlated with certain political developments. The theological stage is reflected in the divine right of kings. The metaphysical stage is reflected in the liberal politics of the Enlightenment. The positive stage entails a scientific or sociological approach to political organization. Critical of the democratic institutions of his day, Comte envisioned a future society governed by a scientific elite who would use Bacon's scientific method to solve human problems and improve social conditions. Although he rejected God and Christianity, Comte recognized the value of religion in contributing to the moral fabric of society. Mimicking Christianity, in his *System of Positive Polity* (1851-54), Comte proposed a secular humanistic religion of his own, complete with secular observances and saints, like Frederick the Great and Adam Smith. Comte's inane religion, of course, never caught on with the general public, but his empirical views on the evolution of science and his theories on positive knowledge increased the number of positivist scientists in the nineteenth and twentieth centuries.

Naturalism: Nature Is All There Is, God Does Not Exist

Perhaps the most widely accepted philosophy of today is the godless human philosophy of naturalism. Naturalists believe that nothing exists outside of nature: no gods, no spirits, nothing. Nature is the whole of reality. Since there is no world outside of the material universe, metaphysics or religion should be de-emphasized, or more preferably done away with altogether. Any reports of supernatural miracles or paranormal phenomenon can be explained by science because all universal phenomena are reducible to the interplay of atomic particles in empty space, as revealed to us by physics and chemistry. There is no intelligent design or purpose to the universe; everything that exists is the result of random chance over eons of time. Since there is no god, ethics and morality must come from within society. Man must determine for himself what is right and wrong. Morals and values should be based on needs, customs, tradition, consensus, inclination, or any other form of utilitarianism that says what is useful is good. Right and wrong evolves just as man evolved. What is wrong yesterday may not be wrong today, and what is true for you may not necessarily be true for me. It all depends on the situation. Like the material world, everything is relative and evolving; nothing stays the same.

The atheistic philosophy of naturalism is not new; it has been around since the ancient Greeks and Romans. It declined in popularity with the advent of Christianity, but rose again in power and prestige during the Scientific Revolution of the seventeenth and eighteenth centuries, reaching its apogee in the nineteenth century in the persons of Charles Darwin and Karl Marx. Marx's dialectical materialism has already been discussed. Darwin, as you remember from your biology classes, popularized the naturalistic theory of evolution, which says that man evolved from monkeys instead of being purposely created by God to inherit everlasting life. Darwin's atheistic theory of evolution, which is taught in practically all of our public schools as infallible truth, claims that all life evolved by chance from some sort of primordial soup on the primitive earth eons ago. Every living thing on earth arose from this soup by random chance: plants, fish, animals, birds, insects, everything. And since all life arose by chance from muddy soup, there is no purpose to life except to eat and replicate. Those who reproduce the most are the "fittest," for they survive to perpetuate themselves. Those who fail to reproduce become extinct. The blind forces of nature, therefore, select who lives or dies by weeding out the weak and infirm by means of what Darwin called "natural selection." Mercy, compassion, morals, and ethics have no place in Darwin's dog-eat-dog world, only survival and frenzied copulation. Eliminating the "unfit," therefore, makes perfect evolutionary sense, as it did with the Nazis.

Both the Nazis and the Marxist communists hailed the cruelty and heartlessness of Darwinian evolution, where only the strong and powerful survive, as a revolutionary scientific truth. Friedrich Engels and the Communist League praised Karl Marx for his Darwinian-like theory of the "evolution" of economic history and his advocacy of a violent revolution by the proletariats. Adolf Hitler used Darwinian evolution as justification for his attempt to engineer a master race of Aryan supermen to populate and rule the world by force. The millions of Jews and Slavs whom Hitler killed in the holocaust were considered by the Nazis as subhuman and unfit to live. Hitler saw nothing wrong with slaughtering millions of innocent people who were not Aryan Germans. He was simply putting into practice Darwin's theory of natural selection and survival of the fittest.

Every sensible American knows that the Nazi's forced sterilization and murder of millions of Jews was an abominable crime against humanity, but what most Americans don't know is that there was a government-backed forced sterilization program right here in the United States at the beginning of the twentieth century that lasted clear up to the 1970s, which resulted in the involuntary sterilization of forty thousand to sixty thousand Americans in thirty-five states, mostly in Virginia and California. At the forefront of America's "eugenics" (good birth) movement, which was social Darwinism at its worst, were Harry Laughlin and Charles Davenport. Davenport founded the Eugenics Record Office at Cold Spring Harbor, New York. It was in operation from 1910 to 1939. This spurious, Darwinian-based organization sought through legal means in the American courts to forcibly prevent the physically and mentally "unfit" from having children.[44] (Laughlin and Davenport's description of "unfit," of course, usually meant European immigrants of Mediterranean descent, blacks, Jews, orphans, paupers, criminals, prostitutes, the "feeble minded," alcoholics, hobos, hemophiliacs, diabetics, epileptics, the blind, the deaf, and the physically deformed.) On May 2, 1927, in an eight to one decision, the United States Supreme Court affirmed in Buck vs. Bell Virginia's involuntary sterilization of Carrie Buck, who was seventeen, while she was forcibly committed to Lynchburg's State Colony for Epileptics and Feeble-Minded. Buck's only faults, it was later found out, were being the daughter of a prostitute, to have been brutally raped, and to have had a baby out of wedlock. Chief Justice Oliver Wendell Holmes Jr., writing for the majority, wrote the following opinion in support of Virginia's forced sterilization of Buck, which echoed the heartless Darwinian sentiment of survival of the fittest: "It is better for the world if instead of waiting to execute degenerate offspring for crimes, or to let them starve for their imbecility, society can prevent those who are manifestly unfit for continuing their kind. . . . Three generations of imbeciles are enough."[45]

Harry Laughlin of Kirksville, Missouri, who wrote the infamous Model Eugenical Sterilization Law, on which most of the thirty-five states based their forced sterilization laws upon, was the so-called scientific "expert" on eugenics when he appeared before Congress in 1924 to urge them to pass the Immigration Restriction Act to prevent, as Laughlin said, the "inferior stock" of "lesser immigrants" from diminishing America's collective "germ plasm." Thanks to Laughlin, an ardent Darwinist and rabid anti-Semite, the law was passed and wasn't repealed until the 1960s. For his contributions to "the science of race cleansing," Harry Laughlin was granted an honorary degree from Nazi Germany in 1936. In an ironic twist of fate, Laughlin developed late-onset epilepsy, one of the diseases requiring sterilization under his own "model law," and died in 1943 without children.

In the early twentieth century, many highly respected people were in favor of Laughlin's eugenics program of forced sterilization. Among them were Presidents Calvin Coolidge and Theodore Roosevelt, Alexander Graham Bell, and Margaret Sanger, the birth control champion and founder of Planned Parenthood, the infamous organization founded in 1942 that has caused the destruction of billions of human lives. But it wasn't until the horrors of the Nazi death camps and the hideous experiments of SS doctors were made public in 1945 that the evil science of eugenics finally fell from grace in America. The heartless theories of Darwin, however, have stubbornly persisted and are still being perpetuated in academia. It is only a matter of time when another megalomaniac emerges and tries to commit mass murder in the name of Darwinian evolution.

The Social Darwinism that contributed to the Nazi holocaust and the communist pogroms of the Soviet Union should actually be called "Social Spencerism," for it was an iniquitous man named Herbert Spencer who popularized the atheistic theories of Darwinian evolution in America, Germany, Russia, and elsewhere. Darwin was actually not very well known in his own country of England when he died in 1882. Herbert Spencer (1820-1903) was an English social philosopher who applied the principles of Darwinian evolution to practically everything under the sun, including philosophy, sociology, law, and justice, you name it. It was Herbert Spencer and not Charles Darwin who coined the term "survival of the fittest." Darwin, in fact, never used his theory of evolution to explain anything outside the realm of biology. In the late 1890s and early 1900s, Spencer was being read much more than Darwin, and people got their views of evolution and social Darwinism primarily through him, especially in America. Some very important intellectuals in America thought that Herbert Spencer was the greatest intellectual since Aristotle. A famous American professor at Yale in the 1880s named William Graham Sumner was against the government providing any relief to the poor, taking a literal interpretation of Spencer's "survival of the fittest."

The terrible consequences of Darwinian evolution are so widespread and so important to understand that Darwinism deserves a chapter of its own, so refer to chapter 2 of this book for a detailed discussion on that deadly human philosophy.

Nihilism: Negate Everything, for God Is Dead

Nihilism (from the Latin word *nihil,* which means "nothing") is a godless human philosophy that rejects all positive values and beliefs.

God is dead, there are no moral absolutes, and there is no purpose or meaning to life except negation and rebellion. Christians first used the word nihilism as a philosophical term during the Middle Ages to describe heretics. In the 1850s and 1860s, it was used in Russia to designate godless radicals and revolutionaries who considered Russian society backward and oppressive. Ivan Turgenev popularized nihilism as a philosophical doctrine in his novel *Fathers and Sons* (1862), where he names, defines, and analyzes nihilism through his main character, Bazarov, an idealistic young revolutionary. Towards the latter half of the nineteenth century, the godless philosophy of nihilism was epitomized in the irreverent writings and teachings of the famous German philosopher Friedrich Nietzsche (1844-1900), who loved to equate himself with the Antichrist. Nietzsche hated Christianity more than anything else, and was the first major philosopher to triumphantly proclaim the death of God. "God is dead. God remains dead. And we have killed him," he said.[46] He also said that "The greatest recent event—[is] that 'God is dead,' that the belief in the Christian God has ceased to be believable. . . . "[47]

Nietzsche's unabashed hatred of Christianity was formed while he was in college by reading the atheistic writings of Spinoza, Rousseau, Hume, Feuerbach, Darwin, and a host of other unbelievers. Never one to parse his words, Nietzsche called Christianity a decadent, effeminate, life-negating anachronism; deeming it a "world of pure fiction" and the "greatest misfortune" inflicted upon mankind to date. "I condemn Christianity. I raise against the Christian church the most terrible of all accusations that any accuser ever uttered. It is to me the highest of all conceivable corruptions," he said.[48] Jesus Christ, he said, was an "idiot" and a "sneak," and the Church was "hostile to life." The Church's notion of morality was "anti-nature," he said, and their concept of "sin" was "an invention against science." Christianity, Nietzsche wrote in his *The Antichrist* (1888), was the religion of the "weak" and the "pitiful," which he condemned with the most terrible of accusations. "What is more harmful than any vice? Active pity for all the failures and the weak: Christianity," he wrote.[49] He also said that "Christianity is called the religion of pity . . . Pity is the *practice* of nihilism."[50]

Like Feuerbach before him, Nietzsche wanted to raise man to the level of God. Inspired by Darwin, Nietzsche advocated the breeding of a master race of Aryan supermen, which he called the *Ubermensch*. Aryan supermen, Nietzsche believed, would become earth's master race and would overthrow Christianity, which, he said, has waged deadly war against the *Ubermensch*. Said Nietzsche: "Christianity . . . has waged deadly war against this higher type of man."[51]

Using Darwinian philosophy and principles, selective breeding would bring about Nietzsche's dream of creating a master race of supermen, whom, he said, were more worthy of life. "The problem I thus pose is not what shall succeed mankind in the sequence of living things (man is an *end*), but what type of man shall be bred, shall be *willed,* for higher value, worthier of a life, more certain of a future."[52]

Christianity, which preached love for one's neighbor, was one of the main obstacles to Aryan man, so it had to be eliminated. "Christianity . . . represents the counter-movement to any morality of breeding, of race, of privilege: it is the *anti-Aryan* religion par excellence," said Nietzsche.[53]

Only with the coming of a leader rich in will, the one who is "beyond good and evil," would Aryan man triumph over Christianity. "He shall be the greatest who can be the loneliest, the most hidden, the most deviating, the human being beyond good and evil, the master of his virtues, he that is overrich in will."[54]

Nietzsche's nihilistic writings were a favorite of fascists, radicals, anarchists, and bohemians of every stripe—particularly the Nazis. Adolf Hitler saw in himself that human being overrich in will who would realize Nietzsche's dream of creating a genetically-superior master race of Aryan supermen. But we all know how Hitler's mad dream ended: eleven million dead in Nazi concentration camps and tens of millions more killed in fighting during WWII, the most horrible war in history. To the Nazis, however, Nietzche was a prophet. Hitler enjoyed handing out copies of Nietzsche to his officer corps.

Although a madman, Friedrich Nietzsche was an incredible prophet of evil and doom. In his *Beyond Good and Evil* (1886), Nietzsche prophesied the coming of horrible wars in the twentieth century as the result of his nihilistic philosophies. "There will be wars such as there have never been on earth."[55]

In *Ecce Homo* (1888) and "Why I am a Destiny," Nietzsche correctly prophesied that only after his death would he be appreciated. He believed that "My time has not yet come. . . . some are born posthumously."[56] He also wrote, "One day my name will be associated with something catastrophic—a crisis such as there has never been on earth, the most profound collision of conscience . . . I am not a man, I am dynamite."[57]

As if he knew it would happen, Nietzsche's life was cut short at an early age. Friedrich Nietzsche went insane in 1889, one hundred years after the beginning of the French Revolution and the same year Hitler was born, and died on August 25, 1900, at the beginning of the century he predicted would be so catastrophic. (Nietzsche, by the way, died on the same date as David Hume.) Because of his madness, Nietzsche was infirm for the last eleven years of his life, and had to be taken care of by his mother and sister. But his sister saw to it that his works were published and circulated, and only after his death did they gain a huge following, exactly as Nietzsche predicted.

As if God were trying to warn him, Nietzsche suffered from terrible health and chronic pain his entire adult life. He was frightfully myopic and nearly blind, and had to wear thick glasses to see. His eyes were buggy like Hitler's and he constantly suffered from severe migraines and convulsions. Nietzsche finally had to quit his position as a philosopher because of poor health, and traveled around Europe trying to find the right climate to ease his constant pain, eventually settling in Italy, the home of the Church he hated so much. It was in the later years of his life that Nietzsche wrote his most infamous antichristian works, which included *Human, All Too Human* (1878), *The Gay Science* (1882), *Thus Spoke Zarathustra* (1883-85), *Beyond Good and Evil* (1886), *Toward a Genealogy of Morals* (1887), *Twilight of the Idols* (1888), *The Antichrist* (1888), *The Will to Power* (published posthumously in 1901), and *Ecce Homo (Behold the Man)*, posthumously published in 1908.

The depressing and evil life of Friedrich Nietzsche is an interesting one to note. Born Friedrich Wilhelm Nietzsche in a small German village on October 15, 1844, the son of a Lutheran minister, Nietzsche was raised in a devout Christian home; his father, his uncles, and both his grandfathers were pastors. As a youth, Nietzsche planned on becoming a Lutheran minister until he lost his faith in college. While attending Bonn and Leipzig universities, Nietzsche was introduced for the first time to the atheistic writings of Darwin, Schopenhauer, Feuerbach, and others. After renouncing the faith of his fathers, Nietzsche decided to become a professor of linguistics. As a professor of philology at Basel University, Nietzsche excelled in the language of sarcasm. His biting satire, satanic diatribe, and antichristian blasphemy quickly gained him a popular following amongst the ungodly German youth, by heartlessly encouraging the elimination of the weak, the poor, the disinherited, and the physically impaired from the face of the earth, especially in Germany.

But in a bit of ironic justice, Nietzsche's own physical infirmities eventually forced him to quit his chair at Basel in 1870, and for the remainder of his life he was a drug-addicted, sickly pariah. Nietzsche collapsed and nearly died on Christmas day in 1875, and on another occasion he nearly choked to death on a fish bone. He saw both events as full of symbolism and meaning. A loner and a sickly recluse, Nietzsche wandered about Europe searching for the right cure to relieve his chronic pain and illnesses, but he never found rest from his many tortures. As Nietzsche's physical and mental health grew worse, his hatred for Christianity likewise worsened and became more and more evident in his writings.

Like his namesake on whose birthday he was born, Friedrich Wilhelm IV, King of Prussia, and his favorite poet, Holdërin, Friedrich Nietzsche went insane. It happened in early January 1889, in Turin, Italy, home of the burial shroud of Jesus, when Nietzsche tried stopping a man from beating his horse. If only Nietzsche had cared as much for human beings. Just months before his insanity, Nietzsche was calling the Catholic Church a "madhouse" which needs "sick" people to exist, and the sickest of all are the saints and holy ones. "Christianity needs sickness . . . to make sick is the true, secret purpose of the whole system of redemptive procedures construed by the church. And the church itself, is it not the catholic madhouse as the ultimate ideal?" he asked.[58] He also said that "'the highest' states that Christianity has hung over mankind as the value of all values have pronounced holy by the church in *maiorem dei honorem* . . . Nobody is free to become a Christian; one is not 'converted' to Christianity—one has to be sick enough for it."[59]

Nietzsche should have heeded God's warning that you must watch what you say because you will be judged according to your words: "On judgment day you will be held accountable for every unguarded word spoken. By your words you will be acquitted, and by your words you will be condemned."[60]

Pessimism: Life Is Hopeless, Death and Extinction Are Preferable

Pessimism is a faithless philosophical belief that life is miserable and hopeless. It is extremely negative and gloomy in attitude; its followers believe that the dream of finding happiness in this life is all an illusion. The material world, pessimists say, is steeped in misery and pain, and an endless struggle of trying to find peace and contentment that never comes. It is the antithesis of faith and optimism, which accepts the struggles of life as they come without foregoing the hope for peace and joy in this life and the life of the world to come.

Pessimism has its roots in Eastern religions and an early antichristian heresy known as gnosticism, the philosophy that the spiritual world is good but the material world in which we live in is evil. A good god, Gnostics say, created the spiritual world, while a bad god created the material world, and both are equally powerful. Gnostics believe that matter in general is evil, that certain foods are evil, that marriage is evil, that the body is evil, and that Christianity is evil. The pleasure of sex, however, is good. Ancient Gnostics often partook in profane sexual practices in the hopes of achieving a higher level of spiritual awareness. Gnostics of the first century A.D. were inspired by an apostate heretic named Carpocratian, who believed that his "gnosis," or secret knowledge, was the true way to find happiness and could only be experienced by a very select few. The practice of gnosticism, Carpocratian said, provided a truer insight into the spiritual world than Christianity, which he vehemently denounced. In his denunciation of Christianity, Carpocratian denied that Jesus was God, denied that he was truly a man, denied his atonement for sins, and denied the Ten Commandments and the God of the Old Testament. The Apostle Paul addressed the severe ascetic practices and false beliefs of gnosticism in his first letter to Timothy, warning him to be wary of the liars who spread its false doctrines:

> The Spirit distinctly says that in later times some will turn away from the faith and will heed deceitful spirits and things taught by demons through plausible liarsBmen with seared consciences who forbid marriage and require abstinence from foods which God created to be received with thanksgiving by believers who know the truth. Everything God created is good; nothing is to be rejected when it is received with thanksgiving, for it is made holy by God's word and prayer.[61]

In the second century A.D., pessimism about the evils of the material world surfaced again in the heresy of Manichaeanism. Manichaean heretics were followers of Mani the Persian, who wandered aimlessly for forty years throughout the Orient proclaiming himself to be the "Last Messenger of the True God," and the "Paraclete" that Jesus had promised to send. Manichaeans believed that Satan was an eternally evil god and was the equally powerful rival of the good god. Satan, they said, came forth from the darkness and wasn't a fallen angel as Christians believe. The devil is the creator of the material universe, which is evil. Mani rejected the authority of the Old Testament, claimed that Jesus and the God of the Old Testament were two separate gods, that Christ was pure spirit and didn't become flesh, that human souls have been cast down to earth into material bodies because they had sinned in heaven, and that souls must undergo reincarnation before they are liberated from the flesh. Manichaeanism rejected the doctrinal authority of the Church, and adhered to a rigorous moral code similar to gnosticism. The supreme objection of mankind, Manichaeans believed, was liberation from evil matter.

The pessimistic spirit of Manichaeanism reemerged in the eleventh century in the heresy of Albigensianism. Centered in the southern

French town of Albi, Albigensians, or Cathars as they were also known, denied the Trinity and the human birth and resurrection of Jesus Christ, believed in a New Testament good god and an Old Testament evil god, believed that Satan, the evil god of the Old Testament, created matter which was evil, rejected the sacraments, disdained marriage, adopted sexual promiscuity and vegetarianism, condemned infant baptism and the forgiveness of the fallen away, denied the Holy Spirit's inspiration in the New Testament, repudiated the right of the state to punish criminals, and discarded the role of the clergy. Pope Innocent III (r.1198-1216) sanctioned new preaching missions to convert Albigensians to Christianity until one of his representatives in the region, Peter of Castelnau, was assassinated by them in 1208. Thereupon, Innocent III authorized the weapons of the Crusades as a means of fighting the Cathars, and by the end of the fourteenth century the Catholic Church had successfully eradicated Albigensianism from Europe.

Like the aforementioned Western heresies, the Eastern religions of Hinduism and Buddhism are extremely negative in attitude about the material world. Hinduism, which began in India about 3500 years ago, is the belief in a multitude of gods and goddesses, both good and bad, and a rigid social caste system where the upper castes like the Brahman (priestly caste) or Sakya (warrior caste) do not associate with or help the inferior lower castes; a violation of the golden rule Jesus gave us to love our neighbors as ourselves. The poorest of the poor in Hinduism are called the Harijan caste, or the "untouchables," and are shown very little charity and compassion by their neighbors because of the Hindus' belief that they are working out their bad karma actions from a previous life. Cows, which are sacred in Hinduism, roam the streets of India and are well taken care of, but not the untouchables, who often lie in the streets in abject poverty and misery. Another reason the Hindus don't care much about the destitute in India is because they believe that the material world is all an illusion (maya), wrought by Brahman, the universal world soul, in a spirit of play (lila). This extremely negative and merciless belief about ultimate reality is one of the main reasons why there is so much suffering and misery in that region of the world, and that is why belief in reincarnation is a sin.[62]

The four sacred texts of the Hindus, which teach a pessimistic, negative outlook on life, include the Vedas, the Brahmanas, the Bhagavad-Gita, and the Upanishads. According to the Upanishads, the material universe is in an eternal cycle of creation and destruction, of death and reincarnation, of good and evil. If you have bad karma you may be reborn as a bug, or an animal, or a demon, or worse, as an untouchable. If you have good karma you may be reborn as one of the infinite numbers of minor gods in Hinduism, or an upper caste member with wealth, beauty, longevity, and power. It all depends on fulfilling one's *dharma* (duty) in a particular caste. That is why those in a lower caste don't complain about the unfair caste system in which they are born—because they are afraid of being reborn in the same or worse circumstances. The perpetual cycle of birth and rebirth, of having and not having, only ends when one renounces worldly desires. This happens when the individual soul or *atman* forswears the material world and becomes united with Brahman, the ultimate reality in Hinduism. Human beings, thereafter, become one in being with this world spirit. Therein lies one of the paradoxes of Hinduism: loving a God with whom one is identical, which is self-worship. We are God.

One of the fastest growing religions in the West is Buddhism. Buddhism is a branch of Hinduism that was founded in India around the sixth century B.C. by Siddhartha Gautama, the Buddha or "Enlightened One." According to legend, Siddhartha, who was raised in pampered and luxurious seclusion by his powerful warlord father, the head of the Sakya warrior caste, left home one day against his father's wishes in search of spiritual enlightenment. Around 533 B.C., according to tradition, Siddhartha, now twenty-nine and out in the real world for the first time, come upon an old man, a sick man, and a corpse. Never having seen sickness, poverty, old people, or death, Siddhartha suddenly realized that the common fate of all of us is misery and suffering. After that traumatic experience, Siddhartha encountered a monk who was full of calmness and serenity, whereupon he determined to adopt his way of life and forsake family, riches, and power in search for the truth (the Great Renunciation). Wandering about northern India as a mendicant monk, Siddhartha took lessons in Hinduism from Brahman priests, but felt that the Hindu caste system was too unfair and only benefited those in power. Siddhartha then decided to become an ascetic, undergoing the most severe forms of self-denial and yoga meditation. Still, Siddhartha didn't find the enlightenment he was looking for. While meditating underneath a fig tree, or bo tree, one day in Buddh Gaya, Siddhartha was suddenly struck with the idea that the true way to salvation from the pain and misery of the world was the "middle way," somewhere in between privileged Brahmanism and rigorous asceticism. Upon this revelation Siddhartha became the Buddha or the "Enlightened One." Immediately afterward, evil demons began to assault the Buddha with temptations of fear, lust, pride, and doubt; but the Buddha, remaining strong, withstood them all. After overcoming his trials, the Buddha set about gathering his disciples, organizing them into a monastic community known as the *sangha,* which practiced Siddhartha's Buddhism, the "true" way of salvation.

As an ideology rather than a theology, Buddhism incorporates the main doctrines of Hinduism, such as karma and reincarnation, but rejects the Hindu priesthood and caste system, the Vedic scriptures, and the Hindu sacrificial cult. Buddhism has four noble truths. The first truth is that life is full of pain and misery. From the moment of birth to the end of life we encounter nothing but pain and suffering. Even death doesn't deliver us from distress because we are reincarnated. The second truth of Buddhism is that all suffering is the result of our attachment to worldly desires and ignorance regarding the understanding of ultimate reality. The third truth is that suffering can only be overcome by detachment from worldly desires and from enlightenment. The fourth truth is that the only path to this understanding is the noble eightfold path, which consists of right views, right intention, right speech, right action, right livelihood, right effort, right-mindedness, and right meditation. These eight paths are usually divided into three categories that form the cornerstone of Buddhist philosophy: morality, wisdom, and concentration.

Buddhism is a human philosophy on how to live rather than a theology on God's existence because it does not recognize a personal deity or an immortal soul as the three monotheistic religions of Judaism, Christianity, and Islam do. Buddhist doctrine teaches that there is no such thing as an immortal soul. Instead, Buddhists believe that man is composed of five distinct "bundles" or *skandhas*: the material body, feelings, perceptions, karmic tendencies, and consciousness. These five bundles are continuously changing; no one has the same bundles for more than a few moments at a time. It is wrong to consider any of these bundles as permanent or eternal; that would lead to egoism, craving for worldly desires, and continued suffering. All existence, therefore, is characterized by three realities: the anatman (no soul), anitya (impermanence), and dukkha (suffering).

Rather than an eternal life after death, Buddhism teaches the doctrine of *pratityasamutpada,* or dependent origination. Dependent origination says that ignorance in a previous life causes the mind and the senses to operate after death, which leads to craving and the clinging to existence, which leads to reincarnation. As in Hinduism, one's good or bad karma leads to rewards or punishments in the next cycle of rebirth. Pleasures or sufferings in this life are the result of one's karma in a previous life, so there are no such things as social injustices: you get what you deserve as a result of some kind of moral natural law. Bad karma can lead to reincarnation as a lowly animal, a poor leper, a demon, a god, a ghost, or whatever.

On the nature and reality of gods, Buddhists believe that they exist but don't assign them any special attributes other than long life and pleasures. Gods are not eternal and have to die and be reborn just as humans do. They did not create the universe and do not respond to prayer or sacrifices. In fact, it is preferable to be reborn as a human being because only humans can reach nirvana, the end of the birth and rebirth cycle (*samsura*). The state of nirvana, Buddhists believe, is the final state of existence and transcends all suffering, bad karma, and samsura. Nirvana is achievable only through the extinction of worldly desires and individual consciousness, and can be achieved while one is still alive. Reaching enlightenment or nirvana in life burns off any residual bad karma and prepares one for final nirvana after death (*parinirvana*). Nirvana, in theory, is achievable for everyone, although it is a realistic goal in Buddhism only for members of the Buddhist monastic community. Faithful Buddhists, therefore, strive to be reborn as a Buddhist monk.

The introduction of Hinduism and Buddhism into Western European philosophy is attributable to the famous German philosopher Arthur Schopenhauer (1788-1860). Born in Danzig on February 22, 1788, Schopenhauer was educated at the universities of Göttingen, Berlin, and Jena, where he studied philosophy and Eastern religion. Renouncing Christianity and belief in God in favor of Eastern mysticism, Schopenhauer became the first Western philosopher to incorporate Hindu and Buddhist doctrines and beliefs into a philosophical epistemology. After reading Schopenhauer's glorification of Eastern religions, Nietzsche wrote in his *Antichrist* that "Buddhism is a hundred times more realistic than Christianity. . . . Buddhism is the only genuinely positivistic religion in history. . . . In my terms, it stands beyond good and evil."[63] He also believed that "Buddhism does not promise but fulfills; Christianity promises everything but fulfills nothing."[64]

Schopenhauer led a solitary existence and was well known for his hatred of women and his pessimistic personality, which deeply upset his mother. At his residence in Frankfurt am Main, Schopenhauer became intensely involved in the study of Buddhist and Hindu philosophies. From them, Schopenhauer formulated his atheistic theory of the universal Will in his two most important books, *The World as Will and Representation* (1818) and *On the Will and Nature* (1836). According to Schopenhauer, universal Will, like the ultimate reality in Hinduism and Buddhism, is the unconscious and inanimate existence in all living organisms. The universe, in fact, is one vast cosmic will to exist, which experiences itself through nature. According to Schopenhauer, will is more basic than thought in both man and nature. It is the inner being of all organisms. Unfortunately, the world in which Will exists is one of constant pain and misery, from which Will constantly seeks to escape, which it can only do so by paradoxically forfeiting its ultimate goal—existence. Pleasure, to Schopenhauer, is only the temporary absence of suffering. Only when people become aware of the misery of their lives as part of the cosmic will to live, will they lose all wishes for existence and self-gratification, as the Buddhist monk does. To Schopenhauer, only the saints are capable of extinguishing their will to live, thereby reaching liberation from matter and its inherent suffering, something akin to nirvana in Buddhism. But the Will, while extinguished from the lives of saints, continues to live on in others and in nature, because it is impossible for all of us to be saints.

Schopenhauer's doctrine of the universal Will maintains that life is but an illusion, which is clearly borrowed from the Hindu concept of *maya,* and that the world is but a reflection of our own will, which is adopted from the Hindu/Buddha concepts of karma and dharma. Nietzsche and Hitler adopted Schopenhauer's Eastern belief that the world is merely a reflection of our own will to justify their demonic craving for power and to create an Aryan race of supermen who would rule the world from Germany. A strong-willed man who rises to power, said Nietzsche, is someone to be praised, not feared. "What is good?" Nietzche asked. "Everything that heightens the feeling of power in man, the *will* to power, power itself."[65] Said Nietzche: "Life itself is to my mind the instinct for growth, for durability, for accumulation of forces, for *power:* where the *will* to power is lacking there is decline."[66]

Reading Schopenhauer's writings on the universal Will and Nietzsche's glorification of the strong-willed leader who rises to power, Adolf Hitler was convinced that he was "the one" prophesied by Nietzsche to lead Germany out of the darkness and despair of WWI.[67] This is a perfect example of the Second Beast of Revelation (human science) making an idol to the First Beast (human power).

Pragmatism: All Truth Is Relative, There Are No Moral Absolutes

In this troubled, post-modern age we live in, society has become so overwhelmed by conflicting philosophical and religious ideas regarding ultimate reality and right and wrong that the vast majority of people today no longer believe that there is such a thing as absolute truth. Our secularized public schools and higher institutions of learning, which are primarily to blame for this tremendous increase in moral relativism, teach that truths, ethics, and morality are all relative to the situation, and that there are no absolute right or wrong answers. What is true for you may not necessarily be true for me, say the moral relativists, so don't judge. Even though moral relativists don't believe in Christianity, they love to tell you, "Didn't Jesus say 'Judge not lest ye be judged?'" Concerned Christian parents know the frustration of trying to teach their children absolute moral truths when their kids have been exposed to this kind of twisted moral logic in public schools or in the media. Role models in high positions of authority haven't helped much either. Pontius Pilate's absurd question "What is truth?" of two millennia ago has been surpassed in ridiculousness by former President Clinton's famous remark during his testimony about the Monica Lewinsky affair: "It depends on what the meaning of the word is *is*." How can our kids be expected to be moral and honest if the former president of the United States argues the meaning of *is*? *Is* is truth. Period.

Moral relativism and its denial of absolute truths is rooted in a mid-nineteenth-century philosophical doctrine that developed in America called pragmatism. Pragmatism, which is closely related to utilitarianism, is the belief that the truth of a proposition lies in its practical utility or usefulness. Truth, therefore, is relative to time, place, and purpose. What is useful or practical at the time is what is true. Truth is based not on moral absolutes, but on how we feel about something at the time. The beliefs of pragmatists are like the lyrics of the once popular song, "It can't be wrong if it feels so right." Reality, however, proves otherwise. Pleasures and good feelings are poor indicators that certain actions are morally right. Actions that bring pleasure at the moment often have terrible consequences in the end. Take, for instance, having sex outside of marriage, which is no longer seen as a taboo by the vast majority of teenagers and adults. Having sex outside of marriage can results in terrible consequences, such as unwanted pregnancies, abortions, single parenthood, poverty, child abuse, and sexually transmitted diseases like AIDS. In South Africa, for instance, there is an epidemic of AIDS-related deaths because of that country's sexual promiscuity, with upwards of 10 percent or more of the adult population testing positive for HIV. A whole generation in Africa may die just because sex feels good. Condoms, despite what our leaders say, are not the answer. Condoms fail to prevent pregnancies 14 percent of the time, so how can we trust them to prevent the spread of the AIDS virus, which is thousands of times smaller than a sperm cell? The only sure way to stop the spread of sexually-transmitted diseases like AIDS is to practice abstinence until marriage, like God commanded. This is an absolute truth that is true for everybody, whether we like it or not. Abstinence works 100 percent of the time.

The three most notable advocates of American pragmatism were Charles Sanders Peirce, William James, and John Dewey. C. S. Peirce (1839-1914), son of Benjamin Peirce, was the "Father of American Pragmatism." He was an American physicist, chemist, mathematician, and philosopher of logic and Boolean algebra (the mathematical logic behind computer languages). Educated at Harvard University, Peirce was appointed to the U.S. Coast Survey in 1861, and between 1864 and 1884 he lectured intermittently on logic, mathematics, and philosophy at Johns Hopkins and Harvard universities. In 1877, he became the first American representative to the International Geodetic Congress. Influenced by the philosophies of Francis Bacon, Immanuel Kant, Jeremy Bentham, and Charles Darwin, Peirce was a die-hard idealist who believed in the superiority of the scientific method of empirical investigations to arrive at what he believed to be the "truth." Peirce's version of the truth, however, got him into trouble at Johns Hopkins. Known for his immoral, libertine lifestyle, rapid mood swings, paranoia, and delusions of grandeur, Peirce's unorthodox beliefs eventually got him removed from his teaching post at Johns Hopkins.

After his dismissal from Johns Hopkins, Peirce delved into philosophical writing.[68] Peirce's new philosophy of pragmatism was first outlined in a paper contributed to *Popular Science Monthly* (1878). According to this new philosophy, no object or concept possesses inherent validity or importance. Its significance or truthfulness lies only in the practical effects resulting from its use or application. Usefulness, therefore, is truth. Peirce rejected all orthodox religions and their authoritarian teachings, and in their place he put science as the only authority. Calling himself a "laboratory philosopher," Peirce abandoned all metaphysics, rationalism, and individual reason in favor of objective scientific inquiry, the purpose of which he said was to settle all doubt and establish a new belief system based upon Bacon's scientific method. To Peirce, truth is what a handpicked scientific elite agrees upon after their completion of thorough scientific investigations, not some religious authority or Church magisterium. In fact, Peirce hoped that his scientific pragmatism would someday eliminate people's belief in metaphysics and religion altogether.[69]

Another prominent figure in philosophical pragmatism was William James (1842-1910), son of a theologian and brother of the great novelist Henry James. William James attended private schools in the U.S. and Europe, the Lawrence Scientific School at Harvard University, and Harvard Medical School, from which he received a medical degree in1869. After a brief retirement from an illness, James taught anatomy, physiology, and hygiene at Harvard from 1872, and was professor of philosophy and psychology from 1881 on. In 1907, James left Harvard and lectured at Columbia University and the University of Oxford. James' first book, the immensely popular *Principles of Psychology* (1890), established his reputation as one of the most influential thinkers of his time. James' *Principles* removed psychology from its traditional place as a branch of philosophy and established it among the laboratory sciences. In fact, twentieth-century psychoanalysis and medical psychiatry owe their existence to William James. After his publication of *Principles,* James applied the empirical scientific methods he used in psychology to philosophical and religious issues concerning the existence of God, the immortality of the soul, and free will. James' views on these subjects were presented in lectures and essays published in such books as *The Will to Believe and Other Essays in Popular Philosophy* (1897), *Human Immortality* (1898), and *The Varieties of Religious Experience* (1902). Theologically, James, like Peirce, was opposed to any absolute moral truths contained within metaphysics or religion. Instead James defined truth as the capacity of a belief to guide one to "successful action." All beliefs should be evaluated in terms of their usefulness in solving problems. Success, therefore, determines the value of our beliefs. Unlike Peirce, James justified the continued existence of religion and metaphysics within society for their pragmatic value in teaching morals and ethics, but he didn't believe in the transcendent, Almighty God of the Bible. God, to James, was the psychic energy of nature, which he called the "mother sea of consciousness."[70] Here again we see the tremendous influence that ancient Eastern religions had on modern Western philosophy and education.

In James' later works, which included *Pragmatism: A New Name for Old Ways of Thinking* (1907), *The Meaning of Truth* (1909), *Some Problems in Philosophy* (1911), and *Essays in Radical Empiricism* (1912), James redefined his pragmatic philosophical beliefs about truth. According to his new definition, truth is that which "works." One determines what works by testing propositions in experience. In so doing, one finds that certain propositions become true and others false. A particular hypothesis can be considered true if the predicted events take place. Truth is something that happens to an idea in the process of its verification. Religious beliefs, therefore, are meaningless because they entail no testable predictions. Since no moral absolutes given to us by the religious authorities can be tested, truth is relative to what works through individual, subjective experience. In this regard, James' truth markedly differs from Peirce's truth. Peirce often angrily denounced James' subjective, individualistic version of philosophical pragmatism as a perversion of his true, objective, scientific pragmatism. This fallout between Peirce and James begs the question: "Since two of the leading pragmatists couldn't agree on what truth is, why should we believe any of them?" We shouldn't.

By far, the most influential pragmatist of the late-nineteenth and early-twentieth centuries was the controversial American educator and secular humanist John Dewey. Dewey is often called the "Father of American Education" and the single greatest influence on our public educational system by those who revere him the most—liberal progressives who hate Christianity and want it outlawed in school. As an unbeliever in Christianity, or any traditional religion for that matter (Dewey was a signer of the 1933 Humanist Manifesto[71] that renounced God and elevated man in his place), John Dewey fought throughout his entire career to take discipline and authority away from our teachers and to keep religion out of our public (government) schools. In this he succeeded: state-mandated prayer in public school was outlawed by the Supreme Court in 1962, mandatory Bible readings banned from public schools in 1963, the Ten Commandments taken down from classroom walls in 1980, and, most recently, the Pledge of Allegiance was banned in June of 2002 by the Ninth Circuit Court of Appeals in San Francisco because it mentions "under God." Nowadays, students and teachers in many school districts, for fear of lawsuits, are reluctant to mention the name of Jesus in class or sing Christian songs during Christmas, which school boards refer to as the Winter Pageant or Winter Break (they want to keep the holiday but get rid of the holy day).

God has been taken out of our public schools and replaced with a hostile, pagan, antichristian, secular, humanistic ideology not unlike communism, thanks in large part to John Dewey, who was also known to be a communist sympathizer.[72] The fruit of Dewey's godless education has been catastrophic. One only has to look at the recent spate of crime and violence in schools to see it. Metal detectors, explosive-sniffing dogs, and uniformed police officers are employed in most of our public schools to prevent another occurrence of Columbine High School or Red Lake, Minnesota. But it is only a matter of time until another mass murder will be committed by some deranged teenager. Where God is not welcome, Satan always is.

John Dewey (1859-1952) was born in Burlington, Vermont, and attended the University of Vermont, where he was introduced to the evolutionary and naturalistic theories of Charles Darwin and T. H. Huxley, and the pragmatic theories of C. S. Peirce and William James, which deeply influenced his philosophy on life and his evolutionary theories of education. Dewey graduated from Vermont with a B.A. in 1879, and was accepted as a graduate student in philosophy and psychology at Johns Hopkins University, where he came under the influence of G. Stanley Hall, one of the most prominent experimental psychologists of his time, and George Sylvester Morris, a German-trained Hegelian philosopher (Dewey's doctrinal dissertation was on the philosophies of the German idealist Immanuel Kant). Upon obtaining his doctorate in philosophy in 1884, Dewey accepted a teaching post at the University of Michigan, which would last for ten years, with the exception of a teaching post at the University of Minnesota in 1888. While at the University of Michigan, Dewey wrote his first two books, *Psychology* (1887) and *Leibniz's New Essays Concerning the Human Understanding* (1888). Both these works reflected Dewey's early training in Hegelian idealism, Jamesian psychology, and evolutionary naturalism, which he tried to synthesize, however difficult it is to do, into one secular philosophy. In 1889, Dewey became head of the philosophy department at Michigan. Influenced by one of his mentors, James Hayden Tufts, Dewey in 1894 joined the recently founded University of Chicago, where he founded and directed the experimental Laboratory School, which implemented his liberal and progressive ideas in elementary education.

Around 1892, Dewey renounced his former belief in Christianity and his earlier adherence to Hegelian idealism and metaphysics. Now totally freed from any influence of religious or metaphysical beliefs, Dewey became an avowed atheist like his role model Charles Darwin, whom he patterned his educational theories upon. Dewey believed that education, like mankind, was evolving towards a higher goal or destiny, which had no room for archaic metaphysical ideas like the existence of God. In fact, Dewey's criticisms of religion were very similar to Marx and Engels'. Dewey said religion "has been petrified into a slavery of thought and sentiment, as intolerant superiority on the part of the few and an intolerant burden on the many."[73]

Dewey was totally against the traditional subject-matter type of education where authoritarian teachers teach a classroom of students the three Rs and the Bible by drills, memorization, and standardized testing. Instead, Dewey advocated a new way of educating children through their "direct involvement" and "experience."[74] In this radical new method of education, kids would teach themselves by "doing" rather than by passively sitting and learning. The seated classroom with the teacher at the front of the class teaching students was out of date, he believed, and had to be done away with. Studies and the classroom, therefore, had to be "psychologized" by using the "scientific method" of "child-centered involvement."

Dewey's ultimate goal of revolutionizing pedagogical learning was to make children functioning members of a democratic society who shared the same ideals and interests of the community as a whole, a progressive theory of education that was also popular in communistic countries. Dewey believed that children should become functional members of the society while in school, rather than be taught to be functional members of society after graduation. Particularly abhorrent to Dewey was vocational education schools where students were taught a trade that they would work at the rest of their lives. Dewey's progressive laboratory school at the University of Chicago, which emphasized teaching social reform and early community involvement, was in large part based upon Darwin's theory of evolution that nothing is constant but that everything is subject to change. Change to Dewey was good, especially getting rid of religion in school. Democratic ideals rather than dogmatic morals should be taught to children early in life. Dogmas, doctrines, religious authoritarianism, and any claims of having knowledge about absolute truths are bad. The moral absolutes given to us by the Bible don't exist anymore. Truth is ultimately based on the pragmatic consequences of one's actions, not any revelations by God on what is good or bad. All ideas about moral truths are open to criticism. Nothing is absolutely true; morals are evolving. Truth is what works, what succeeds, what leads to a solution.

Dewey was adamantly opposed to the teaching of Christian doctrines in school. In fact, he said the teaching of biblical values and moral absolutes in school were detrimental to the progressive goals of liberal educators like him, and that "A narrow and moralistic view of morals is responsible for the failure to recognize that all the aims and values which are desirable in education are themselves moral."[75]

Human thought, believed Dewey, is practical problem solving achieved by testing rival hypotheses against experience, through the empirical methods of scientific inquiry. Something is true if it leads to a pragmatic solution of a problem. All knowledge, therefore, is the product of human activity directed to the fulfillment of human purposes, not something that is revealed to us by God.

Besides his atheistic pragmatism, Dewey was a materialist who believed that nothing exists apart from matter. Dewey criticized the old mind-body theory of philosophical dualism, claiming that human thought is the result of the material processes that take place within the brain, not a result of some incorporeal mind or soul that is temporarily seated within the brain and ascends to heaven or hell after death.

Disagreement in the administration of his progressive laboratory school at the University of Chicago finally led to Dewey's resignation in 1904. Dewey then joined Columbia University where he spent the remainder of his career lecturing and writing. A prolific writer, Dewey wrote close to a hundred books and papers, which included such popular works as *The School and Society* (1899), *The Child and the Curriculum* (1901), *Studies in Logical Theory* (1903), *Ethics* (1905), *The Influence of Darwin on Philosophy and Other Essays in Contemporary Thought* (1910), *How We Think* (1910), *Essays in Experimental Logic* (1916), *Democracy and Education* (1916), *Reconstruction in Philosophy* (1920), *Human Nature and Conduct* (1922), *Experience and Nature* (1925), *The Public and its Problems* (1927), *The Quest for Certainty* (1929), *A Common Faith* (1934), *Art as Experience* (1934), *Liberalism and Social Action* (1935), *Logic: The Theory of Inquiry* (1938), *The Theory of Valuation* (1938), *Experience and Education* (1938), *Freedom and Culture* (1939), and *The Knower and the Known* (1949).

However popular pragmatism was in America, it was criticized by European philosophers as a purely American philosophy that had its roots in the American entrepreneurial spirit of success. The practice-oriented philosophy of pragmatism, Europeans said, is a poor criterion of truth. Something may work but can plainly be false or illegal. And they were right. Take, for instance, corporate greed and the devastating consequences of the recent accounting scandals at Enron, Global Crossings, and WorldCom. The greed that benefited the top CEO's of these companies financially devastated their employees and shareholders. Their actions were morally wrong, despite that the investment icon Ivan Boesky said in a lecture to business students that "greed is good."

Pragmatism's consequences on American education have been devastating. Poor test scores, illiteracy, teenage drug abuse, unwanted teen pregnancies, school violence, and a general lack of respect for authority are all results of the American educational system adopting the progressive, pragmatic ideals of John Dewey. Today, outcome-based education no longer requires kids to study and make good grades. Scholastic aptitude testing and grading is seen as too discriminatory towards minorities, who have the lowest scores. Value clarification programs now teach that morals are relative and ethics are situational. Self-esteem programs make the excuse that the reason Johnny can't read is because of his low opinion of himself, not because of the failure by teachers to make Johnny read. There is a deliberate dumming-down of our educational system so that kids can pass and the schools can graduate them. Core subjects like English, reading, writing, history, mathematics, foreign languages, and science, which are all necessary to successfully compete in the modern business world, are being neglected in favor of unnecessary subjects like environmentalism, human sexuality, gay and lesbian diversity training, gender and minority studies, multiculturalism and the like.

The American educational system is entirely to blame because they have allowed this to happen to our kids. They revise our history books to deny the impact or importance of hard work and conservative moral values that were characteristic of preceding generations. The importance of Christianity and the impact it has had on history is hardly ever mentioned in school textbooks nowadays. There is an intentional effort to keep religion and the truth away from our children. The myth of separation of church and state, which is not in the Constitution, is constantly being used as an excuse by judges and school administrators for taking God out of our textbooks and Pledge of Allegiance. There is a widespread hostility towards Christianity by the liberals in charge of our educational system. The teaching of Islam and Eastern religions is far more acceptable in our public schools today than the teaching of Christianity. In fact, some California school districts have recently made the study and practice of Islam a mandatory subject, despite the protests of outraged parents and the violation of the law, which forbids mandatory religious worship in public schools. The University of North Carolina, which won the NCAA men's basketball tournament in 2005, even made the Koran a mandatory reading assignment for incoming freshmen in 2002. Where is the outrage from the ACLU about this? If the Bible were mandatory reading they would be screaming all the way to the Supreme Court.

Logicism: Human Science Is Logical, Divine Science Illogical

Mankind has fallen in love with his own intelligence and knowledge and worships at the altar of human science instead of divine science. Truth today comes not from the Word of God but from what science tells us is true. We were all taught in school that Darwinian evolution, which says that the universe was created by random chance from nonliving cosmic substances that randomly evolved human beings from seawater for no apparent reason is the ultimate truth. This widely held belief, however, is totally illogical. Random chance has no being, no mind, no intelligence, no foresight, no power, no purpose, no spirit, and no love. It is only the mathematical likelihood of the occurrence of a given event, like rolling a seven in craps. A living being still has to roll the dice to cause the event to happen; a "Prime Mover" if you will. As painfully obvious as this truth is, most liberal educators, philosophers, and scientists still believe in evolution rather than divine creation, even though the mathematical statistics are on the side of creation and intelligent design. The statistical likelihood that random chance created the universe is about as likely as a tornado creating a 747 out of a junkyard, or a monkey accidentally typing out a complex computer algorithm. It just won't happen no matter how much time elapses. Even though they know this, because there are many criticisms being leveled today at Darwinian evolution by highly intelligent people, hard-core evolutionists in our educational system still try to convince us that given enough time, chance can create anything. Do not be deceived. Chance did not create the world.

The real reason that atheists and Darwinists put their faith in random chance instead of God is that they don't want to obey God's commandments. They want total freedom to do as they please, especially when it comes to having lots of sex. One famous twentieth-century philosopher and sexual libertine who renounced God in favor of random chance and evolution was Bertrand Russell, the brilliant mathematician from England who made up the absurd philosophy of logicism. Logicism is an atheistic philosophy that human science alone, especially the science of mathematics, represents the purest logical truth. Unlike religion, mathematical symbols, such as classes, sets, and members, represent undeniable truths, which can all be proven by logic. However, it was later shown in 1931 by the mathematician Gödel, much to Russell's discouragement, that not all mathematical truths can be logically proven. The claims of divine science and the Bible, Russell said, cannot be proven by science, so they must be rejected as untrue and illogical.

Russell's atheistic philosophy of mathematical logicism gave rise to a similar philosophy called logical positivism (a.k.a. logical empiricism), the belief that philosophical truths exist only when their linguistic symbols are verifiable by human science. Logical positivists reject the traditional positivist ideas of Auguste Comte, who held that personal experience is the basis of true knowledge, and instead emphasize the importance of scientific verification by a select panel of scientific elites.

Logical positivism, an extremely popular movement within philosophy during the 1930s and 1940s, greatly influenced the rise of today's modernism within Christianity, which more and more comes to see the stories within the Bible as mere symbols for teaching ethics and morality rather than literal truths.

One of the leading logical positivists of the twentieth century was Ludwig Wittgenstein (1889-1951) from Austria, a philosophy and mathematics student of Russell's at Cambridge. Wittgenstein's first major work, *Tractatus Logico-Philosophicus* (1921), was a philosophical treatise on the use of language and semantics as tools for conveying thought. Wittgenstein believed that only concepts and propositions that represent facts, such as the languages used by scientists, are to be considered logical, meaningful, and true. Like his mentor Russell, Wittgenstein believed that metaphysical and ethical statements by theologians and clergy are not verifiable by modern science, and should therefore be discarded by intellectuals as meaningless.

Bertrand Russell (1872-1970), godson of John Stuart Mill and the founder of philosophical logicism, was a British philosopher, third earl, brilliant mathematician, secular humanist, and Nobel laureate. He became a world-famous lecturer on philosophy and mathematics at Cambridge University, in addition to being a social and political activist for peace and nuclear disarmament and an outspoken critic of religion. Born in Trelleck, Wales, May 18, 1872, Russell was educated at Trinity College, University of Cambridge. After graduating in 1894, he traveled the lecture circuit in France, Germany, and the U.S. and was made a fellow of Trinity College. Russell's first major work on mathematics was *The Principles of Mathematics* (1902), in which he attempted to remove mathematics from the realm of abstract philosophical notions and to give it a precise scientific framework. Russell's second major work was *Principia Mathematica* (3 vol., 1910-13), in which he collaborated with Alfred North Whitehead to try to show that mathematics can be stated in terms of the concepts of general logic, which he believed represent the only undeniable, provable truths. *Principia Mathematica* became a monument of rationalistic science and spawned other atheistic philosophies in science, such as logical positivism.

In addition to being an outspoken atheist and critic of Christianity, Russell was a pacifist and critic of both sides during WWI. For his outspoken criticism of the British and their allies, Russell was imprisoned and deprived of his teaching post at Cambridge. While in prison, Russell wrote *Introduction to Mathematical Philosophy* (1919), which combined the two areas of knowledge he regarded as indivisible: mathematics and philosophy. An ardent admirer of the antireligious and communist tenets of Marx and Lenin, Russell visited the Soviet Union after WWI, but became disillusioned with Lenin's tyrannical rule. In his book *Practice and Theory of Bolshevism* (1920), Russell criticized the brutal tactics used by Lenin to achieve communism's redistribution of wealth. Although critical of Lenin's practices, Russell continued to be a diehard and unapologetic socialist throughout his entire life.

Russell taught mathematics and philosophy at Peking University in China from 1921 to 1922, and from 1928 to 1932 he taught at the highly progressive Beacon Hill School for young children in Britain, which he and his wife Dora founded (Russell had four wives altogether). From 1938 to 1944, he taught at various educational institutions in the U.S.

He was barred, however, from teaching at the College of the City of New York (now City College of the City University of New York) by the state supreme court because of his attacks on Christianity in such works as *What I Believe* (1925), and his advocacy of free sex, divorce, adultery, homosexuality, and adult-child nudity in *Manners and Morals* (1929). An outspoken critic of conventional sexual morality, Russell believed that all human beings should be allowed to express their natural sexual desires, children as well as adults. Some impulses, of course, must be restrained in the best interests of society, but it is the individual's relatively unimpeded natural growth and self-realization of his sexuality that makes for the good life and a harmonious society. The detached sage and ascetic saint who practice celibacy, he said, are poor role models for our children because they are sexually incomplete human beings. Likewise, the Christian doctrine of hell for committing sexual immorality and other mortal sins is an invention of "moralists" (a common charge you hear nowadays by liberals leveled at conservative Christians), who delight in inflicting punishment on those who don't believe as they do.

The concept of hell is one of the main reasons why Russell hated Christianity. He had a guilty conscience and was inwardly afraid of going there. In his famous *Why I am Not a Christian: And Other Essays on Religion and Related Subjects,* Russell publicly denounced Jesus Christ for teaching about the existence of hell, declaring that "There is one very serious defect to my mind in Christ's moral character, and that is he believed in hell, I do not myself feel that any person who is really profoundly humane can believe in everlasting punishment."[76]

Russell believed that Christianity was against all moral progress in society and, therefore, an evil institution. "I say quite deliberately that the Christian religion, as organized in its churches, has been and still is the principal enemy of moral progress in the world," he said.[77]

Russell compared the Christian Church to Lenin's oppressive communism, and blamed Christianity for fomenting WWI.Not just content to bash Christianity, Russell said that all religions are harmful and dangerous because they are all based on lies, fear, superstition, cruelty, and greed. "I think all the great religions of the world—Buddhism, Hinduism, Christianity, Islam and Communism—both untrue and harmful," he said.[78]

Russell believed that Darwin had debunked the false teachings of religion once and for all, so no one should be afraid of going to hell anymore. Therefore, scientists should not take the Bible and the teachings of Christianity seriously. "It is evident that a man with a scientific outlook on life cannot let himself be intimidated by texts of Scripture or by the teaching of the church," he said.[79]

Bertrand Russell returned to England in 1944 and was reinstated as a fellow of Trinity College. By then he had rethought his pacifism and publicly supported the Allies against Nazi Germany and Japan during WWII. In 1949, Russell was awarded the Order of Merit by King George VI, and a year later he received the Nobel Prize in literature and was cited as "the champion of humanity and freedom of thought." He returned to his pacifist ways later in life. At the ripe old age of eighty-nine, Russell was an ardent opponent of the Vietnam War, and was imprisoned again for participating in an antinuclear demonstration.

A radical secular humanist and antichristian until the day he died, Bertrand Russell died at ninety-seven on February 2, 1970, the Feast of the Presentation of Christ (Candlemas), a holy day for Christians that is also celebrated by Satanists, occultists, and New Agers for purely pagan and evil reasons. Russell's teaching career and published works had a tremendous influence on the secularization of Western society and the disbelief in Christianity. Among Russell's major works were *The Problems of Philosophy* (1912), from which he refuted the tenets of Kantian idealism, *The Analysis of Mind* (1921), *The ABC of Relativity* (1925), *The Analysis of Matter* (1927), *Education and the Social Order* (1932), *In Praise of Idleness* (1935), *History of Western Philosophy* (1945), *Human Knowledge: Its Scope and Limits* (1948), *My Philosophical Development* (1959), *Political Ideals* (1963), *War Crimes in Vietnam* (1967), and his *Autobiography* (1967-69).

Existentialism: Life Is Meaningless

The last major philosophical school of the twentieth century was the atheistic school of existentialism, which was popular in Europe around the 1930s and 1940s, during the time of the Nazi holocaust. Existentialism is a totally nihilistic and pessimistic philosophy, that maintains that since there is no God or heaven, life has no meaning, and it is pointless to live. Human beings wander throughout life aimlessly without purpose, as in an endless void. Man is not made in the image and likeness of God and is no different than the clouds, which exist one moment in time and are then gone forever. Only when we realize the absurdity and meaninglessness of our lives can we overcome the angst of living, and die a peaceful death, which finally ends our suffering and existence forever. It is foolish to hope for life after death, and even more foolish to keep the Ten Commandments to merit an eternal reward. Death ends everything and we are not judged by some supreme deity for the good or bad we've done in life.

The first roots of atheistic existentialism go all the way back to the famous Christian writer and social critic from Denmark, Søren Kierkegaard (1813-1855). Kierkegaard criticized Hegel's idea of the all-encompassing Absolute Spirit or universal consciousness, and instead stressed the irreducibility of the personal, subjective aspect of human life—the unique "existence" of the individual. Kierkegaard's first use of the word "existence" struck a chord with the secular German philosopher and Nazi sympathizer Martin Heidegger (1889-1976), the actual founder of existentialism.

As the founder of existentialism, the most important philosophical question to Heidegger was what it means to "be" or what is the meaning of our *existenz,* as it is spelled in German. In his most famous existential work, *Being and Time* (1927), Heidegger said that man's essence (which he called *Dasein*) lies in "its always having its being to be, and having it as its own." *Dasein's* existence and finitude is an issue for each of us, Heidegger said, and how we face up to this determines our individual *existenz.* All of us are thrown into a world we have no choice in. We live, we suffer, and we die. Because our existence in a certain place and time is out of our control, our passions and emotions in life are likewise not willed. The most debilitating of these unwilled emotions is our universal fear or angst about the uncertainty of life and death. A correct understanding of the inevitable and eternal end of one's personal existence, according to Heidegger, will alleviate our universal angst over death. Only when *Dasein* realizes that death is the end of everything can it obtain true freedom, a freedom that understands the meaninglessness of life and the finitude of one's own personal existence. There is no God, no heaven, and no hell. Death ends everything. Heidegger didn't believe in God, but did hold out the possibility that he could appear in the future.

Heidegger's successor, Jean-Paul Sartre (1905-1980), the highly-acclaimed French existentialist with Marty-Feldman eyes, coined the term "existentialism." He was famous for saying "hell is other people."[80] Sartre agreed with Heidegger about the meaninglessness of man's existence and the finality of his death, but disagreed with Heidegger's belief that our passions and emotions in life are unwilled. Going back to Kierkegaard, Sartre believed we choose our emotions and feelings like anything else. Choice is "the fundamental act of freedom," Sartre said, a phenomenon of our "imagination" contained within our "self-consciousness." To Sartre, the freedom to choose whatever we wish, to do as we damn well please (while still respecting the freedom of others), is the most important thing in life. There is, however, one catch to being free: our freedom and the choices we make in life, Sartre said, are "absurd" because our self-estrangement and powerlessness over our destinies is an inevitable part of the wretched human condition. Existence in this world is absolutely pointless and meaningless because human beings are "nothing."

Nihilism, fatalism, and self-negation are at the heart of Sartre's bewildering concept of the nothingness of human beings who, he believed, come from nothing and go to nothing. Nothingness therefore is the true essence of our being. Sartre's strange tenet that our individual nothingness comes from some sort of other Nothingness makes absolutely no sense philosophically or ontologically or epistemologically. If something can't come from nothing all by itself, then our individual nothingness can't come from some other Nothingness either. Nothingness can't become being, since by definition it doesn't exist. Sartre, therefore, was illogical in saying that "The being which Nothingness comes to the world must be its own Nothingness"[81] and that "Things are entirely what they appear to be, and behind them . . . there is nothing."[82]

What Sartre is really implying in these illogical statements is that human beings, who foolishly think they are made in the image and likeness of God, who is Everything, are really nothing because God does not exist. Sartre did not believe in the existence of God and held out no possibility of him revealing himself in the future. Since there is no God, human beings should create their own world by rebelling against the prevailing civil and moral authorities of the day. The freedom to rebel is good, and the freedom to negate and do away with is even better. Sartre reveled in the ideas of nothingness and negation, which was exemplified in his two most famous works, *Being and Nothingness* (1943) and *Nausea* (1938), and his two most popular dramas, *Flies* (1943) and *No Exit* (1944).[83]

Most of Sartre's depressing works were written during the time of the holocaust of Nazi Germany, when he was a secret member of the French resistance, and reflect the despair and hopelessness of many Europeans under Nazi occupation, a time when hope and trust in God had all but vanished. The prevailing question of Sartre's time, especially in the Nazi death camps, was how could a good and loving God, if he existed, allow so much evil in the world? The answer to this question is the same as it's always been: because man wants evil to exist in the world. God is not to blame for the existence of evil. He made man free to choose between evil and good, and many times man chooses evil. People like Sartre are hypocritical when they glorify in their own freedom but rail against God when evil men choose to exercise their God-given freedom and wreak havoc on the world. Don't blame God for the catastrophes that befall mankind. How can we truly be free if God didn't allow us to choose to love and obey him or not? Love can never be forced, or it is not love. Love must always be given freely from the heart.

An impassioned Marxist revolutionary and antichristian atheist, Jean-Paul Sartre, Europe's most celebrated existential philosopher, dramatist, novelist, and political journalist during the middle of the twentieth century, was awarded the Nobel Prize in literature in 1964 but rejected it, explaining that to accept such an award would be to compromise his integrity as a writer. This, of course, infuriated the Nobel Prize committee, which had it coming for honoring so many ingrates and atheists in the past, but endeared Sartre amongst his radical followers.

The rapid rise of absurdist theater in Europe during the 1950s was in large part attributable to the popularity of Sartre's existentialism. The two most notable absurdist dramatists were the Romanian-French playwright Eugene Ionesco, whose play *Rhinoceros* (1959) saw "man as lost in the world," where "all his actions become senseless, absurd, useless," and the Irish-born writer Samuel Beckett, whose famous play *Waiting for Godot* satirizes the absurd lives of two hapless drifters who patiently and faithfully wait for a man named Godot to show up and help them out, but he never does. Godot, of course, represents God, and the absurd lives of the faithful drifters represent naive believers who blindly put their faith and trust in God only to be disappointed in the end.

Jean-Paul Sartre died in 1980, long after having gone blind. As he lived his life in spiritual darkness, so he died in physical darkness. Like the blind leading the blind, Sartre influenced the direction of many Western philosophers, writers, and playwrights of his time, which was usually away from a belief in the one true God. Those greatly influenced by Sartre include the Algerian French philosopher Albert Camus (1913-1960), whose absurdist ideas were reflected in his existential works *The Stranger* (1942), *The Plague* (1947), *The Fall* (1956), and *The Rebel* (1951); Martin Esslin and his *Theater of the Absurd* (1968); the absurdist playwrights Arthur Adamov, Fernando Arrabal, Jean Genet, Jean Tardieu, Samuel Beckett, and Eugene Ionesco; and absurdist writers such as Harold Pinter, Tom Stoppard, Gunter Grass, Peter Weiss, Edward Albee, Israel Horovitz, Sam Shepherd, and Vaclav Havel of Czechoslovakia.

Chapter Two

Darwinian Evolution: A Fairytale for Unbelievers

"Darwinian man, though well behaved, at best is only a monkey shaved."
—W. S. Gilbert, *Princess Ida*

The Rise of Atheistic Darwinism

If you are an atheist and don't believe in God, chances are that you became one because you have a psychological hatred towards your father or because you have been taught in school that Darwinian evolution is a proven scientific fact; that man evolved from monkeys instead of being purposely created by God in his likeness and image. Since most children naturally love their father if he is good, one can safely conclude that the teaching of Darwinian evolution is the primary reason people convert to atheism in the modern world. I say convert to atheism because every child has a natural, inborn belief in God, and has to be converted to atheism by the negative influence of evil people in their lives, such as an abusive father or father-figure or an unbelieving teacher or mentor. Atheists by and large are converted in their youth, when their young minds are impressionable and easily persuaded by authority figures such as liberal high school teachers and college professors, many of whom teach the godless human philosophies that I outlined in the previous chapter and the lie of Darwinian evolution.

The godless human science of Darwinian evolution is the most insidious and evil atheistic philosophy that has ever been taught in school because it sounds so logical to the scientifically uneducated that it has to be true, but it is not. It is an insult and calumny against Almighty God, the Creator of all life. Within the last twenty or so years, with the tremendous advances in the art of gene sequencing, many scientists have come to realize that life is just too complicated to have arisen all by itself from muddy water without an intelligent designer, and are beginning to ridicule Darwinian evolution as an impossibility and an absurdity. Regrettably, even though confronted with mountains of scientific evidence that suggests intelligent design, liberal elitists in academia stubbornly continue to teach Darwinian evolution as scientific fact rather than theory because they don't want to believe in God. To do so would mean that they would have to submit to God's authority and repent of their wicked lifestyles, so they willfully refuse to admit the obvious truth.

The majority of people at the beginning of the eighteenth century, when Charles Darwin was a boy, believed that God had created mankind and all the species inhabiting the earth in just a few thousand years time. They took their belief, of course, from a strict, literal interpretation of the Book of Genesis and the biblical story of the creation of Adam and Eve. Slowly but surely, however, the literalists' theory of a young earth was challenged when the fossilized remains of extinct species not mentioned in the Bible were discovered in Europe, Russia, and North America. The first fossil discovered was a mastodon tooth in the late 1600s. By around 1730, Russian woolly mammoths, which only differ from the mastodons by the shape of their molars, were being unearthed in abundance and brought back from Russia to Paris for exhibition and study. Within the next thirty years or so, mastodon fossils were likewise being found in abundance in North America, and the ideas of animal extinction and a much older earth were now taken seriously.

In addition to the discovery of the molars, bones, and tusks of extinct ice-age animals in the 1700s, British geologists and surveyors such as William Smith discovered in the early 1800s that the different geological strata or layers of rock have evidence of multitudes of different types of extinct life forms, prompting other scientists like the French paleontologist and comparative anatomist Georges Cuvier (1769-1832) to conclude that the earth goes through periodic cycles or "revolutions" of creation and extinction. Fossils, as scientists discovered, can be used to date the different geological strata and give us some idea of the age of the earth, which is much older than people originally thought.[1]

Around the time Charles Darwin set sail on his famous voyage with the H.M.S. *Beagle* in the 1830s to study the biological and geological diversity of the southern hemisphere, other scientists like the Swiss geologist and amateur fossil collector Louis Agassiz, the inventor of the ice age theory, and Britain's comparative anatomist Lord Richard Owen, the inventor of the term "dinosaur," which means "terrible lizard," were just beginning to categorize the fossils of dinosaurs and prehistoric mammals. Darwin was well aware of the existence of prehistoric animals but not yet aware of the existence of prehistoric cavemen, whose remains were first discovered while Darwin was at sea from 1831 to 1836.

The remains of human-like bones were first discovered alongside the bones of fossilized animals in the Engis Cave in Europe in 1834, which further indicated to scientists that our human species was much older than previously thought. A second discovery of the existence of prehistoric cavemen was made not long afterwards in Boucher de Perthes, Abbeyville, in northwest France, where ancient stone tools were discovered alongside the fossilized remains of extinct animals. The third discovery of evidence of prehistoric cavemen came in 1856 when the fossilized skull of a Neanderthal specimen was found in the Neander valley near Dusseldorf, Germany. The fourth and perhaps most important discovery of ancient cavemen came in 1868 in southern France in an area just south of Toulouse, where a landowner named Edouard Lartet, who published papers on Paleolithic cave art, luckily found on his own land the remains of Cro-Magnon man, whose bones were practically identical to those of modern man but at least fifteen thousand years older. Yet another important discovery was made in Brikham Cave in 1859, where the bones of woolly mammoths and man-made tools were discovered alongside each other.

With the discovery of the fossils of ancient dinosaurs and mammals, and the tools and bones and artwork of prehistoric cavemen lying near the animals they hunted, by the 1860s human antiquity older than the Bible and the existence of ancient mammals and dinosaurs that had somehow become extinct was now accepted by a majority of people. Helping to bring about the general public's acceptance of a much older earth than what biblical fundamentalists believed was the very popular book *Prehistoric Times,* published in 1865 by the premier anthropologist from Britain, John Lubbock, who coined the terms Stone Age, Bronze Age, Iron Age, Neolithic, and Paleolithic.

By the mid nineteenth century, the theory of evolution (all species evolving from a single common ancestor from the sea all by itself without the help of God millions of years ago) was still a very novel and contentious idea. It stirred up a lot of emotion and heated criticism, especially with members of the clergy, who held tremendous influence over their members who still believed in a fundamentalist interpretation of the Bible. It wasn't until the late-nineteenth and early-twentieth centuries when several key scientific discoveries were made that evolution finally began to be accepted by the general public.

Acceptance of Darwinian evolution took many years because of critics' ability to raise serious objections to Darwin's theory of evolution without God. One of the first obstacles Darwin faced in defending evolution was the debate over the age of the earth. For Darwin's theory of evolution to be feasible, the age of the earth had to be hundreds of millions, if not billions of years old. Darwin believed that the earth was that old, but there was no way for science to measure its age in his time so he could defend himself against his critics who believed in a much younger earth. During the 1870s, around the time Darwin published his controversial book *Descent of Man,*[2] which claimed that man evolved from monkeys and was not specially created by God, the age of the earth was estimated to be much younger than what Darwin assumed, based on the British physicist's Lord Kelvin's incorrect theory of the cooling of the earth. However, with Henri Becquerel's discovery of radioactivity in 1896, and the discovery of radioactive dating in 1907 by Bertram Boltwood,[3] the age of the earth was determined to be much older than previously thought: about 4.5 billion years old. So, with the discovery that the earth is much older than originally thought, the theory of evolution gained credibility.

The blending of traits theory, which was the most commonly accepted model of inheritance in the eighteenth and nineteenth centuries, was another obstacle Darwin faced. The blending of traits model was the incorrect theory that inheritance of traits is like the mixing of paints. A tiny drop of modified color would be blended into a bucket of paint, and when thoroughly mixed, become totally imperceptible. For example, if a white man were to travel to a tropical island and marry a dark-skinned islander and they had children and grandchildren, his white color would eventually be blended totally out in succeeding generations. Thus, the critics of Darwin said, any slight modifications in a species as seen in Darwinian evolution would be so diluted out by the tremendously large population of non-modified species that evolution could not possibly occur. For the blending of traits objection, Darwin had no real answer, but tried his best to defend his theory of evolution in six subsequent editions of his famous first book, *Origin of Species,* by using Lamarck's incorrect theory of the inheritance of acquired characteristics.

Ironically, it would be a man of God, a Catholic monk from Austria whom Darwin never even heard of, who disproved the incorrect blending of traits theory and helped Darwinian evolution become that much more acceptable to the scientific community. With the discovery of genes in 1866 by the Austrian monk Gregor Mendel,[4] and how genes are passed on to succeeding generations in dominant and recessive pairs of alleles and not blended away in successive generations as previously thought, the blending of traits model was finally proven wrong in

1900 when Mendel's work was rediscovered and made public. As a result, Darwin's theory of evolution, where slight modifications in anatomy could survive and be passed along without being totally lost in the gene pool, was given even more consideration and credibility.[5]

But how could evolution of similar species take place on isolated islands and landmasses that were not connected to the main continents where the original species were thought to have evolved. This objection to evolution was finally solved by a scientist named Alfred Wegener (1880-1930). Wegener was a German astronomer and father of the continental drift theory. In his 1915 book, *The Origin of Continents and Oceans,* Wegener theorized that sideways forces uplifting subterranean rock formed mountains, and that continents were at one time all connected but were separated over eons of time by tectonic plate movement. Thus, it was entirely possible for similar but modified species to be located on isolated landmasses as Darwin observed in the Galapagos Islands and elsewhere. Wegener's continental drift theory, therefore, became another important ally for Darwinists.

To summarize, with the discovery of fossilized dinosaurs, ice age mammals, prehistoric cavemen, a much older earth, Mendelian genetics, and continental drift, Darwinian evolutionists of the early-twentieth century were able to piece together all the scientific evidence in very convincing fashion to convince the general public that man evolved from monkeys and was not specially created by God to have dominion over the earth and be with him in heaven after he dies. Evolutionists of the twentieth century had their scientific proof that Darwin never had, so they thought, and were able to persuade billions of men, women, and children through the government-run schools and liberal media that God doesn't exist and that all life evolved by itself from the sea through random chance, just as Darwin had theorized. Recall the Scopes Monkey trial of 1925 and the movie *Inherit the Wind,* which made a laughing stock out of biblical fundamentalists.

Darwinian evolution is now the most widely accepted model of how life came to be on this planet, and as a result, religion and the biblical story of creation has been legally forced out of our classrooms as an intolerant, bigoted anachronism of the Middle Ages. The very mention of the name of Jesus is avoided in most public schools for fear of lawsuits, and any attempt to offer creationism as an alternative theory to evolution is met with open hostility and lawsuits as it was in Kansas during the 1990s by such liberal groups as the ACLU, the People for the American Way, Americans United for the Separation of Church and State, the National Education Association, and others who hold the keys to power in our public educational system.[6]

But after all these years of open hostility to Christianity and the outlawing of its teaching in our government-run schools, is godless Darwinian evolution still supported by the scientific evidence as people originally believed it was, especially with the tremendous explosion of new scientific knowledge since the discovery of DNA in the 1950s[7] and genetic mapping in the 1990s? The answer to this very important question is absolutely not. As one prominent scientist from Harvard Medical School by the name of Dr. Isaac Manly, M.D. said about Darwin's theory of evolution, "What I have learned in the past ten years of review of recent scientific knowledge of cellular morphology and physiology, the code of life [DNA], and the lack of supporting evidence for evolution in the light of recent scientific evidence is a shocking rebuttal to the theory of evolution. . . . There is no evidence of any kind for this theory."[8]

Charles Darwin: A Man without God

To expose the theory of evolution for what it is—junk science meant to destroy people's faith in God—I must first give a little history about the one who started it all. Charles Darwin was born in Shrewsbury, England, on February 12, 1809, the fifth child of a wealthy and sophisticated family. Charles's paternal grandfather was Erasmus Darwin, a well-known physician, naturalist, and freethinker who had a big influence on Darwin's future atheism and interest in evolutionary theory. On his mother's side, Charles was related to the famous Wedgewoods, who made their fortune in fine china and pottery. Born with a silver spoon in his mouth and fine china on the table, Charles Darwin never had to worry about money and was free to do with his life as he pleased. As a youngster, Darwin was fond of collecting minerals, plants, and sea creatures, and showed great interest in a career in science. After graduating from the prestigious school at Shrewsbury in 1825, young Darwin enrolled at the University of Edinburgh to study medicine. Darwin, however, hated medicine and dropped out in 1827, whereupon he entered Christ's College at the University of Cambridge in 1828 to study for the ministry, which was more his dad's wishes than his own. While at Cambridge, Darwin met two stellar figures in science: Adam Sedgewick, a geologist, and John Stevens Henslow, a botanist, both of whom greatly rekindled his boyhood passion in the natural sciences.

Before studying for the ministry at Christ's College, Darwin had already been exposed to the theory of evolution, and he weighed the theological arguments for creation vs. evolution based on his observance of the obvious design in nature. To prove that evolution was more than a fanciful supposition of unbelievers, Darwin knew that he would have to somehow answer the question of intelligent design in the universe. At this point in his life, Darwin had no answers. As his luck would have it though, in 1831 Darwin, through his friend Henslow, won an appointment to be the unpaid naturalist on the H.M.S. *Beagle,* which was scheduled to begin a five-year scientific expedition around the world.

Darwin's visits to South America, the Pacific Islands, Australia, New Zealand, the Indian Ocean islands, and Africa allowed him to meticulously observe the diverse flora and fauna of these areas, as well as the various geological formations found on different continents and islands along the way. In his geological observations, Darwin noted the tremendous effects that natural forces such as earthquakes and volcanoes have on shaping the earth's surface, and used these observations to theorize that natural forces could likewise shape the formation of the many different species found on earth over eons of time. He also noted that with his discovery of coral reefs thousands of feet deep, that the earth was much older than what biblical fundamentalists maintained, which he thought would give the theory of evolution more than enough time to work.

At the time of Darwin's voyage in the early 1830s, the prevalent theory amongst geologists and scientists on the origin of life was the catastrophic formation theory, which stated that the earth had undergone several successions of creations of plant and animal life, as insinuated in the Book of Genesis, and that each creation had been destroyed by a sudden geological catastrophe.[9] In the opinion of the catastrophists of the 1830s, species were individually created by God and, therefore, were immutable or unchangeable for all time. Geological catastrophes wiped out extinct species, with the last catastrophe being Noah's flood. One of the main opponents of the catastrophic theory (but not the immutability of species) during Darwin's time was the English geologist Sir Charles Lyell, author of the three volume work *Principles of Geology* (1830-1833). In his *Principles,* Lyell argued that the geologic variations of the earth are the result of constant changes over long periods of time, not sudden catastrophes, and that the earth is much older than most people think.

On his voyage with the Beagle, Darwin had access to the ship's scientific library, which contained Lyell's books on geology. Lyell's theory of constant change over long periods of time struck a chord with Darwin, who was observing that the fossils of extinct species were very similar in structure to the living species he came across (the "succession of types"). He also observed that different geological areas, some very close to each other and separated only by water, have unique variations among common species such as iguanas, giant tortoises, mockingbirds, and finches, particularly in the newly-formed Galapagos Islands off the coast of Ecuador. From these two observations, Darwin concluded that Lyell was right about the active forces of nature being in constant change, but was wrong about the immutability of species, since he noticed such wide variation within species in the different places he explored. Darwin was drawing closer to formulating his own theory of evolution, but had not yet put it all together. During his trip, Darwin sent back his diary to his sister in installments and some of his papers to his friend Henslow, who printed and circulated some of them. When Darwin returned to England, he had already become a well-respected geologist.

Not long after returning to England in October of 1836, Darwin recorded his findings of his voyage on the Beagle in his *Notebooks on the Transmutation of Species,* which included his observation of the variations within species. But it wasn't until October of 1838, when Darwin first read *An Essay on the Principle of Population* (1798) by Thomas Malthus (1766-1834), that he was suddenly struck with an idea of evolution (which he simply called "my theory") by means of environmental pressures and the "struggle for existence," which, he said, weeds out the less fit, leaving only the favorable variations. Malthus, who so influenced Darwin, was a British economist and graduate of Jesus College, University of Cambridge, who wrote in his *Essay* that "population increases geometrically while food supplies increase only arithmetically." Thus, according to Malthus, population growth inevitably outstrips food supplies, which results in a struggle for survival, and when severe enough a natural decrease in population by means of war, famine, and disease.[10] Borrowing from Malthus's theory of the dynamics of human population growth linked to the struggle over food and shelter, Darwin formulated his now famous theory of "natural selection," which says that in nature, there is a constant struggle for survival and only the strongest survive. Said Darwin in his autobiography about Malthus:

> In October of 1838 I happened to read for amusement 'Malthus on Population,' and being well prepared to appreciate the struggle for existence which everywhere goes on from long-continued observation of the habits of animals and plants, it at once struck me that under these circumstances favorable variations would tend to be preserved, and unfavorable ones destroyed. The result of this would be the formation of new species.[11]

So, according to Darwin, any species that has an advantage over its competitor, however slight, will be much better able to survive and reproduce its own kind. These advantages, which are the result of favorable mutations in anatomy, are passed on to succeeding generations, and over long periods of time result in new species formation.[12] Nature, therefore, selects the most fit and weeds out the unfit, and this is how new species evolve from preexisting ones and how extinct species die out. Species are continually changing and new species evolving, said Darwin, because—like Sir Charles Lyell had observed—the environment is constantly changing. For example, when environmental conditions favor plants that produce large seeds with hard shells, finches with big strong beaks will prevail. When environmental changes take place that favor plants that produce smaller, softer seeds that fall into tight spaces, then finches with smaller beaks will prevail. Like man, who has the ability to breed any type of dog or horse he wants, so the more nature has the ability to produce newer and better species on her own without the help of God.

What Darwin and his followers didn't understand was that there is a huge difference between macroevolution, which Darwinism implies, and microevolution, which is what is observed in real life. Macroevolution is the unproven theory that totally new species arise from preexisting ones by means of natural selection, such as amphibians arising from fish, or birds from lizards and dinosaurs. Microevolution, on the other hand, is the scientifically proven theory that individual species have inheritable variability already encoded within their DNA, and under certain environmental conditions or selective breeding, a particular variation will be prominent. It is very important to understand the difference between macro and microevolution. Macroevolution says that nature creates new species on her own without the need for special creation by any deity. Microevolution says that individually-created species have tremendous variability already encoded within their DNA, but individual species cannot and do not evolve into new species.[13] Animal husbandry, which has been practiced for thousands of years and proves the existence of microevolution, has shown that you can selectively breed a dog or horse into all kinds of varieties or breeds, but a dog can never be bred into a pig, or a horse into a cow. A dog will always be a dog and a horse will always be a horse because species are immutable or fixed, as Lyell maintained, but they do show wide ranges of anatomic variability, as Darwin observed in finches.

Darwin's theory of evolution by means of natural selection was first published in 1859 in his *On the Origin of Species by Means of Natural Selection or the Preservation of Favored Races in the Struggle for Life,* the so-called "book that shook the world." Darwin had waited that long to publish his theory because of the harsh criticism leveled against Richard

Chambers' anonymously-published book, *Vestiges of the Natural History of Creation* (1844), which likewise suggested that somehow evolution gives rise to new and different species. When Darwin found out that his friend and colleague Alfred Russell Wallace, a British naturalist, specimen collector, and explorer of South America and Southeast Asia, was about to compile his own book on evolution, Darwin quickly finished *Origin* and beat him to press. Wallace had actually sent Darwin a copy of his manuscript for review, and Darwin felt bad that he betrayed him, but some historians say that Darwin's primary motive for going to press was that ever since he returned to England in 1836 he had become increasingly sick and was afraid of dying before he published his theory.[14]

Darwin's *Origin* was much more thorough and evidentiary than Chambers' *Vestiges* and sold out immediately. Nevertheless, it was initially met with extreme hostility and criticism by many respected members of the scientific community (including Agassiz, Lyell, Cuvier, and Owen), and was roundly attacked in the highly-respected publications *The Quarterly Review* and *Edinburgh Review*. Darwin's biggest critics were the comparative anatomist Richard Owen, the geologist Adam Sedgewick, and Bishop Samuel Wilberforce, who had earlier attacked Chamber's *Vestiges*.

A scientific meeting was convened in Oxford in May of 1860 to debate Darwin's new theory, which became the most important scientific meeting in British history. Darwin, however, was absent due to illness and his lack of desire to take on the popular cleric Wilberforce. The most vocal opponents against Darwin at the Oxford meeting were Bishop Wilberforce and Captain FitzRoy of the *Beagle*, and the most ardent supporters of Darwin were Thomas Huxley and Joseph Hooker. American John Draper also gave a talk at that meeting entitled "On the Intellectual Development of Europe, Considered with Reference to the Views of Mr. Darwin and Others, that the Progression of Organisms is Determined by Law." As history records, the Darwinists won out the day, and the theory of evolution without God was given the blessing of the scientific community.

The theory of evolution, of course, did not originate with Charles Darwin, as it can be traced all the way back to ancient Greece, but he was the first to try to explain it with what he thought was overwhelming scientific evidence, which was a thimbleful of nothing compared to what we know today.[15] One of the first scientists to suggest the theory of evolution was the nineteenth-century invertebrate zoologist Jean Baptiste de Lamarck (1744-1829), who argued that the anatomical variation amongst species such as tigers and lions and gazelles and giraffes were the result of environmental processes which act to modify a common lineage. Lamarck believed that "acquired characteristics," such as a giraffe stretching its neck for higher leaves, were passed on to the next generation, who would thereby have longer necks. Lamarckism, the theory of passing on acquired characteristics through the influences of striving and the environment, has since been proven to be totally untrue.

Lamarck was one of the first scientists to suggest evolution, but had no working model or mechanism to prove it. To him, species spontaneously generated and remained fixed, but showed wide variations due to their striving in the environment. Darwin's theory of evolution, however, went much further than Lamarck's theory of the inheritance of acquired characteristics. He argued that all living organisms, from the largest elephant to the smallest insect, evolved from a primitive common ancestor over millions, if not billions, of years. The accumulation of infinitesimally small favorable changes or variations in the morphology of organisms over long periods of time, said Darwin, caused all new species to emerge from a single common ancestor. For instance, the favorable mutations that led to fish developing legs from fins allowed them to walk out of the ocean one day and populate dry land. Land-roving fish, having abundant sources of food and no enemies, gave rise to an entirely new species called amphibians, such as frogs and salamanders. Amphibians, of course, could live inside and outside of water, but had to have a source of water nearby to reproduce their progeny, since they still lay non-shelled eggs in water like fish.

As time went on, said Darwin, favorable new mutations in amphibians allowed them to lay eggs encased in hard shells, which gave rise to a completely new species called reptiles. Reptiles, such as turtles and lizards, have an advantage over amphibians because they don't have to depend on a nearby water source to lay their eggs,[16] and can roam further inland on dry land where there is less competition for food and unoccupied habitat. As more time went by, favorable mutations in reptiles or dinosaurs allowed them to grow feathers and fly, which gave rise to birds. Birds have an advantage in survival because they can cover more territory and find more food and shelter in trees.

About the same time as birds evolved, different mutations in the same primitive reptile stock that gave rise to birds allowed reptiles to grow hair, become warm-blooded, and conceive and carry their young inside their bodies without the need for laying a stationary nest of eggs, which gave rise to another new species of animals called mammals. Mammals have an advantage over reptiles because they can cover more territory while pregnant, survive in colder climates because they are endothermic rather than ectothermic, nurse their progeny with their own milk, and have more time to teach and raise their young, which allowed for the development of intelligence and socialization with other members of their kind, which, in turn, increased their survival rates.

From the primitive mammal stock, such as tree-dwelling lemurs that have a prehensile tail, which allows them to grasp limbs and live in tall trees away from ground-dwelling predators such as lions and tigers, came the monkeys and great apes. Primates, in turn, evolved an opposable thumb to grasp objects and fashion instruments into tools, and learned to stand and walk erect. The evolution of an opposable thumb in lower primates allowed higher primates to start fires to cook food, and fashion weapons to defend themselves and hunt with. The ability to walk erect allowed them to see long distances over tall grass and navigate their way to unpopulated lands, increasing their survival. As time went by and these higher primates reproduced and began to socialize in larger groups, primitive ape-men evolved. And from ape-men evolved the primitive cavemen, with their larger brain case and social gregariousness. And, finally, from primitive cavemen evolved modern man, or *Homo sapiens.*

This in a nutshell is Darwin's theory of the evolution of man from monkeys; a degrading theory that is universally taught to our school children as unquestionable truth. God didn't purposely create man and the roughly six million different species of plants and animals on earth; it was random chance over 4.5 billion years. To the scientifically uneducated, Darwinian evolution sounds so logical and easy to understand that it must be true. Wrong! Darwinian evolution is a fairytale for unbelievers invented by a man who tried to use science to disprove God's existence because he lost his faith in God and tried to ease his troubled conscience about how life arose on earth without the need for a Creator. The early death of his daughter was the final straw for Darwin's belief in God. How could a good and loving God allow such a tragedy as this.

But the enormous number of scientific discoveries since the time of Darwin and the absence of the theorized transitional species in the fossil records, which Darwin desperately hoped someday would be found so it would vindicate him against his critics, have shot holes in his theory of evolution so large that you could drive a space station through it. During Darwin's time, genetics, statistical probabilities studies, molecular biology, cell biology, cytology, biochemistry, thermodynamics, electron microscopy, X-ray crystallography, and nuclear magnetic resonance (NMR) hadn't even been invented yet; and vertebrate embryology, comparative anatomy, animal physiology, paleontology, physics, and chemistry were in their infancy. Darwin had no idea about the astronomical complexity of living organisms and the way individual species pass their genetic code down to successive generations. In fact, genes, the basic unit of inheritance contained within all DNA and RNA, hadn't even been discovered when Darwin first wrote down his theory of evolution in a science paper in 1844.[17] Because of his deficiency of scientific understanding, the simple theory of descent from a common ancestor by numerous slight modifications in anatomy over millions of years seemed quite logical to Darwin and his most ardent followers, such as Thomas Huxley (Darwin's "bulldog"), Herbert Spencer, and Alfred Wallace. But if Darwin knew what scientists know today, how extremely complex life really is and what goes on inside and outside anatomy, I'm sure he'd never have invented his ridiculous theory of macroevolution in the first place, and probably would have become a minister like his dad wanted instead of an atheist who tried to relieve people of their faith in God because God had disappointed him.

In spite of the mounting scientific evidence against Darwinian evolution, it is still being promoted today by a majority of scientific elites as fact rather than fallacy, because without Darwinism one would have to believe in God, and there are far too many elitists and liberals in academia and public life who don't want to do that, so they'd rather perpetuate and defend the obvious lies of evolution than believe in the truth of creation. Another reason that Darwinian evolution is still being taught today is that those in positions of power in the colleges and universities who doggedly promote and teach evolution as science fact rather than fallacy have tremendous influence on gullible young students and aspiring teachers. No one who wants a career in science or education wants to be seen as an obscurantist, or worse, a biblical fundamentalist, so many aspiring young scientists and teachers who secretly have doubts about evolution readily buy into its lies to keep from ruining their careers. They blindly accept what they've been told since elementary school because of the tremendous influence of authority figures. But even the most ardent supporters of Darwinism, like the late Stephen Jay Gould, know that it is getting harder and harder to defend evolution. The monumental complexity in the recently completed human genome project, for instance, which mapped the entire genetic code of human DNA and showed that our DNA contains tens of thousands of pairs of genes with over three billion bits of information, enough information to fill one thousand telephone books of a thousand pages each, have prompted those involved with the project to admit that the evidence suggests that intelligent design rather than random chance created life on earth.

The First Fallacy of Darwinism: Life Originated All by Itself from the Sea for No Purpose

> "All appearances to the contrary, the only watchmaker in nature is the blind forces of physics . . . Natural selection, the blind, unconscious, automatic process which Darwin discovered, and which we now know is the existence and apparently purposeful form of all life, has no purpose in mind. It has no mind and no mind's eye. It has no vision, no foresight, no sight at all. If it can be said to play the role of watchmaker in nature, it is the *blind* watchmaker."
>
> —Richard Dawkins, *The Blind Watchmaker*

Darwinian evolution maintains that God does not exist and that he did not create the universe. Rather, the universe created itself from eternally existing atoms over eons of time by random chance without reason or purpose. Time plus chance plus matter created all things. This modern evolutionary theory that all living organisms originated gradually by chance through natural forces from non-living elements, with older species evolving into newer species, did not, however, originate with Darwin. It can be traced as far back as the ancient Greek philosophers Anaximander of Miletus (550 B.C.) and Empedocles (495-435 B.C.), who, like Darwin, believed that all life originated from some kind of primordial sea slime that arose from preexisting matter, which they, like Darwin, believed is eternal and uncreated.

It is difficult for the human mind, however, to believe that matter has always existed by itself for no apparent reason. It's even harder to envision atoms just milling around for an infinite amount of time and then one day start arranging themselves into a very complex and orderly universe for no apparent reason. We don't even know yet how small matter is. Physicists like Albert Einstein have shown that the atoms of Democritus and Leucippus, the ancient Greeks who first invented the theory of indivisible atoms, are not indivisible as previously thought, but can be split into much smaller particles of protons, electrons, and neutrons, which can be further divided into Lord knows how many smaller parts. So where did all these smaller particles come from, and how and why did these smaller particles come together to form all the atoms and elements of the different types that make up the universe? Is matter infinitely divisible and infinitely small? Given enough time, could matter come into existence from nothing all by itself? The answer to these important questions comes not from science but from logic.

Logic says that "from nothing, nothing comes" (*ex nihilo, nihlo fit*).

Nothing, therefore, can't create a "something." Matter, which is "something," must necessarily come from some other "something," which has no beginning or end because an infinite regression of created things coming from other created things is illogical. There has to be an original source from which all things in the universe derive, which has no beginning or end, which is self-sustaining and uncreated, which created everything there is. That source is the one we humans call God, who is the "Eternal and Uncreated One." When explaining this to skeptics and unbelievers they, of course, will challenge you with "If God created the universe and everything in it, than who created God?" The answer to this often-heard question is that God is not matter but pure spirit, eternal and uncreated. Skeptics, in turn, will reply, "If matter can't be eternal and uncreated, than how can God's spirit be eternal and uncreated?" The answer to this question is itself a question: When did love begin, or truth begin, or goodness begin, or mercy begin, or justice begin? They exist and never had a beginning and will never have an end. Spiritual things are totally different from material things, and you can't equate the two as to origins and kind. If skeptics then say that they can't imagine anything without a beginning or end, which at first they tried to say was a property of matter, then ask them what is the highest number which no other number exceeds, or the lowest number which no other number can be lower. There are, of course, no such things: numbers are infinite. So it is with God, who is the "Infinite Being," who can't be understood by mere human imagination. It helps to think of it this way: being can never come from non-being, which by definition is nothingness, so life can never come from non-life. All beings and all life, therefore, must come from God, the "Original Being" who is "Life Itself."

Of course, don't-confuse-me-with-the-facts skeptics who refuse to believe in God no matter how much logic and historical and scientific evidence you provide must still answer for themselves how the universe originated on its own. This is where Darwinian evolution comes in. Darwinian evolution says that the universe arranged itself into an orderly system from eternally existing, infinitesimally small chaotic matter of infinitesimal density by random chance alone for no apparent purpose or end. As their bedtime story goes, one time in infinity past, all the matter in the universe was compacted to less than the size of a pin's head. Then there was a big bang somewhere in infinity past (15 billion years ago according to the world-renowned atheistic and physicist Stephen Hawking) that dispersed all of this eternal matter into the infinite void like a tremendous shotgun blast. How this big bang happened all by itself or where the infinite void came from, scientists can't explain.

As the story continues, at the time of the big bang the laws of physics and time did not yet exist but came to be on their own, which common logic says is impossible because the sudden existence of laws implies the existence of a lawgiver who formulates the laws. As the primordial gases, liquids, and solids of the big bang slowed down and gathered together by gravitational forces, they began to form the trillions of planets, stars, solar systems, and galaxies we now see, according to the newly-formed laws of physics and time. The earth cooled and formed a solid outer crust, or mantle, that surrounded a much-hotter inner core formed by various molten rocks and radioactive elements that steadily decay and give off heat. On the surface of the primitive earth some 4.5 billion years ago were gaseous atoms and molecules, primarily of inorganic type, such as hydrogen, oxygen, and nitrogen.

By chance one day, gaseous hydrogen atoms on the earth's surface combined with oxygen atoms in a 2:1 ratio that formed a very stable bond called water, the most abundant molecule on earth. Liquid water, being heavier than air, collected on the earth's surface due to the gravitational forces of planetary rotation, and in time covered the globe. Soon, according to the evolutionists, elements of all kinds from the earth's crust were mixed into the seas, and by chance one day, perhaps with the help of an outside energy source such as lightning or the sun, the first simple organic molecules were formed. These simple organic molecules, which Darwinists say were formed because of the stable bonds they made, were in ceaseless competition for other atoms to form longer, more complex molecules known as "polymers." For reasons evolutionists can't give, these polymers suddenly liked being together in highly complex arrangements more than they liked being in their elemental state.

So out of this "primordial soup" of non-living polymers, which sounds so good and warm and motherly that it must be true, came the first precursor to all life forms: a primitive organic polymer that found a way to replicate itself.[18] Of course, there is no geological evidence that there was ever any such thing as a primordial soup on the earth's surface, just water.[19] Now this primitive "replicator," which modern evolutionists say was either the first RNA or DNA molecules, didn't like being eaten up by its cruel and hungry competitors (other replicators formed from the original replicator as a result of mis-copies say the Darwinists), or zapped apart by radiation from the sun (thermolysis), or dissolved by the churning action of the oceans (hydrolysis), so it cleverly learned one day how to encase itself in a protective wall or membrane, forming the first living cell. It could breathe, eat, excrete, move, respond to stimuli, grow, and reproduce—the seven characteristics necessary for life.

Being very intelligent about its identity and surroundings, this primitive cell knew that it wouldn't be around forever, so it found a very clever way to divide itself into like copies that would perpetuate its genetic code, guaranteeing that its DNA (or RNA) would survive to populate the seas and inherit all that was good. Again, the billions of copies that the first primitive cell made, however, couldn't get along with each other because of their increased competition for other molecules to bond with, so some of them got together and joined forces, forming the first multicellular organism. Primitive multicellular organisms were bigger and stronger and smarter than the original one-celled replicator, and were better able to survive and produce like copies of themselves, so they flourished throughout the oceans and evolved into all kinds of life forms. Some evolved the capability to synthesize their own food from the sun's energy working on chlorophyll and carbon dioxide, and became the first members of the plant kingdom. Some were predators that lived off the energy of other living organisms, and became the first members of the animal kingdom.

This is the most widely accepted evolutionary fairytale of how life evolved on earth all by itself from the very first replicator. For the rest of this chapter I will not distinguish between twentieth-century Neo-Darwinists, who incorporate modern genetics into their theory of evolution, and nineteenth-century Darwinists, who were not aware of genetics, and just call them all Darwinists because none of them believe in God. Here is the rest of their story on how the first sea-slime replicator evolved into all the known plants and animals of the earth.

According to the fossil records, the earliest signs of plant and animal life on earth were bacteria/blue-green-algae-like creatures called cyanobacteria, which date back some 3.8 billion years ago. Anatomically and physiologically, cyanobacteria are part plant and part animal, being capable of making their own food yet capable of living off other molecules. About 2.7 billion years ago cyanobacteria, as evolutionists maintain, evolved into the first true unicellular oceanic plants and animals, such as algae and protozoans. Plants, by definition, make their own food through photosynthesis, while animals must get their energy source by feeding off other living organisms. About 1.8 billion years ago, the first multicellular organisms appeared in the seas, such as red algae. Somewhere between 700 and 500 million years ago, the earliest known communities of animals, the Ediacarans, evolved. Ediacarans are called animals but scientists really don't know what they really are. The first fossils of animal embryos were discovered in 1998 and have been dated to around 570 million years ago. These embryos were triploblastic in structure, having three germ layers that develop into the muscles, the digestive tract, and the nervous system.

The pre-Cambrian Period of 700 to 550 million years ago comprises 85 percent of the history of life on earth, and is characterized by the presence of bacteria, protozoa, algae, Ediacarans, and embryos of early animals. For reasons unknown, around 530 million years ago, the Ediacarans began to decline in number while the triploblasts began to explode. In the following Cambrian Period of around 535 to 500 million years ago, the period of the developing chordates that have a primitive neural tubes or notochords, many new life forms emerged, such as starfish, lamp shells, corals, jellyfish, sponges, acorn worms, sea squirts, and the primitive sessile filter feeders, which supposedly evolved into free-swimming gill filter feeders like the amphioxus, the supposed ancestor of modern-day fish. The "Cambrian explosion" took place entirely under water, with the continents being bare except for scattered bacterial crusts.

Signs of life on dry land first began to appear around 500 million years ago with the appearance of green algae, the first multicellular inland plants. The next plants to appear on land looked very similar to moss and liverworts, and were located alongside riverbanks and coastlines. Around 500 million years ago, in the Ordovician Period, the first vertebrate fish known as the ostracoderms began appearing in the oceans. Fifty million years later, in the Silurian Period, upright plants appeared on dry land; various insects, centipedes, and spiders evolved from aquatic worms and migrated inland; and jawless fish known as the agnathas predominated the oceans. The Devonian Period of 400 million years ago saw the rapid evolution of bony and cartilaginous fish with true jaws, the appearance of land-dwelling amphibians such as frogs and salamanders, which supposedly evolved from fish, and the first trees. For the next fifty million years, in the Carboniferous Period, ferns and fern coal forests dominated the land, sharks were abundant in the oceans, and the first reptiles.

In the Permian Period some 280 million years ago, reptiles became the dominant land species and displaced the amphibians in importance. During the Triassic Period 230 million years ago, the first dinosaurs appeared, along with the first primitive mammals. Dinosaurs were actually huge reptiles, according to evolutionary taxonomists. In the Jurassic Period 180 million years ago, dinosaurs ruled the land. Birds began to evolve from reptiles and mammals increased their numbers on land. By the end of the Cretaceous Period some 66 to 65 million years ago, the large dinosaurs suddenly became extinct, perhaps due to a catastrophic event such as an asteroid. During the Tertiary Period of about 65 to 2 million years ago, many species of grazing mammals evolved and flourished on the developing grassy plains, with the evolution of large mammalian predators such as the saber-tooth tiger. The first hominid ancestors of human beings appeared about five million years ago as the common ancestor of humans and chimps diverged. About two million years ago, in the Pleistocene Epoch, the mammoths, the the saber-tooth tigers, and other large mammals began to die out as glaciation became increasingly widespread and living conditions and hunting pressures from cavemen worsened for them during the ice age. The first true human beings or *Homo sapiens* appeared about 600,000 years ago in Africa. With an abundance of game animals and the invention of fire to cook with and caves for warmth and shelter, Neanderthal cavemen of a couple hundred thousand years ago were the dominant hominid in Europe. The subsequent appearance of Cro-Magnons, with their larger braincase and greater intelligence, probably caused the extinction of Neanderthals. Modern man, as their story goes, is a relative of the Cro-Magnons, whose archaeological records date back about ten to fifteen thousand years ago.

This is how all life evolved on earth according to the evolutionists of the twentieth century. From the simplest and most primitive of proteinaceous life forms all the species of the earth evolved, including monkeys and men. Darwin speculatively wrote in 1871 that if this original aquatic life form were ever discovered, it would be formless, diffuse, and inchoate, and it probably developed in warm, sunny water that was free of predators and rich in inorganic elements electrically charged by frequent lightning:

> It is often said that all the conditions for the first production of a living organism are now present which could ever have been present. But if (and oh! what a big if!) we could conceive in some warm little pond with all sorts of ammonia and phosphoric salts, light, heat, electricity present that a protein was formed ready to undergo still more complex changes at the present day such matter would be instantly devoured or absorbed which would not have been the case before living creatures were formed.[20]

Darwin's supporters thought that this original primitive life form could be found on the ocean floors, so in 1872 the British scientific expedition ship H.M.S. *Challenger* set sail to collect deep ocean specimens in hopes of finding it. But after four years and 69,000 miles of unsuccessful dredging, the scientific expedition was forced to give up. What originally prompted the expedition was this: Thomas Huxley, Darwin's "bulldog" and presiding scientist on the *Challenger*, had in 1868 studied some mud samples from the bottom of the sea that were preserved in alcohol. When he noticed a gelatinous, inchoate substance he immediately announced to the scientific community that he had discovered a species called *Bathybius haeckelii,* Darwin's long-anticipated formless, non-nucleated, homogenous albumin capable of nourishment and reproduction. So the *Challenger* expedition was commissioned in 1872 to search out and find Bathybius on the ocean floors. But much to Huxley's disappointment, he never found Bathybius. In fact, the expedition's chemist later discovered that *Bathybius* was just a colloid precipitate of calcium sulfate, formed when alcohol was added to sea mud, and not a primitive life form after all. Modern supporters of Darwinian evolution don't like to be reminded of this embarrassing episode of mistaken identity, and try to keep it out of the biological text books, but it happened nonetheless.

Another embarrassing episode for evolutionists came in 1894 when a different so-called primitive life, *Eezoon canadense,* was found to be nothing more than an inorganic byproduct of volcanoes. *Eezoon canadense,* discovered by a collector of rocks for the Geological Survey of Canada in 1858, were thin, concentric layers of alternating serpentine (a silicate) and calcium carbonate originally thought to be fossils of a giant, living foraminifer (marine rhizopod), a primitive relative of *Bathybius* that had somehow evolved a seashell millions of years ago to protect itself from the environment. Canada's leading paleontologist of the time was elated at this find and named the supposed new species *Eezoon canadense.* But upon further investigation, scientists rejected *Eezoon canadense* as a primitive living organism when the same thing was found in the highly transfigured blocks of limestone ejected from Mount Vesuvius in 1894, proving that *Eezoon canadense* was an inorganic byproduct of high heat and pressure in metamorphosed rock and not an ancient life form of evolution. Evolutionists don't like to talk about this mistake either.

Today, no scientist will put his reputation and career on the line and claim that they have discovered Darwin's original life form, from which all other life forms evolved, because every time someone has done so they have been proven wrong. Evolutionists, however, still hold tenaciously to Darwin's theory of the original, living protein that produced all life on earth, because their theory of a universe without God would fall apart without it.

As evidence of their theory, modern Darwinists invariably point to the outdated experiment done by Stanley Miller at the University of Chicago in 1952. As an undergraduate student working in the laboratory of Nobel laureate Harold Urey, Miller sought to prove that around four billion years ago, the common, inanimate molecules on the lifeless earth—such as water, methane, and ammonia gases—recombined in a reducing environment (without the presence of oxygen) with the aid of an external power source such as lightning, to produce the first amino acids, the building blocks of proteins, which are vital for life. In a simple experiment, Miller mixed a combination of boiling water vapor, methane, and ammonia gases in a large cylindrical flask and gave it a spark of electricity to simulate lightning bolts, and lo and behold, in about a week's time, amino acids formed in the mixture. Taking a tremendous leap of faith from a simple experimental observation to a highly speculative conclusion, Darwinists around the world hailed the Urey-Miller experiment as proof of the origin of life in dirty water all by itself through the random, natural processes of evolution.

Being able to synthesize amino acids from an artificially controlled mixture of water vapor, ammonia, and methane charged with electricity, however, in no way proves the origin of life; it just shows how to synthesize amino acids. To reach any other conclusion would be like saying that the sun, which can turn water, mud, and straw into hardened bricks, built the pyramids of Egypt. Today's biochemists can synthesize all kinds of organic and inorganic molecules, and can separate living cells into their many essential component parts, such as phosphorous, lipids, proteins, sugars, amino acids, nucleic acids and the like. Yet with all their scientific knowledge and laboratory capabilities, scientists can't even create a simple blade of grass. That's because only God can create life. Show me one scientist who has created life from non-life and I will become an evolutionist. Scientists can manipulate life through genetics and cloning, but they can't create it, even thought they claim they can.

The general consensus amongst today's credible scientists is that the earth's early atmosphere did not contain a reducing environment of methane, ammonia, and hydrogen like in the Urey-Miller experiment,[21] and that any polypeptides that would have been built up by amino acids through the power of lightning would have been immediately dissolved by the hydrolytic churning motion of ocean currents and waves. Another scientific fact that discredits the Urey-Miller experiment is the formation of D-amino acids in Miller's flask. Let me explain. Artificially-generated amino acids (such as in the Urey-Miller experiment) have chiral forms of dextrorotary (D) and levorotary (L) acids, otherwise known as right-handed and left-handed racemates, which are mirror images of each other in a fifty-fifty ratio. Proteins that make up real life, however, contain only left-handed or L-amino acids. Artificially or randomly generated amino acids would, therefore, by the laws of probability and chance, produce equal amounts of L-amino acids and D-amino acids. But all proteins in living organisms are made up of only L-amino acids, and any incorporation of D-amino acids into a protein would render that protein nonfunctional. Therefore, it is mathematically impossible for random chance to have created life because the laws of probability say that D-amino acids would have been necessarily incorporated into proteins along with L-amino acids, making proteins non-viable. It is impossible to randomly produce 100 percent optically pure L-amino acids from inorganic molecules as seen as the Urey-Miller experiment, and if someone could, that person would immediately be a candidate for the Nobel Prize in chemistry. Thus, evolutionary scientists and teachers who say that the Urey-Miller experiment proves that life originated from non-life are either woefully ignorant of chemistry or are willfully misleading the general public, who know nothing about science.

The Second Fallacy of Darwinism: Mutations Are Good Because They Drive Evolution

A single cell, the basic structure of all living organisms, was thought to be in Darwin's time just a simple blob of protoplasm surrounded by a primitive cell wall or membrane. Modern science, however, with the aid of electron microscopy, has determined that this so-called "primitive cell" of Darwin's time is actually a machine so complex that it dwarfs the most complex of human factories. The eukaryotic (nucleated) cells of plants and animals contain dozens upon dozens of individual structures that help the organism carry on the vital processes of protein formation and cell division. At any one time, there are literally hundreds of actions and reactions taking place within a living cell in the cellular components so far identified by scientists.

At the center of the human cell is the nucleus, which contains twenty-three pairs of chromosomes (the nuclear material which contains our DNA), half coming from the mother and half from the father. On these chromosomes are the tens of thousands of precisely arranged genes that encode how we are made. Genes are actually individual units of triplet code (codons) made up of purine and pyrimidine bases called A, T, G, and C (adenine, thymine, guanine, and cytosine) which are precisely arranged on cross-linked, double-stranded helices of DNA (deoxyribonucleic acid). Each double-stranded helix of DNA is composed of alternating links of phosphate and deoxyribose. The two strands of DNA are cross-linked by pairs of purine and pyrimidine bases projecting inward from the deoxyribose sugars and joined by hydrogen bonds, with adenine linked to thymine and cytosine to guanine. Double-stranded DNA is arranged in such a way as to be separated and read by incoming single-stranded messenger RNA (m-RNA), whose primary function is to read the genetic code on DNA and then link together specific amino acids outside of the nucleus, based on the genetic information of DNA. The linked amino acids, in turn, constitute the proteins that are specific for every cell. How this is executed is as follows. Individual amino acids are first linked to groups of small, single-stranded transfer RNA (t-RNA) residing outside of the nucleus, which contain a specific triplet codon for a particular amino acid. The individual amino acids are then assembled together as chains on cellular structures called ribosomes outside of the nucleus by the messenger RNA to produce specialized proteins, according to our DNA code. Proteins are the building blocks of life and regulate the millions of biochemical processes going on within our cells, such as highly specialized gene replicating mechanisms, enzymatic production and activation, complex metabolic pathways, internal and external on-off controls, feedback loops, recognition systems, cascading systems, cell receptor binding/release systems, stabilizing factors, metabolic precursors, promoters, exchangers, etc.

The millions of biochemical processes that take place simultaneously in our bodies and cells at any given moment, such as blood clotting, antibody formation, respiration, or insulin production, are said by scientists to be "irreducibly complex" because to change or omit even one of these biochemical processes will most likely result in the organism's death. Thus, to survive, the genetic code of all organisms must be very precise and highly accurate. Mistakes or mutations in our genetic code, which mess up the way proteins are made and assembled from m-RNA, are almost always deleterious, resulting in decreased viability and sooner or later death. In the highly sophisticated world of molecular biology, which Darwin knew absolutely nothing about, accidents or mutations in genetic codes are not favorable to organisms; they are deadly. This is the second fallacy of Darwinian evolution: infinitesimally small mutations over long periods of time are favorable to an organism's survivability; and, therefore, perpetuate evolution. Scientists and medical doctors who are honest with themselves know that just the opposite is true: mutations are counterproductive because God has made each individual species viable in every way with no need for improvement to survive. In fact, changes in our genetic sequences are known to cause an estimated three thousand to four thousand inheritable human diseases; such as Huntington's disease, cystic fibrosis, muscular dystrophy, and sickle cell anemia. Try telling sufferers of these diseases that mutations are good for evolution.

Darwinian evolution rests on the false premise that ever-so-slight favorable changes in anatomy via genetic mutations, which supposedly increases an organism's ability to survive and reproduce, drive the beneficial evolution of the lower species to the higher. Thus, by its very own definition, evolution infers that the higher species are better and more successful, or else they wouldn't have evolved in the first place. But in reality, simple creatures survive very well just as they are and have no need to become more complex. In fact, nowadays it is the simpler creatures that are threatening to wipe out the lives of the more complex, as seen by the viruses that cause AIDS and SARS and West Nile fever, not to mention the ever-increasing number of strains of deadly bacteria that are becoming more and more antibiotic resistant. Bigger and more complex, when it comes to evolution's theory of fitness and survival, is not always better. The Black Plague of the Middle Ages wiped out nearly half the population in some parts of Europe, and another plague far worse than the Black Plague could happen again if man is not careful. It could be AIDS or some far deadlier disease we haven't yet discovered, or which we thought we have conquered, such as smallpox. Stock piles of deadly bacteria and viruses, such as anthrax and smallpox, could be let loose at any moment by terrorists, so it might be man himself who unleashes the fury of the simple yet deadly microbe.

The Third Fallacy of Darwinism: The Only Purpose of Life Is Replication

In Darwin's world there is no God or heaven or afterlife to look forward to when we die. There is no reward for leading a virtuous life, nor punishment for being wicked. Too bad if you suffer in this life because there is no one to make amends or wipe away your tears in the next life. This life, which arose by accident on a watery globe in the Milky Way Galaxy far out in some distant universe, has no meaning or purpose except to replicate.

Richard Dawkins, the pre-eminent twentieth-century Darwinist from Oxford University, wrote in his best-selling book in 1989, *The Selfish Gene,* that human beings are nothing more than fleshy shells or vehicles, which DNA has randomly evolved to protect and transport itself prior to replicating. Deoxyribonucleic acid or DNA, according to Dawkins, evolved millions of years ago by random chance in Darwin's primordial soup, and this primitive molecule, which he called the first replicator, found a way to first copy itself and then enclose itself in a protective shell. As time went by, DNA replicators evolved more and more complex shells until DNA finally ended up inside a shell called man.

Human beings don't realize it, but we are nothing but a shell and are unconsciously acting out the will of our primitive DNA replicators, which drives us to replicate more DNA. That is the ultimate and only reason for our being, to replicate deoxyribonucleic acids. Quoted Dawkins: "Genes [contained within DNA] created us and their 'preservation' is the ultimate reason for our being."[22]

To Dawkins, if anything approaches immortality it is DNA. The problem with this patently ridiculous and extremely naive form of philosophical reductionism is that it ascribes intelligence and reason to atoms and molecules, which don't have any. Why should DNA molecules care if they are passed on to the next generation? What's in it for them? DNA has no feelings, no self-awareness, no self-consciousness, no memory, no judgment, no hope, no goals, and no happiness. It is nothing but phosphorous, sugar, and nucleotide bases held together by extremely fragile hydrogen bonds. It can be split, dissolved, dissected, recombined, manipulated, extracted, implanted, frozen, and thawed by scientists without uttering a single protest. It dies when we die and is born when we're born.

In reality, it is not DNA that is directing us but we who are perpetuating them by our own choosing. What about the many people who opt for celibacy? Celibacy and abstinence contradict Dawkins' inane theory that we are unconsciously obeying the higher will of DNA to replicate itself. If DNA is so powerful in its will to survive that it drives us to robotically reproduce our DNA, why then did man work so hard to invent contraceptions and legalize abortion? Wouldn't human DNA have found a way to prevent these roadblocks to its own replication? Of course it wouldn't because Dawkins' theory of the willful, intelligent, and immortal DNA replicator is ludicrous and inane.

As every sensible person knows, our lives have much more meaning than just making copies of DNA. God has given us truth, direction, and purpose to our lives: to love and serve him on earth and be with him when we die; to be caretakers and custodians of this earth he created for us; to love our neighbors as ourselves; and not to fight and kill one another in never-ending struggles to survive and pass on our genetic code as Darwinism suggests. We are not just mere physical receptacles for housing DNA. We are human beings made in the likeness and image of God. Besides, DNA never remains the same from one generation to the next. Geneticists have shown that with meiotic division, gene recombination, chromosome crossovers, point xmutations, inversions, linkages, etc., that the original set of one parent's genetic information is virtually gone within a few generations.

One of the biggest questions that extreme reductionists like Dawkins can't answer is why the idea of God is so ingrained within the human soul if all we are is a vehicle for DNA. Dawkins tried to account for this variance in evolution by saying that somewhere along the line genes for cultural and spiritual ideas evolved in our DNA to comfort and strengthen us psychologically. Dawkins called these genes "memes." According to Dawkins, memes are genetic codes that reassure us when we are troubled, but are not at all truthful. They tell our brains comforting bedtime stories so that we will not be afraid of the dark or the bogeyman under our bed. Memes are no different than a doctor's placebo. Said Dawkins about memes:

> Consider the idea of God. We do not know how it arose in the meme pool. Probably it originated many times by independent 'mutation.' In any case, it is very old indeed. How does it replicate itself? By the spoken and written word, aided by great music and great art. Why does it have such high survival value? . . . The survival value of the god meme in the meme pool results from its great psychological appeal. It provides a superficially plausible answer to deep and troubling questions about existence. It suggests that injustices in this world may be rectified in the next. The 'everlasting arms' hold out a cushion against our own inadequacies which, like a doctor's placebo, is none the less effective for being imaginary. These are some of the reasons why the idea of God is copied so readily by successive generations of individual brains.[23]

A paradoxical problem arises with Dawkin's wacky theory that we are simply material vehicles for passing on DNA, which is somewhat like the old chicken-and-the-egg debate. In order for the first DNA molecule to replicate itself, there must have been present a proteinaceous enzyme called DNA-replicase, which splits the double-stranded helix of DNA and allows it to open up and be read by incoming single-stranded messenger RNA (m-RNA). DNA-replicase, therefore, had to have been present before the evolution of DNA for DNA to replicate, yet DNA-replicase could only have been made after DNA had been split and read by incoming m-RNA, which, in turn, requires DNA-replicase to open up the DNA strands before m-RNA can read it and decode the genes for the production of DNA-replicase. This a paradox of life that evolutionists can't explain: all components of the DNA replicating process had to have been in existence before the first DNA molecules evolved, which is impossible since DNA is the only thing that makes them. Instead of the old question "Which came first, the chicken or the egg?" evolutionists are faced with the unanswerable question, "Which came first, DNA or DNA-replicase?" Evolutionists have tried to skirt this problem by saying that single stranded RNA probably evolved before DNA. But there is a problem here too. Organisms that have single-stranded RNA as their genetic material, such as viruses, are dependent on other organisms with double-stranded DNA because as a parasite, they inject their RNA into their host's DNA, which then become production factories for the replication of RNA viruses. It is a very ingenious yet simple way of viral replication and is how the HIV virus that causes AIDS replicates. DNA, therefore, had to have been present before RNA or else these types of organisms would never have survived.[24]

The Fourth Fallacy of Darwinism: Those Who Produce the Most Offspring Are the Most Fit

Charles Darwin defined evolutionary fitness as one's capacity to survive and produce more offspring than one's competitor. The more offspring one produces, the more successful one is. But here again is another problem with Darwinism: if the number of offspring were the most important thing in evolution, then the whole world should be made up of nothing but microbes, since they have extremely high reproduction rates. This, of course, is not what's seen in the real world.

Bacteria, viruses, protozoa, and other single-celled microbes, all of which have reproduction rates millions of times faster than higher animals, are very successful at producing progeny and passing on their genetic information. So if all life forms descended from a single-celled microbe as Darwinists say, and if high reproduction rates assure the most progeny, why then did whales and elephants, which have very long gestation periods and very few progeny over their lifetime, evolve? Their continued existence makes no sense according to Darwinian evolution. They reproduce too slowly and too few numbers, yet they have been around for millions of years and continue to perpetuate their own kind.

Which brings up another question: If homosexuals have sex with there own gender and can't reproduce new human beings, why are their numbers and influence increasing in the general population? Genes for homosexuality can't be passed on to succeeding generations if homosexuals don't have children. The only plausible answer to this politically incorrect question is that there is no such thing as a gene for homosexuality. Homosexuality is a moral choice, not an inheritable genetic trait.

Despite what homosexual activists say, no one is born a homosexual. If they were then all identical twins who share the exact same genetic code would be either 100 percent homosexual or 100 percent heterosexual.

But that is not what's seen in real life. There are many examples of identical twins where one twin is gay and the other straight. Similarly, if homosexuals were born that way then they would never be able to change their sexual identity. Homosexuality would be like eye color. Again, that is not what's seen in real life. Homosexuals can and do give up their homosexual lifestyles and revert back to heterosexuality. Just ask the many thousands of members of Exodus International, a popular support group for former gays who have given up their homosexual or bisexual lifestyle.

If not genes, what then causes homosexuality? Honest psychologists and psychiatrists know that true homosexuality (not the faddish, media-glorified homosexual experimentation seen these days in college dormitories) is a result of one's environment and the way children are brought up, rather than one's DNA. In the case of girls, rejection by boys who choose the prettier and more popular girls often results in disheartened adolescent females bonding together in common interests such as sports or women's groups, which many times leads to lesbian experimentation and relations. In boys, poor or absent father figures, being sexually abused by an adult male, and/or being raised by a dominant mother and effeminate father are the primary causes of boys becoming gay men. Young boys naturally long for male companionship and mentoring from their father, and when they don't get it they are often tempted to seek out sexual relationships with older boys and men to fill the masculine void in their hearts. Sex becomes confused with love and longing for dad, whereupon many young homosexual boys begin having tremendous guilt feelings over their lifestyle, which more often than not results in self-destructive behavior such as drug and alcohol abuse, promiscuous sex with multiple partners, AIDS, and premature death. Statistics have shown that there is a much higher incidence of suicide and premature death in the homosexual population than in the heterosexual population, with the average life span of a gay man being about eight to twenty years less than heterosexuals, according to some studies.

Again, by using Darwin's definition of fitness as a longer life and producing the most offspring, natural selection should have weeded out a gene for homosexuality a long time ago. And don't try to tell me that a gene for homosexuality persisted in the gene pool because their was a need for an effeminate male to baby-sit the cavekids or cook the brontosaurus burgers while the burly cavemen were out hunting dinosaurs and conking cavewomen on the heads with their clubs. That often-heard theory from homosexuals is a joke and has no scientific basis whatsoever. You don't have genes for other people.

In today's decadent and materialistic society where practically every human action is reduced to the interplay of matter and chemicals within the human body and brain, genes are said by the "scientific" experts to be responsible for all the wrong choices we make in life. There is a gene that makes us gay, a gene that makes us violent, a gene that makes us greedy, a gene that makes us alcoholics, a gene that makes us fat and lazy, and so on and so on. The idea of free will and individual responsibility has been practically eliminated from our belief system, especially when it comes to homosexual desires. In the 1990s, several articles came out in prestigious scientific journals linking homosexuality to ones's genes. In August of 1991, a San Francisco anatomist by the name of Dr. Simon Le Vay published a report in the highly respected journal *Science* stating that the hypothalamic nuclei in the brains of dead gays were twice as large as in the brains of dead straights. La Vay's inference, of course, was that the brains of gay men are different than the brains of straights, which explains their different sexual orientations. Never mind the fact that these men were dead and couldn't tell anyone whether or not they were really gay or how they felt about their own sexuality. And never mind the fact that their brains were not studied while they were in a functioning state. As chairman of the Institute of Gay and Lesbian studies in West Hollywood, California, Dr. Le Vay, himself a homosexual, was obviously biased and sympathetic towards gays from the start, and had a political agenda to find a genetic link for homosexuality. Le Vay's research was never taken very seriously by the scientific community, but was picked up by the pro-gay media and run with, convincing scores of scientifically-naïve people that homosexuality is a genetic trait not an immoral choice.

Another futile attempt to link homosexuality to genes occurred in July of 1993 when Dr. Dean Hamer of the National Cancer Institute in Bethesda, Maryland, himself a gay, published a report in the journal *Science* that there was a genetic marker on the X chromosome of men, the Xq28 region, that appeared different than in straights, based on his genetic studies of dozens of gay men and their gay brothers and maternal gay relatives who likewise had a higher incidence of this chromosomal aberration. The news of Dr. Hamer's discovery of the "gay gene" was trumpeted as a scientific breakthrough by such liberal news outlets as National Public Radio, the *New York Times,* and *Newsweek* magazine, which ate the story up. Six years later, however, in April of 1999, a Canadian team of researchers at the University of Western Ontario headed by Dr. George Rice came out with their own scientific report stating that in a much larger study of gays and their maternal relatives they could not duplicate Dr. Hamer's findings and, therefore, his report had no credence. Hamer's "gay gene" got a rude slap in the face in the Canadian and American press in 1999, but his influence has forever shaped public opinion in favor of the idea that homosexuality has some sort of genetic component. Serious researchers today, however, are far too reluctant to claim that genes alone are the root cause of any and all behavioral abnormalities because of the high risk of being discredited, so they prefer to theorize that a combination of genetics and the environment are responsible for our immoral actions. Never is it a simple matter of free will and incorrect choices on our part, which we used to call sin. Our genes or someone else are to blame. Tell that to God come judgment day.

The Fifth Fallacy of Darwinism: Given Enough Time, Anything Is Possible

Charles Darwin maintained that infinitesimally small mutations in primitive species over eons of time allowed for the evolution of lower species into higher species. Therefore, it was the great lengths of time between species formations that allowed for higher species to evolve. Time is always the faithful ally of evolutionists, who believe that given enough of it, anything is possible. Richard Dawkins, the famous zoologist and Darwinist from Oxford and author of the best-selling book on evolution *The Blind Watchmaker: Why the Evidence of Evolution Reveals a Universe Without Design,* was one of those Darwinists who believed that the seemingly impossible can happen if given enough time: "Given infinite time, or infinite opportunities, anything is possible."[25]

In *Blind Watchmaker,* Dawkins maintained that an infinite amount of time could produce an infinite array of possibilities. What would seem to be a miraculous event to the casual observer is nothing more than random chance acting over long periods of time. For instance, if a statue of the Virgin Mary suddenly waved its hand, we should not be surprised or alarmed because it was going to happen sooner or later, given that the molecules within the marble are continually vibrating back and forth. Thus, according to Dawkins, given enough time, miracles should be expected to happen. Said Dawkins:

> So what do we mean by a miracle? A miracle is something that happens, but which is exceedingly surprising. If a marble statue of the Virgin Mary suddenly waved its hand at us we should treat it as a miracle, because all of our experience and knowledge tells us that marble doesn't behave like that. I have just uttered the words 'May

> I be struck by lightning this minute.' If lightning did strike me in the same minute, it would be treated as a miracle. But actually neither of these two occurrences would be classified by science as utterly impossible. . . . In the case of the marble statue, molecules in solid marble are continuously jostling against one another in random directions. The jostlings of the different molecules cancel one another out, so the whole hand of the statue stays still. But if, by sheer coincidence, all the molecules just happened to move in the same direction at the same moment, the hand would move. If they then all reversed direction at the same moment the hand would move back. In this way it is *possible* for a marble statue to wave at us.[26]

Despite what Richard Dawkins said about anything being possible given enough time, the science of statistical probabilities doesn't agree with him on time giving rise to evolution because the earth has not been around for that long a time. The earth has been around for only 4.5 billion years or so, which, statistically speaking, is not nearly enough time for elemental atoms to organize themselves into even the simplest of living cells. To understand the impossibility of the evolution of life on earth in such a short period of time without God, consider the odds of winning the lottery. To win the lottery you have to pick six numbers in a row, with dozens of numbers in a tumbler that randomly spits out one number at a time. The odds of winning the Powerball lottery are about 120 million to one, much more than your chance of being struck by lightning. Now consider a much harder lottery containing eighty-four numbered balls, with new rules saying that in order to win you must have the correct number of balls all in a row, starting with the number one ball first, the number two ball second, the number three ball third, all the way up to the eighty-fourth ball. According to the experts in probabilities, your chances of the tumbler randomly spitting out the correct sequence of eighty-four numbered balls in precise order from one to eighty-four would be one chance in 4.8×10^{50}, which when written out on paper without the exponential notation is one chance in 4,800,000,000,000,000,000,000,000,000,000,000,000,000,000,000,000,000, a probability so small as to be virtually non-existent. Even if you tried non-stop for 4.5 billion years, the approximate age of the earth, you would never get a one through eighty-four in perfect sequence. No one would win this lottery.

Now consider what is seen in real life. The DNA that comprises the human genome contains billions of precisely arranged nucleotides of A, T, G, and C, where a single mistake in one DNA nucleotide sequence could mean the death of that individual. Statistical experts have calculated that the odds of a tumbler correctly arranging the entire human genome by itself one nucleotide at a time to be one chance in $10^{1,000,000,000,000}$ (a one followed by a trillion zeroes), a number so large that it would take 300,000 miles just to write it out on paper.[27] All the DNA of all the species on earth have much more than eighty-four nucleotides in perfect sequence in their DNA or RNA. A bacterium, for instance, has over 3,000,000 nucleotides in its genetic code, and a simple amoeba has as much information in its DNA as one thousand *Encyclopedia Britannicas*. The mathematical probabilities, therefore, of non-living molecules precisely forming and encoding complex DNA nucleotide sequences by random chance out of the elements of the earth, randomly organizing themselves into all the components of a living cell, building proteins to structure and protect themselves, taking in oxygen and respiring by mitochondrial oxidation, and executing all the other functions that a living cell requires is so astronomically large that it would be a waste of time to even try to calculate them.[28] Four and half billion years is not nearly enough time for even a single-celled life form to have arisen on earth all by itself. And that is the evolution of just a single cell. The human body contains approximately ten trillion cells working in unison to sustain life, raising the odds for evolution to an incalculable level.

The French probability expert Emile Borel said that there is a single law of chance, that beyond which there is no possibility of its occurrence no matter how much time elapses. Borel calculated that number to be one chance in 10^{50}, beyond which things will never happen, even with an eternity of time (one chance in a billion is 1×10^{9}). To understand just how large one chance in 10^{50} is, it would take you millions of years just to count that high. If you recall that the odds of a human DNA sequence encoding itself by random chance is one chance in $10^{1,000,000,000,000}$, the evolution of man could never have happened in 4.5 billion years, or an eternity for that matter. Thus, the evolutionists' main argument, that an infinite amount of time allows for the impossible to happen, has been disproved by statistical probability studies using Borel's Law.

Believing in the infinite creative power of time and chance as evolutionists do is absurd, and is like believing that hurricanes can create elaborate multi-story sand castles on Miami Beach if just given enough time. Don't you believe it; elaborate design points to intelligence and an intelligent designer, not to time and chance. No one is foolish enough to believe that time and chance can create something as complicated as a computer, so why believe that evolution created life all by itself, which is infinitely more complex?

Speaking of computers, the genetic code of DNA is very similar to a computer program's electronic code of ones and zeroes, except that DNA's code is tertiary rather than binary, in the sense that one nucleotide triplet of A, T, G, or C codes for one amino acid (our genetic dictionary has sixty-fourDNA words of three letters each and more than one set of DNA triplets can code for the same amino acid). Genetic codes are also very similar to computer programs in that they have start and stop points, and they must be accurately written by a programmer in order to execute, which is also evidence of intelligent design. Any comma or period or semicolon or space or typo or command out of place in a computer program can cause it to abort and fail. So it is with DNA, where a single nucleotide out of place can cause the death of the organism, such as seen in sickle cell anemia where a single change in one nucleotide (adenine for uracil in this case) results in the substitution of the amino acid valine for glutamic acid and the physical disease of sickle cell anemia.

To further illustrate the ridiculousness of life evolving on its own by accident, let's substitute Darwin's ancestral monkey for random chance. No one believes that a monkey sitting at a keyboard can randomly type out a complex computer program for a NASA supercomputer, yet evolutionists maintain that random chance, like a monkey at NASA's computer, created DNA out of the earth's preexisting elements. The statistical probablility is one in ten thousand million million million million million million (1×10^{40}) for a monkey to correctly type out just one sentence like Hamlet's "Methinks it is like a weasel." If the same monkey typed non-stop without breaks, it would take him about a million million million million million years to type out just that one sentence, which is about a million million million times as long as the universe has been in existence.[29] Even Richard Dawkins, the most publicized Darwinist of the twentieth century, knew that Darwin's single-step theory of natural selection, where evolution occurs gradually one infinitesimally small mutation at a time, is mathematically impossible because of the extremely low probability of biochemical events occurring by random chance (the statistical probability, for instance, of the evolution of a single molecule of hemoglobin all by itself is one in 10^{190}). Dawkins, therefore, concocted a new theory of evolution he called "cumulative selection," whereby each new random mutation that is an improvement is used as a basis or template for the next improvement, sort of like the *Wheel of Fortune* game show. Said Dawkins:

> If evolutionary progress had had to rely on single-step [natural] selection, it would never have gotten anywhere. If, however, there was any way in which the necessary conditions for *cumulative* selection could have been set up by the blind forces of nature, strange and wonderful might have been the consequences. As a matter of fact, that is exactly what happened on this planet.[30]

Dawkins' theory of cumulative selection implies that the blind forces of evolution weren't really blind at all, but were watching what happened and guiding themselves along a predetermined path towards an end goal. This suggests a purpose or intelligence to life, just the opposite of what Dawkins and other Darwinian evolutionists believe. Realizing he had contradicted himself in his beliefs, Dawkins tried dismissing his insinuation that cumulative selection was proceeding towards a purposeful, predetermined goal by later saying in the *Blind Watchmaker* that "cumulative natural selection is blind to the future and has no long-term goal."[31]

Neo-Darwinists like Dawkins know that the scientific evidence points to intelligent design in the universe, yet they refuse to admit to an intelligent designer. They tenaciously hold on to random chance as the creator of the universe; but as I mentioned in the previous chapter, random chance can do nothing on its own without a Prime Mover. Chance is only the statistical probability of the occurrence of some event, like rolling a seven or eleven in craps. In Dawkins' theory of cumulative selection, he admits that the "necessary conditions" must have already been "set up" for evolution to work. But just what these necessary conditions were or how they were set up Dawkins has no answer for except some silly-looking branched lines he created on his computer that he called "biomorphs," which looked to him like spiders and bugs and were supposed to explain the creation of life.

Darwinists love to say that given enough time, anything can happen—except, of course, something they don't want to happen, like evolution going backwards. Darwinists believe that life evolved from the simple to the complex by means of natural selection over millions of years, but that it can't evolve backward from the complex to the simple by the same route. And why not? Haven't evolutionists said that given enough time, anything can happen, and that nature selects the fittest? I would think that if a statue of the Virgin Mary could wave its hand, then a mud puppy could evolve back into a fish if the world suddenly became deluged with water again. To support the theory that evolution only moves forward, Darwinists have come up with something they call Dollo's Law, which says that the odds of evolution reversing itself are just too improbable to go back along the same path twice. Evolution, therefore, according to the Darwinsts, is irreversible. Excuse me, but aren't the odds of evolution going forward just about as ridiculous as they are going backward? Dollo's Law, therefore, contradicts what evolutionists say about given enough time the miraculous and the statistically impossible can happen. Said Dawkins about Dollo's Law:

> Dollo's Law is really just a statement about the statistical improbability of following exactly the same evolutionary trajectory twice (or, indeed, any *particular* trajectory), in either direction. . . . There is nothing mysterious or mystical about Dollo's Law, nor is it something that we go out and 'test' in nature. It follows simply from the elementary laws of probability. . . . For just the same reason, it is vanishingly improbable that exactly the same evolutionary pathway should ever be traveled twice.[32]

If the truth be known, Darwinian evolutionists are hypocrites. When they want something to happen they say given enough time it will happen. When they don't want something to happen they revert back to statistical probabilities and say that it can't happen. But they can't have it both ways: either time can do all things or it can't. Time actually hinders evolution because evolution is all a matter of proper timing, and if favorable changes don't take place at the exact same time in the exact same place in the exact same species, evolution is impossible. For example, for Darwinian evolution to have occurred, many organisms of the same species had to have followed the same evolutionary trajectory at the same time, because the random, so-called "favorable" mutation of just one organism would have been lost in the gene pool to successive generations, especially with fish who have an enormously large breeding populations. Gene flow or mixing of the gene pool in large breeding populations like fish so dilutes genetic variations that the population as a whole remains virtually unchanged.

Time also tends to disprove evolution because Darwinism violates the Second Law of Thermodynamics, which says that the universe tends towards maximum disorder or entropy. The universe was at its maximum entropy at the time of the big bang, and has been moving back that way ever since the world was created. We know from the Second Law of Thermodynamics that things get old, wear out, and return to a state of disorder with the passing of time. We are created from the dust of the earth and to the dust we shall return. If it were not for God who gives order to the universe and sustains it with his power, the world would have disintegrated into chaos eons ago, or would never have been formed in the first place. Darwinian evolution, on the other hand, says that the world ordered itself out of maximum disorder for no apparent reason, somewhat like dissolved sugar water in a hummingbird feeder miraculously coming together to form a perfect sugar cube. The idea that nature can order itself and sustain itself without a higher power contradicts the Second Law of Thermodynamics. Evolution, therefore, is impossible without God.

Furthermore, Charles Darwin was positive that life evolved by means of natural selection over eons of time, yet he didn't even know the most simple of facts about how babies are made. When formulating his theory of evolution by means of natural selection, Darwin was totally ignorant of the method of genetic inheritance and the reproduction of offspring in animals, that from the meiotic division of male and female germ cells and the subsequent fertilization of the ovum by the sperm, half of the genetic information comes from the father and half from the mother. Instead, Darwin thought that tiny pieces of every adult cell in the body, which he called "gemmules," broke off in the bloodstream and made their way to each spermatozoa and ova, which then reformed inside the mother's womb after mating—a childish theory Darwin called "pangenesis" which is laughed at today.[33]

Just as ridiculous as his theory of pangenesis was Darwin's "disuse atrophy" theory, the idea that if you don't use a part of your anatomy it will diminish and eventually disappear in your progeny. For example, someone who never exercises will give birth to scrawny, couch-potato kids, or if mice have their tails cut off their progeny will eventually have no tails. Darwin's belief in the inheritance of acquired characteristics, commonly referred to as "Lamarckism," has been proven to be totally untrue.[34] So, if Darwin was wrong about these basic facts of life and modes of inheritance, why do so many people still believe in his inane theory of evolution through random chance and natural selection? Because they don't want to believe in God, that's why. It never ceases to amaze me how people will risk their eternal souls on such a highly improbable theory as Darwinian evolution yet refuse to believe in the one God who has revealed himself to us time after time. The odds for atheists going to hell for not believing in God are like the flip of a coin, fifty-fifty, in that hell either exists or it doesn't. On the other hand, the odds that evolution really happened, according to statistical probability studies, are so infinitesimally small as to be uncountable. You wouldn't bet your eternal soul on the flip of a coin, so why believe in the impossible odds of Darwinian evolution? It makes no sense to me.

The Sixth Fallacy of Darwinism: Ontogeny Recapitulates Phylogeny

Although it has been proven untrue many times over by experts in comparative anatomy and embryology, many Darwinists still believe in and promote the bogus theory of the "recapitulation of species" during embryological development. The phony evolutionary theory of the recapitulation of species says that the embryonic development of a species (the ontogeny) goes through all the embryonic stages of its evolutionary ancestors (the phylogeny) from start to finish. For instance, a human embryo starts out as a worm, then proceeds to a fish, then proceeds to an amphibian, then proceeds to a reptile, then proceeds to a bird, then proceeds to an animal, then proceeds to a man. The main promulgator of this totally untrue theory was a German biologist and professor from the University of Jena in Germany by the name of Ernst Haeckel (1834-1919). Haeckel was the first evolutionist to illustrate the ever-popular evolutionary tree you see so often in science books and from whom the term "missing link"[35] originates. As an avowed Darwinist, Haeckel studied the embryological development of several species of fish, amphibians, reptiles, birds, animals, and humans, and concluded that all these embryos were the same during their early embryonic development. Haeckel's conclusion had tremendous implications for the scientific community: we all evolved from a single common ancestor like Darwin had said. Haeckel formalized this theory of the recapitulation of embryos into a scientific law in 1866, his "Biogenesis Law" as he called it, which he published in his infamous pro-evolution book entitled *Natural History of Creation* in 1868.

Haeckel's bogus theory of the recapitulation of species, however, was based on misinterpretation of anatomy and outright deceit. It is a classic case of junk science. It turned out that Haeckel so wanted to prove Darwin's theory of evolution that he intentionally manipulated his research to fool the general public into believing that human embryos develop from one species then the next. In his *Natural History of Creation,* Haeckel printed a variety of woodcut illustrations depicting the embryonic development of several different species of fish, dogs, birds, tortoises, and humans, in which he attempted to show that the early embryos of these species were identical in form and appearance. From these woodcuts, in which the different embryos appear exactly alike to the casual observer, Haeckel explained to his readers that all these species evolved from a single common ancestor as Darwin maintained and, therefore, Darwinian evolution is true. But upon closer examination of Haeckel's illustrations by other scientists, it was quickly discovered that the woodcuts of the different embryos were actually the same woodcut printed over and over again. In 1868, Ludwig Rutimeyer, a professor of zoology and comparative anatomy at the University of Basel in Germany was one of the first scientists to refute Haeckel's phony illustrations in an article he entitled "Referate," which appeared in the anthropological publication *Archiv fur Anthropologie.*

In addition to Rutimeyer, in the 1880s a leading German embryologist of the time by the name of Wilhelm His, Sr. printed numerous illustrations of his own showing the true appearance of the different embryos, which were not at all identical in appearance and development, as Haeckel had fraudulently depicted back in 1868. As a result of their investigations, Haeckel was charged with fraud by five of his peers and convicted by a university court at Jena. Haeckel tried defending himself by saying that he was only filling in and reconstructing the missing links in his illustrations when the physical evidence was thin and the embryological specimens were poor.

Extensive research in comparative vertebrate anatomy and vertebrate embryology in the twentieth century has definitively proven that Haeckel's theory of embryonic recapitulation is untrue, yet it continues to live on in the hearts and minds of modern-day Darwinian evolutionists who desperately want science to substantiate their belief in godless evolution. Even though proven untrue, modern science textbooks and college professors still repeat Haeckel's false maxim that "ontogeny recapitulates phylogeny." In teaching Darwinian evolution and the false theory of recapitulation, some biology teachers still continue to use pictures and illustrations of human embryos that appear to show fish gill slits, bird yolk sacs, and animal tails which they say is evidence that we evolved from fish, birds, and lower mammals. But nothing could be farther from the truth. I know this to be true because I have taken courses in comparative vertebrate anatomy and embryology and saw firsthand the deceit that professors use to convince their students of evolution. Let me take a moment to scientifically explain why Haeckel's theory of recapitulation is nothing more than smoke and mirrors.

What appears to the casual observer to be fish gills slits in human embryos are actually five pharyngeal pouches or ridges (a.k.a. branchial arches) separated by grooves, not open slits, on the ventrolateral surface of the embryonic head and neck during the fourth week of life. These five pouches develop into the neck, jaw, face, and external ears of human beings. The tip of the first pharyngeal pouch in human beings becomes the auditory tube and the tympanic cavity of the middle ear. The tip of the second pharyngeal pouch becomes the fossa and covering epithelium of the palatine tonsil. The tips of the third, fourth, and fifth pharyngeal pouches give rise to a series of specialized organs called the parathyroids, thymus, and ultimobranchial bodies respectively. There are no fish gills or gill slits in human embryos. That is simply a lie to perpetuate the myth of evolution.

Regarding the presence of a yolk sac in human embryos supposedly indicating that we evolved from birds, no ovum or egg of any species is totally devoid of yolk. Yolk is a universal nutrient for the development of embryos, and consists of fatty and albuminous substances: the deutoplasm. Bird and reptile eggs have a tremendous amount of yolk because they are formed inside a protective shell and then laid outside the mother's body, while human embryos have much smaller amounts of yolk in their yolk sac because they very quickly acquire nutrition and oxygen from the developing placenta within the mother's uterus (only mammals have placentas, which makes them totally different and unique from all other species and, therefore, serves to discredit the theory of evolution). In fact, the amount of yolk produced for a developing embryo is used in classifying the eggs of the different species. The large, yolk-rich eggs of bony fishes, reptiles, and birds are termed megalecithal (much yolk). The eggs of amphibians have moderate amounts of yolk and are termed medialecithal (medium yolk), while the eggs of the invertebrates and in all but the lowest mammals, which includes humans, have relatively little yolk and are termed isolecithal (equal yolk) or miolecithal (little yolk). In addition to a very minute, early nutritive function, the human yolk sac is the first part of the embryo that provides for the development of blood cells (homopoieisis) at about four weeks after conception, which is vital for the survival of the embryo, since bones and red bone marrow which provides blood in the adult has not yet been formed. The yolk sac quits developing blood cells in the second month, with the formation of blood cells transferred to the body mesenchyme and blood vessels (fifth week), liver (sixth week), spleen, thymus, and lymph glands (second to fourth months), and finally, the bone marrow (fourth month). By the fifth month, all blood cells are being made in the bone marrow, which is now functional. Besides the initial function of the human yolk sac as an aid in the development of blood formation, the human yolk sac also aids in the development of the human digestive system, as the entodermal roof of the yolk sac provides the material from which the tubular gut is fashioned by infolding. In contrast, the yolk sac of birds and reptiles has massive amounts of yolk, and its principal function is nutritive. Thus, to say that humans evolved from birds simply because human embryos have a yolk sac is a ridiculously naive and a scientifically ignorant statement.

Furthermore, what may appear to the scientifically-uneducated to be an animal's tail in the developing human embryo of three to five weeks is actually the elongation of the growing spinal column, which the rest of the developing embryo eventually catches up with. During the sixth to ninth weeks of human embryonic development, this embryonic "tail" disappears from external view through regression. The human coccyx, which represents the remnants of the embryonic "tail," recedes in the fetus to a higher position in relation to the buttocks, and the coccygeal fovea, or postnatal pit, of a newborn baby marks the site where a vestige of the caudal tip of the early spinal cord attaches to the skin. In rare cases, the so-called embryonic "tail" of human fetuses have known to persist and grow rather than regress, with some growing as large as nine inches at twelve years of age. However, these anatomic anomalies are soft and fleshy and do not represent a bony vestigial animal tail from some sort of ancestral monkey, as evolutionists would have you believe.

Another common argument that evolutionists make is that the since the developing human embryo has three types of nephric (urinary) structures in its embryonic development—the pronephros, the mesonephros, and the metanephros, in that order of development—we must have evolved from the lower animals and fishes. Since each of these types of urinary structures is found in the adult form of different species (the adult hagfish has a pronephros kidney, the adult kidney of other fishes and amphibians is mesonephric, and the adult kidney of mammals is metanephric), evolutionists have assumed that the development of human kidneys in three stages reflects our evolution from these lower species. But this is not the case. The three different nephric structures in the developing human embryo are functional and have nothing to do with evolution. A developing human embryo produces waste long before it has a functioning adult kidney, so it has need for a primitive urinary system to handle wastes long before the adult kidneys are formed. Thus, the three-stage system of human urinary development is an ingenious system that God devised so that the developing human embryo would not die from its own waste.

The same can be said with the developing human heart, which appears in its early development to have different numbers of chambers, each consistent with the number of chambers in the heart of an adult worm, a fish, an amphibian, and a bird or reptile, but not necessarily in that order. The development of the human heart, however, has nothing to do with the hearts of these other species. The developing human embryo has a need to pump blood and oxygen to its different parts long before human lungs and an adult heart have appeared. The functional heart of a human embryo first appears as paired tubes which then fuse and twist and develop into three chambers and then four, which is the number of chambers of the adult human heart. The human heart, therefore, is not a phylogenetic representation of its primitive ancestors, as Darwinists would have you believe. If it were it would develop a 1-2-3-4 chambered pattern, but instead it develops a 2-1-3-4 pattern. So much for the idea that ontogeny recapitulates phylogeny.

The Seventh Fallacy of Darwinism: Homologous Features Prove Common Origin

Another persisting myth of evolution is that because we have anatomical features that are similar (homologous) to the lower species, we must have evolved from them. In other words, because fish have eyes, ears, a nose, and a throat, we must have evolved from fish; because amphibians and reptiles have lungs and legs we must have evolved from them; because warm-blooded lemurs and monkeys have hair and maturate their fetuses inside their uteruses we must have evolved from them.

Having the same anatomical features as lower species, however, does not prove anything; it simply points to the economy of intelligent design by an intelligent designer. God sets a workable pattern of anatomical development for each species that provides it with optimum survivability and then fixes it. That's why we only have two eyes, two ears, one nose, one mouth, two arms, two legs, two hands, two feet, and so forth. Any deviation from God's design is a medical anomaly that produces freaks and mutants much less able to survive.

The theory of fixed designs, however, is anathema to evolutionary thinking because evolution says that species are constantly changing and improving in the struggle to survive. This then begs the question: if it were true that species are continually improving and evolving as evolutionists maintain, why don't humans have six eyes and eight legs like spiders so that we can see better and run faster, or four hands so that we can perform more tasks, or tough external shells like turtles to protect us from the violence which is so prevalent in society? We don't because each species is uniquely and wonderfully created by God for a reason and purpose, and external pressures from the environment cannot and will not change God's plan for each of his creatures.

Species are immutable and fixed in that they forever maintain their primary anatomical design, such as bilateral symmetry of the body with one head, neck, thorax, abdomen, dorsal and ventral planes, head and tail ends, and four appendages as is seen in higher animals. Species are, however, capable of showing wide ranges of variability in appearance, such as height, weight, shape, posture, color, etc. Breeders and scientists can manipulate external appearances, but dogs will always be dogs, fruit flies will always be fruit flies, and flowers will always be flowers. While it is true that fruit flies, with their fast maturity and high reproductive rates, can be genetically bred or manipulated by scientists to produce mutants that have deformed or absent wings, or different colors, or dumpy bodies, they can't reproduce an entirely new species of insect because creatures are genetically pre-programmed to only reproduce like kinds. You can interbreed related animals such as a donkey to a horse or a lion to a tiger because they are like kinds, but you can't breed a horse to a lion or a frog to a bird.

Darwinian evolution maintains that a single common ancestor gave rise to all forms of life on earth because it had ample amounts of time to evolve into new and better species. Ample amounts of time were necessary for evolution to occur, Darwinists say, given the long time it takes species to mature and replicate. But most of the lower species such as bacteria, viruses, protozoa, and insects mature tremendously fast and produce incredibly high numbers of progeny in a very short period of time, thousand of times faster than higher organisms. So if Darwinian evolution were really true we should expect to find ample evidence of macroevolution in progress amongst these species because their biological time clocks are so much faster than ours. But we don't. Species continue to produce like members because, as I said before and will say it again, like only begets like. Bacteria have remained bacteria since the day they were created. In fact, bacteria were the first known creatures that appeared 3.5 billion years ago, and they exist today relatively unchanged from the day they were made. So it is with sharks that have existed unchanged for hundreds of millions of years. Sure, bacteria can become antibiotic resistant, and insects can became insecticide resistant, but they are still bacteria and insects. People become immune to certain diseases, so why should it surprise us that bacteria become immune to certain antibiotics?

Let's get back to embryology. During Darwin's time it was mistakenly believed that the homologous anatomical structures between the different species were always derived from the same three germ cell lines (the endoderm, ectoderm, and mesoderm) of developing embryos. However, recent staining techniques of the early blastula (dividing egg) and gastrula (gut-forming egg) of the different species have shown that each species has their own species-specific "fate maps," where particular cells in the dividing ball of cells develop into a particular structure in the adult. As a general embryonic rule, in higher animals the endoderm cells primarily form the lining of the digestive tract and the substance of the organs such as the liver and pancreas, as well as the endocrine glands. The ectoderm forms the skin, mouth, cloaca, hairs, feathers, mammary and sweat glands, sensory epithelia of the nose and internal ear, eye lenses, enamel of the teeth, the nervous system, the eye retina, and other nervous tissue structures. The mesoderm forms the connective tissue, the musculo-skeletal system, the lining of the body cavity, the mesenteries, and the notochord (the primitive supporting structure of the developing spinal column). Homologous structures between species however, do not always originate from the same three germ germ-cell lines, as would have to happen if higher animals evolved from lower animals. For instance, the skin or scales of fish is comprised of epithelial cells in the epidermal layer together with bony elements in the dermal layer, while the skin of mammals have no such bony elements in their dermis. It is also found that although most skeletal material arises in orthodox fashion from mesodermal cells, certain cartilages in the head and throat come from cells of the neural crest, which itself comes from the ectoderm and not the mesoderm as one would expect. It is likewise found that certain cells that produce spinal nerve cord will, if transplanted, produce skin cells. Therefore, it is not always the germ cell layer that determines the future development of cells, but the "potentiality" of each individual cell within a particular germ layer. It has been a real headache for embryologists to understand, but potentiality studies offer great hope in the treatment of diseases. Everyone is aware of the ongoing debate over fetal stem cell research in the treatment of diseases. Stem cell research is a prime example of the potentiality of cells to become more than one organ.

Comparative embryologists have known for a long time that different genes in different species specify homologous structures. They have a term for this called "pleiotropy." In laymen's terms, pleiotropy means that every species has genes that affect more than one structure and, therefore, genes are said to be "species specific" in that almost all genes are pleiotropic for one species only. In other words, the genes that form kidneys in fish are not the same genes that form kidneys in reptiles. Homologous organs within the various species, therefore, are not generated from the same homologous embryonic tissues as Darwin and his colleagues incorrectly believed. Darwin was mistaken when he defined homology as the "relationship between parts which results from their development from corresponding embryonic parts."[36] This one very important mistake is enough to discount Darwin's entire evolutionary theory of descent with modification from homologous embryonic parts.

Let's take kidneys again. Fish and amphibians develop kidneys from an embryonic organ known as the mesonephros, while reptiles and mammals develop kidneys from an embryonic organ known as the metanephros. This development of kidneys in different species from different embryonic parts, or different genes if you want to reduce it even further, is a huge problem for Darwinists to explain.[37] Evolution rests on the premise that since all species evolved from an original species, homologous organs and parts in higher species should come from the same embryonic cell structures in lower species, but they don't.

Another huge problem Darwinists have is metamorphosis, the anatomical and physiological changes that insects such as butterflies undergo during their life cycle from pupae to adult. During metamorphosis, a pupae's internal anatomy virtually melts into an undistinguishable blob, then miraculously reforms into a beautiful butterfly housed in a golden chrysalis or external shell, whereupon it emerges at just the right time and gives color and beauty to the world. How metamorphosis occurs in butterflies no one knows. Darwinian evolution certainly can't explain it.

I have briefly shown how Darwinists incorrectly use the appearance of similar anatomical structures as proof of evolution. Yet another explanation that Darwinists have mistakenly used to try to prove that we all evolved from a primitive common ancestor is the supposed evolution of the anatomical parts of lower animals into different and more complex anatomical structures in higher animals. Take for instance our inner ears. Darwinists claim that the three little auditory ossicles of the human middle ear were in earlier days part of the jaw apparatus of bony fish, and appear still earlier to have been part of the gill supports of the fish's ancestors. This theory, however, is mere speculation and wishful thinking because there is no scientific evidence or transitional fossils to show in stepwise fashion that this really happened.

One can never assume that evolution occurred simply because different species have anatomical features that look alike or act alike. Many different species have similar looking sperm and egg and embryonic cells that only an expert can tell apart, but obviously we didn't descend from all these creatures just because our sex cells and embryonic cells look similar to theirs. Early human and shark embryos look almost identical and most people can't tell them apart, but that doesn't mean our ancestors were sharks, except maybe for trial lawyers. Homology doesn't prove common origin as Darwinists try to maintain; it only points to the economy of intelligent design by the Great Designer himself.

The Eighth Fallacy of Darwinism: The Fossil Record Proves Evolution

Let me be perfectly clear right from the start on the controversial subject of the fossil record: there are no missing links between any species of any kind that prove Darwinian evolution ever happened, especially within the order of primates, which includes man. As the world-famous primatologist John Buettner-Janusch proclaimed, "There is no known living member of the [primate] order which represents the first erect, bipedal hominid. Nor does there exist any form which is transitional between the quadrupedal primates and the bipedal."[38]

Despite what your high school biology instructor taught you in class, there are no missing links between any species of any kind. True, paleontologists, anthropologists, and other scientists and fossil hunters have discovered many different kinds of fossils of many different species, including those of extinct hominids that walked about on two legs and built fires and fashioned primitive tools, but there has been no fossil discovery of any missing link between any species that proves Darwinian evolution. The common picture of the ladder-like ascension of man from monkeys you see so often in biology books and science museums is completely fictitious and is no more real than graphic arts computer software that can morph a picture of a monkey into a man.

Sure, there have been many attempts in the past by Darwinists to link fossils of one species to fossils of another, but they have all been proven wrong in the end. We have no fossils that link the evolution of fish to amphibians, or amphibians to reptiles, or reptiles to dinosaurs, birds, or mammals, or of monkeys and great apes to men. If one were to be found it would be the greatest discovery ever made in the history of science, and that person would receive the Nobel Prize for having disproved Genesis—but don't hold your breath. Here's why you should keep breathing.

If Darwinian evolution is true, in that species gradually evolve over long periods of time from one species to the next, then by simple statistical averages the number of transitional fossils that are buried within the earth should be around 98 percent. That is, if dinosaurs evolved into birds, then we should expect the number of uncovered fossils between the two species to number around 98 percent transitional forms, ranging somewhere between pure dinosaur on one end and pure bird on the other. But do we see that in real life? No way. What we see in the fossil records is only pure dinosaur fossils and pure bird fossils, with no transitional forms in between. We do, however, see fossils of mosaic creatures that contain anatomical features characteristic of more than one species, as in the fossils of the ancient species *Archaeopteryx,*[39] *Ichthyostega,*[40] and *Acanthostega,*[41] but these fossils are no more transitional than the modern-day platypus, which has both a duck's bill and a mammal's fur.

The problem with the absence of transitional forms in the fossil records is usually explained away by the Darwinists as a result of the "incompleteness" of the fossil record. They are down there in the earth, they say, we just haven't found them yet. Darwin was well aware that the fossil record of his time did not support his theory of evolution, yet he hoped that with the passage of enough time his transitional forms would be uncovered and his theory proven right. The lack of transitional forms in the fossil record really bothered Darwin though, and he admitted in his sixth edition of *Origin* that it was a very obvious difficulty for him to explain to his critics. "The distinctness of specific forms and their not being blended together by innumerable transitional links is a very obvious difficulty," he said.[42]

Since evolution is not very amenable to scientific testing in the laboratory because of the tremendous length of time it takes new species to evolve, the only way to prove the theory of evolution is to examine the fossil records. Fossils are like snapshots of history; they tell us what species were around in what periods of geological history. They don't lie or speculate; what you see is what happened. Paleontologists, those who study and search for fossils, have discovered approximately 250 million fossils of some 250,000 species. Since Darwinian evolution supposedly occurs so slowly and is imperceptible to the eye, we should expect to find that 98 percent of these fossils are transitional forms between one species and the next. But that's not what's been discovered. The fossil record shows that species suddenly appear and disappear in geological time without transitional forms. Take again the dinosaurs. Since they are so large and easily recognized, the fossil record should have plenty of transitional forms that diminish in size and change in form until they reach the smaller, prototypical reptile that they supposedly descended from. Again, there is no fossil evidence suggesting that this ever occurred. On the contrary, dinosaurs suddenly appear fully formed in geological history and then suddenly disappear, suggesting that they were created in a very short period of time and then wiped out by some catastrophic event such as an asteroid, which is the currently accepted theory for the extinction of the dinosaurs.

In answer to the lack of transitional forms, modern Darwinists have come up with a new theory of evolution called "punctuated equilibrium," the idea that evolution doesn't occur gradually as Darwin maintained, but occurs suddenly, geologically speaking, in earth's history, which explains why the fossil record doesn't show transitional forms. Species jump from one species to the next in a very short period of time (geologically speaking), leaving a long quiescent period between new species formation. If it took, say, only a hundred thousand years or so for a new species to evolve, which Darwinists say is just the blink of an eye when compared to the age of the earth, then the likelihood of finding transitional species is greatly diminished. This is why we don't find transitional forms in the fossil records, because they are too isolated in location and too few in number.

The Neo-Darwinian theory of punctuated equilibrium was first publicized in the 1970s by the popular evolutionary biologist from Harvard, Stephen Jay Gould, and his colleague Niles Eldredge, a paleontologist from the American Museum of Natural History. Transitional fossils and missing links are out there, the two said, we just can't find them because they are too few and too isolated. The problem, however, with Gould's theory of punctuated equilibrium is twofold. First, since there is much less time for evolution to work with, it increases the odds against random chance creating new species by means of the environment and natural selection preserving favorable mutations. Second, it suggests that there was a biological time clock where the hundreds of thousand of known fossil species were static for very long periods of time, but then for reasons unknown, suddenly and inexplicably evolved precisely at the same time, which further increases the odds against punctuated evolution ever happening. Because of the vastly shortened time frame for random chance to work with, punctuated evolution, therefore, is even more unlikely, statistically speaking, than Darwinian evolution. Species don't just jump from one species to the next in a short period of time.

Other apologetic theories on evolution put forth by Neo-Darwinists have been even more preposterous than punctuated equilibrium. In the 1940s, the German-American geneticist Richard Goldschmidt from Berkeley invented what his critics called the "hopeful monster" theory of immediate, rather than gradual evolutionary change from one species to the next; such as a dinosaur suddenly giving birth to a bird or a lemur giving birth to a monkey.[43] That absurdity doesn't even merit the ink on this paper. Another wacky theory of evolution that attempts to explain how life evolved on earth is the directed panspermia theory that space aliens impregnated the earth with primitive life forms that evolved into our modern-day species. That B-movie theory was invented by of all people the co-discoverer of DNA, Francis Crick of Cambridge University, who received the Nobel Prize for medicine and physiology in 1962. As you can see, brilliant minds don't necessarily have common sense.

Darwin's Ape-Man: Fact or Fiction?

At the beginning of the eighteenth century, all kinds of newly discovered monkeys and apes from Asia and Africa were exported into Europe for scientific study and amusement. As European explorers traveled deeper and deeper into the heart of Africa, the mountain gorilla was discovered for the first time in 1847. With the nearly simultaneous discoveries of gorillas by European explorers and prehistoric cavemen by paleontologists, a cladogram, or family tree, of the evolution of man from monkeys was beginning to be worked out. Although Darwin hinted at it in 1859 in his *Origin of Species,* Thomas Huxley was the first to write, in an 1863 essay entitled "Man's Place in Nature," that man most likely evolved from monkeys, and is probably more related to African chimps and gorillas than the East Asian orangutans. With the release of Huxley's essay and the prodding of Alfred Russell Wallace, the cofounder of the theory of evolution, in 1871 Darwin published *Descent of Man,* in which he agrees with Huxley that man evolved from monkeys and apes, and that man's closest living ancestor is probably the gorilla.[44] Huxley and Darwin, therefore, were the biggest promulgators of the modern-day theory that man evolved from monkeys instead of being especially created by God in his own likeness and image.

Towards the end of the eighteenth century, with Darwin's theory of evolution gaining greater and greater acceptance amongst the scientific community, the hunt was on for man's missing link, Haeckel's *Pithecanthropus,* half-man, half-ape. Here is what scientists have found so far in their futile search for *Pithecanthropus.*

The search for man's missing link has uncovered two genuses of hominids within the family *Hominidae,* which are designated *Australopithecus* and *Homo. Australopithecus* means "southern ape," and represents a diverse group of early, ape-like hominids that existed in Africa between one and four million years ago before suddenly going extinct. They had no culture and no language and are not known for making tools. There are currently seven recognized species in the genus *Australopithecus,* which includes *africanus, robustus, boisei, anamnesis, aethiopicus, afarensis,* and *bahrelghazali.*

Australopithecines are characterized by their small craniums, prominent sagittal crests or ridges on the top of their skulls where huge chewing muscles attached, flat faces with massive protruding jaws, huge cheek bones and orbital ridges, giant molars, a reduction in teeth number compared to man, a small pelvis, curved fingers and toes, a postcranial skeleton with long arms and short legs, an upright gait (indicating some capacity for bipedalism), and a height of about four to five feet in the adult male.

The first *Australopithecus* fossil, a juvenile skull, was found in 1924 by Raymond Dart in Taung, South Africa, and was called the "Taung child." The Taung child, which Dart's critics first thought was a baby chimp, was subsequently named *Australopithecus africanus* for its South Africa location, and has been radiometrically dated to around 2.5 million years old. A second, more adult specimen of *Australopithecus africanus* was found in the Transvaal region of South Africa in 1947 by Robert Broom, which proved to the critics that the Taung child was not just a baby chimp, but a unique and different kind of species. A second australopithecine species was likewise found in South Africa in 1938 by Broom, and was named *Australopithecus robustus* for its robust mandible, massive teeth, and prominent bony crest on top of its skull. It supposedly lived around two million to one million years ago. A third fossilized australopithecine was found by the famous paleoanthropologists Mary and Louis Leakey in Northern Tanzania in 1959, almost exactly one hundred years after Darwin wrote his *Origin,* and was nicknamed "Nutcracker Man" for its huge cheek teeth, flat face, small cranium, and visor-like cheekbones. Nutcracker Man, or *Australopithecus boisei* as it is scientifically named, lived about 2.3 to 1.4 million years ago and was for a time claimed by Louis Leakey to be the long-searched for missing link between man and ape, which it is not. A fourth australopithecine species, *Australopithecus anamneses,* was found in 1965 by Bryan Patterson near Lake Turkana in Kenya, and again in 1994 by Meave Leakey and Peter Nzube. It supposedly dates back four million years. In 1968, a toothless mandible was found in southern Ethiopia by two French paleontologists, and was recognized as an entirely new species called *Australopithecus aethiopicus.* In 1985, Alan Walker in Lake Turkana, Kenya found a blackened skull of *Australopithecus aethiopicus* dating back some 2.7 to 2.5 million years ago. *Australopithecus aethiopicus* is characterized by having a flat face, prominent sagittal crest, huge teeth, and a protruding face. In 1974, Donald Johanson discovered the most famous australopithecine fossil of all in Hadar, Ethiopia, a specimen of the species *Australopithecus afarensis* known as "Lucy," which is estimated to be around 3.2 million years old. Lucy is the best preserved and most complete of any walking hominid ever found, and has an apelike skull, projecting face, small cranium, long arms, and short stature. The species itself supposedly dates back three to four million years ago, and is the oldest of the seven known australopithecines. In 1996, the seventh and last known australopithecine was discovered in central Chad by a French paleontologist named Michel Brunet. It is called *Australopithecus bahrelghazali,* but not much is known about it because only a lower jawbone was found.

The family *Hominidae,* to which the aforementioned australopithecine species belong, also includes the genus *Homo,* which currently has eight recognized species: *habilis, rudolfensis, eraser, erectus, heidelbergensis, neanderthalensis, floresiensis,* and *sapiens* or modern man. *Homos* are distinguished from *Australopithecus* by their larger brains, a more humanlike postcranial skeleton, smaller teeth and jaws, the existence of cultures, and the ability to fashion tools and create elaborate works of art. The first fossilized specimen of the genus *Homo* was discovered in the Neander Valley in Germany in 1856 and was named *Homo neanderthalensis* (neander means "new man" and thal means "valley"). Neanderthals lived in Western and Central Europe and the Near East some 150,000 years ago, but suddenly died out about 35,000 years ago. They were tall with a stocky build, low forehead, flat face, protruding teeth, prominent eyebrows, and thick orbital ridges. They were the prototypical cavemen.

The second specimen of the genus *Homo* was found in 1868 in Abri Cro-Magnon, Les Eyries, France by Louis Lartet and Henry Christy and is called Cro-Magnon. Cro-Magnons have a shorter, squarer face than Neanderthals but a larger brain case. Their eye sockets are more rectangular than Neanderthals and their skulls resemble modern Africans more than they do Europeans. Cro-Magnons existed about thirty thousand years ago and are actually members of our own species *Homo sapiens.* They are famous for their beautiful prehistoric cave drawings and elaborate ivory carvings and Venus statues, which have been found throughout Europe.

The third species of *Homo* were found in the Far East. In 1891, Eugene Dubois, a physician from Holland, set out to find man's missing link in Indonesia, and as his luck would have it, he discovered the skull cap of Java Man in the Dutch East Indies, which has been estimated to be about 1,000,000 to 500,000 years old.[45] Java man was first given the scientific name *Pithecanthropus erectus* for its ability to stand and walk erect, but since the 1950s it has been renamed *Homo erectus.* The obvious reason for the name change from *Pithecanthropus erectus* to *Homo erectus* was because it was once thought that it was the missing link, but later proved not to be. *Homo erectus* specimens have a low forehead, small cranium, thick bones around the eyebrows, and prominent cheekbones. More specimens of *Homo erectus* were found in the Zhoukoudian Cave just south of Beijing from 1928-37, and were named Peking Man for their location. China has yielded more specimens of *Homo erectus* than anywhere else on earth, but the origin of erectines is believed to be in Africa some 1.5 million years ago.

The fourth specimens of the genus *Homo* were found in Germany. In 1908, Otto Schoetensack found a mandible in Mauer, Germany, near Heidelberg, which was identified as a separate species of *Homo* (*Homo heidelbergensis*). *Homo heidelbergensis* lived in Europe and Africa and West Asia some 300,000 to 600,000 years ago, and is thought by many evolutionists to have evolved into the Neanderthals, based on similar-looking facial features.

The fifth discovery of *Homo,* this time outside of Asia and Europe, came in 1964 with the unearthing of the skull bones of *Homo habilis* by Louis Leakey in Olduvai Gorge, Tanzania, in Africa. Raymond Dart, the discoverer of the Taung child, first suggested the name *habilis* to Louis Leakey because it means able, handy, and mentally skillful, which Leakey thought were attributes of *Homo habilis* or Handy Man, as he was called. Leakey believed that *Homo habilis,* which dates back some two million years ago, was the missing link and evolved directly into *Homo sapiens,* and that *Homo erectus* was just a side branch that migrated northward, but in 1986 a partial skeleton of *habilis* was found at Olduvai Ridge which showed that *Homo habilis,* which has a protruding face like the australopithecines, has apelike limbs and long arms like *Australopithecus afarensis,* and, therefore, is more ape than man, so it couldn't be the missing link.

The sixth discovery of *Homo* came in 1969, when Ronald Clark discovered the partial cranium of another species of *Homo* in Swartkrans, South Africa, called *Homo ergaster,* which lived around 1.5 million years ago. *Ergaster* has more humanlike limbs than *habilis,* a more modern body build, a larger cranium, reduced teeth and jaws, and the ability to work with tools (the name ergaster, by the way, is Greek for "workman").

The seventh species of *Homo,* called *Homo rudolfensis,* which dates back around two million years ago, was found in the Koobi Fora, Kenya, in 1972. It has a large braincase and a broad, flat face. The eighth and last known species of *Homo* was just recently discovered in the Flores Islands regions of Indonesia and has been named *Homo floresiensis.* It is a very small hominid with long arms and an upright stature, and disappeared only some twelve thousand years ago. Evolutionists are troubled by the recent discovery of "Hobbit" people, as they are nicknamed, because they have anatomical features that were thought only common to hominids millions of years ago, but are now known to have been in existence just eight thousand years before Abraham was born.

Fossils of the species *Homo sapiens* which includes modern man, have been found on practically every major continent in the world, including Europe in 1868 (Cro-Magnon man, 30,000 years old), the Skhul Cave in Mount Carmel, Israel in 1932 (90,000 years old), Ethiopia in 1967 (130,000 years old), Australia in 1967 (10,000 years old), the Qafzeh Cave in Israel in 1969 (100,000 years old), and Dali, Shaanxi Province, China in1978 (200,000 years old).

The current, most popular evolutionary theory on the origin of man is that we didn't descend from the Neanderthals in Europe as previously thought, because *Homo sapiens,* we've just recently found out, predate Neanderthals by some 170,000 years. We are direct descendants, according to the new theory, of *Homo erectus,* who evolved from *Homo habilis,* who evolved from *Australopithecus,* who evolved from the great apes in Africa around ten million years ago, and so on, and so on. According to the evolutionists, our genus (*Homo*) first appeared in Africa 2.4 to 1.2 million years ago and then migrated throughout the world to all the continents, with the North American continent being the last place *Homo sapiens* migrated to some fifteen thousand years ago. As the evolutionary story goes, about four to ten million years ago hominids of the australopithecine type began walking upright, and then about two million years ago hominids from the genus *Homo* began making tools. In less than one million years time, *Homo* spread about the earth, and around 100,000 years ago *Homo sapiens* began speaking languages and establishing cultures (the oldest human culture is said by anthropologists to have originated some forty thousand years ago).

The generally-accepted evolutionary sequence of monkeys to great apes to australopithecines to habilines to erectines to modern man, however, is by no means agreed upon by the Darwinists, who have come up with many differing accounts regarding the evolution of modern man; which includes the Don Johanson and Tim White *Australopithecus robustus-aethiopicus-habilis-erectus-Homo sapiens* model, the Colin Groves *Australopithecus africanus-Homo ergaster-Homo sapiens* model, the Richard Leakey *Australopithecus robustus-africanus-Homo habilis-erectus-Homo sapiens* model, and the Bernard Wood *Australopithecus robustus-boisei-africanus-Homo ergaster-Homo sapiens* model. But until they have proof that one of these models is correct (which they never will), none of them should be believed.

The only thing, and I repeat, the only thing that scientists know for sure is that all these hominid species existed at one point in time in certain geographic locations. That's it! To claim that all these species are somehow interrelated and descended from a single common ancestor millions of years ago is like saying you know *War and Peace* by reading the first two pages of the book. No one can prove that any of these hominid species ever evolved from any other hominid species, alive or extinct. It's more plausible to believe that all these hominid species were created to be unique and individual by God then it is to believe that they all evolved from a single common monkey or some arboreal lemur.

Actually, it is much more plausible to believe in individual creation than Darwinian evolution because, as we have seen from radiometric dating, some of the species of the genus *Homo* existed at the same time as the australopithecines and the monkeys and the apes from whom they supposedly evolved from. Go back and look at the radiometric dating. Furthermore, the australopithecines from whom man supposedly evolved from would had to have evolved in a much shorter time period (tens of thousands of years instead of tens or hundreds of millions of years), which, as we have already seen, dramatically increases the odds against evolution ever happening because of the shortened amount of time random chance and natural selection have to work with. Fossil records of sharks and crocodiles and fish like the Coelacanth (pronounced see-la-canth) show that they haven't changed at all in the hundreds of millions of years since they first appeared in the fossil record, so why should we believe that modern man evolved so fast in Africa from monkeys and spread throughout the world as Darwinists would have you believe?

I'm sure you've seen the plethora of pictures in science textbooks of primitive African ape-men having to fight off lions, hyenas, and leopards, and vultures who are after their kill, so why haven't all these beasts and scavengers in Africa evolved as fast as hominids? And how do Darwinists explain the sudden appearance of language and writing in *Homo sapiens* only 100,000 years ago, or art and music at about the same time, or sophisticated cultures and societies springing up practically overnight only forty thousand years ago, or all the different races and languages of men? Why is man the only creature gifted with extraordinary intelligence and reason, with self-awareness about his existence, capable of pondering the meaning of his life and death, and where he goes when he dies? Why does every culture on earth recognize the existence of a higher being and has some form of religion or reverence for God?

Some evolutionists, like the late Alfred Russell Wallace, who believe in both God and evolution, say that Darwinian evolution formed the body but that God created the soul or spirit. This is one of the most commonly held beliefs of today among liberal clergy and theologians. Yes, God could have created men and women anyway he wanted to—instantaneously from the dust of the earth, symbolically from the rib of a man, or through gradual changes in preexisting creatures through evolution. The argument is not so much how God created man but whether or not God exists. Darwinists don't believe he does. The scientific evidence, however, suggests otherwise. As I have said repeatedly, Darwinian evolution by itself from a primordial soup by random chance and natural selection working over billions of years on so-called "favorable" mutations without the aid of God is a statistical impossibility. It just couldn't have happened.

Darwinists know that the fossil record just doesn't support their theory of evolution, so some of the more unscrupulous and devious ones have resorted to concocting elaborate hoaxes to fool the general public into believing in evolution. Here are just a few examples of the lies Darwinists were caught telling.

Fossil Fakes and Hoaxes

Atheistic evolution is a faith no different than any religious faith; just that Darwinism teaches that there is no God while religion teaches there is. Both beliefs are based on faith in what constitutes ultimate reality and both ideologies seek proofs for their beliefs. No one wants to believe in something that is a lie, so when one believes deeply in something, it is only natural to seek out evidence to substantiate one's belief. Christians do it and so do evolutionists. To their credit, most Darwinian evolutions are honest people who firmly believe that man evolved from monkeys and that the scientific evidence supports evolution. Be that as it may, there have been far too many times when the so-called experts on evolution have been fooled by hoaxers, or have simply been proven wrong in labeling a fossil a missing link for us to believe everything they say.

Perhaps the biggest fossil hoax of all time came in 1912 with the discovery of *Eoanthropus dawsoni,* commonly known as the Piltdown Man, in a quarry site near Sussex, England, by the amateur archaeologist, geologist, and fossil collector Charles Dawson. It seems that Charles Dawson had been digging around in Piltdown quarry for four years when he first found some ancient-looking bone fragments, and lo and behold he immediately claimed he discovered two prehistoric hominid skulls that were Haeckel's missing links. From 1912 to 1915, Dawson and his highly respectable team of academic supporters—which included Arthur Smith Woodward, Teilhard de Chardin, Arthur Keith, Grafton Elliot Smith, and the author Sir Arthur Canon Doyle of Sherlock Holmes fame, who was also the promulgator of the 1917 "Fairy Hoax" in England—revealed that they had discovered in Piltdown quarry two ancient hominid skulls with the large braincase of a man, a human-like canine tooth, a simian-like jawbone, along with tools carved from elephant tusks and animal teeth. Fitting these bones together, they announced to the world that they were the artifacts of the long-searched-for missing link.

Obviously, the academic world was ecstatic with these discoveries and hundreds of books, papers, articles, pamphlets, and letters were written about Piltdown Man. But there was something odd about the fossils that bothered a few people. The skulls and the jawbone and the teeth had the appearance of ancient fossils, but they just didn't fit together very well. Sadly, it would take forty years for scientists to discover that Piltdown Man was nothing but a gigantic hoax. In the meantime, the lies of Darwinism grew ever more popular. In the early 1950s, with the aid of fluorine absorption dating, Sir Kenneth Oakley discovered that the skulls of Piltdown Man I and II were only about 620 years old, and had been cleverly stained with an iron and chromic acid solution to make them appear ancient. The jawbone was subsequently identified by experts as that of an orangutan of about five hundred years old, and the teeth were identified as an elephant molar, hippopotamus tooth, and the canine of a Pleistocene era chimp. The canine tooth had been painted and filed to show wear, and the jawbone molars filed down to make the teeth fit into the jawbone and hominid skulls to make them appear genuine.

No one knows for sure who originated this elaborate hoax, but several names have been suggested, including members of the Hastings Museum and the British Museum of Natural History. Charles Dawson was probably in on it from the start since he was the first one to discover the fake fossils. It has been theorized that Dawson and his collaborators dreamed up this elaborate hoax because they were jealous of the discoveries of Cro-Magnon Man in France in 1868 and Neanderthal Man in Germany in 1856, and didn't want to be left out of the scientific history books. In any case, *Eoanthropus dawsoni,* which means "Dawson's Dawn Man" in Latin, has been a huge embarrassment for Darwinian evolutionists because so many people accepted uncritically that this was the missing link between apes and man and were made to look like fools.

You would think that because of the Piltdown scandal the scientific community and the media learned their lesson not to believe in everything you read and hear, but they haven't. Just recently, on October 15, 1999, *National Geographic* magazine announced the finding of an important new fossil that was supposedly the missing link between birds and dinosaurs, and were prepared to publish the amazing new discovery in their November 1999 issue. In an article entitled "Feathers for T. Rex" written by senior assistant Christopher Sloan, the *National Geographic Society* announced that a brand new fossil was found in the Liaoning Province of China that was part dinosaur and part bird, which proved Darwin's theory of evolution. The name of this new fossil species was *Archeoraptor liaoningenesis. Archeoraptor* appeared to have the body and head of a dinosaur but the tail of a bird, and was proudly declared by *National Geographic* to be the long-sought-after missing link between the two species.

Not everyone, of course, was ready to accept *National Geographic's* story at face value without doing an investigation of their own. A team of investigators headed by Hans-Dieter Sues of the Royal Ontario Museum in conjunction with an expert Chinese paleontologist by the name of Xu Xing examined the fossil and declared it to be a clever fraud. It seems that a fossil of a dinosaur and a fossil of a bird had each been carefully broken in half by a hoaxer in China, and then the bottom half of the fossil of a bird had been added to the top half of the fossil of a dinosaur to make it look like the dinosaur had a feathered tail. Totally embarrassed about being hoodwinked, the prestigious *National Geographic Society* had to eat crow and admit to the public in April of 2000 that a con artist had fooled them and that there was no such missing-link species as *Archeoraptor liaoningenesis.*

This wasn't the first time that the *National Geographic Society* had been party to a Darwinist's hoax. In 1972, the *National Geographic Society* and the television network NBC flashed a news story about the discovery of a primitive stone-age tribe called the Tasadays on the remote island of Mindanao in the Philippines. It was later discovered that there was no such tribe as the Tasadays, and that it was all a big hoax orchestrated by Manuel Elizade, the presidential assistant on National Minorities under the Marcos government, and a hunter named Dafal from a village of slash-and-burn farmers who persuaded some local villagers to strip and pretend for the cameras that they were stone-age people.

Yet another false claim was made by the *National Geographic Society* in September of 1960, when *National Geographic* published an article by the famous paleontologist Louis Leakey entitled "Finding the World's Earliest Man," in which Leakey supposedly discovered in Africa the remains of *Zinjanthropus (Australopithecus) boisei* or "Nutcracker Man," which was our earliest known human ancestor. It was later determined by scientists that Nutcracker Man was not an ancestor of man at all and *NG* had to retract the story.

Zealous evolutionists have had to eat crow on quite a few occasions. Shortly before 1927, a strange-looking tooth was found in Nebraska, and from that one tooth it was declared to be the missing link between man and ape, the "Nebraska Man" as he was called. That tooth, however, was later identified as belonging to a pig when the rest of the pig's jawbone was discovered in 1927, and the puzzle pieced together. In 1869 the public was fooled again when workmen supposedly dug up a ten-foot, three-thousand-pound fossilized giant in Cardiff, New York. People from everywhere traveled miles to see the fossilized Cardiff giant, which was nothing more than a concrete hoax.

The evolutionists ate crow again when the supposedly extinct species of fish known as the Coelacanth, universally declared by Darwinists as the missing link between fish and land animals, was discovered off the coast of West Africa alive and unchanged after hundreds of millions of years, just two days before Christmas in 1938. Thank you, Jesus, for that Christmas gift. It has done a world of good in discrediting Darwinism.

A Fish Out of Water

The search for the missing link between fish and land animals has been just as important to Darwinists as the search for the missing link between man and apes. If scientists could identify the first fish that crawled out of the oceans it would prove that evolution really exists and that Christianity is just a myth. Needless to say, the stakes are extremely high for both Christins and Darwinists. So high that it would change the course of human history by making religion and Christianity untruthful and obsolete.

By the beginning of the twentieth century, Darwinists had narrowed the list of potential missing links down to three species of fish: the lungfish, the Rhipidistia, and the Coelacanth, with the Coelacanth being the most probable. The lungfish, found recently in Australia, South America, and Africa during the nineteenth century, were still alive and could be thoroughly studied, but the Rhipidistia and the Coelacanth had long ago become extinct and existed only in the fossil records. Many fossils of the Coelacanth had been found throughout the world, so it was assumed by Darwinists that its large, fleshy-lobe fins enabled it to crawl out of water, where it evolved into land animals and eventually disappeared from the oceans and the fossil records.

According to the Darwinists, legged fish evolved from the Coelacanth about 400 to 450 million years ago in the late Silurian or early Devonian periods when "favorable" mutations in fish DNA turned its pectoral and caudal fins into functioning arms and legs, allowing it to support its weight and crawl out of its watery universe onto dry land. As the fairytale story goes, earth's dry land during the Devonian period afforded the first *Ichthyostega,* which literally means "fish with legs," a greater margin of safety from predators, so naturally it found the ability to move back and forth between water and dry land a survival advantage for itself and its offspring.

To impressionable young school children not advanced in comparative science, the story of the evolution of fish to land animals, which is taught in practically all of our public schools, sounds quite convincing: fish simply grew legs and feet from their fins one day and walked onto dry land. However, upper-level students of science who have studied comparative vertebrate anatomy and developmental embryology can find many flaws in this impractical theory. First, let's suppose that in spite of the odds against a fish ever growing appendages from its fins by a cumulative series of infinitesimally-small random mutations passed down from one generation to the next over millions of years, one day a fish did suddenly grow legs and feet and walked out of the ocean. Now let's suppose that this fish survived, and, in turn, its entire offspring had legs and feet too. For a breeding population of legged fish to survive and keep its species going by not having the mutant genes lost in the fish gene pool, which is enormous, it would take a substantial number of similarly mutated fish to have evolved legs and feet at the exact same time and place as the first ones who crawled onto land, which, as common sense and mathematical probabilities tell us, is a statistical impossibility. In all the years that man has been fishing, no fisherman has ever reported catching a fish with anything resembling legs and feet in any of its transitional forms.

Second, it takes much more evolution for a fish to live out of water than simply growing a pair of legs and feet. Take respiration, for example. Except for the three genera of lungfish in Australia, Africa, and South America that can survive in stagnant water and mud during the seasonal drought because they have both gills and lungs, the approximately twenty-six thousand species of modern gilled fish breathe by pumping water through their gills, which directly exchange carbon dioxide in their bloodstream for dissolved oxygen in the water. No gilled fish can survive very long by breathing atmospheric air alone. And no fish, not even the lungfish, can survive without water for very long because of their very sensitive skin. Fish skin is composed of a dermal layer of scales overlaid by a fragile layer of epidermis. Water and mucous-secreting cells in the fish's skin keep their epidermis moist and slimy, which protects it from bacterial invasion, and as most anglers know, a fish's skin can become permanently damaged from the repeated handling of catch and release fishermen. A fish's sensitive skin makes it impossible for it to live on dry land because of the constant abrasion from solid objects and the harmful drying effects of the heat and sun and wind, and the freezing effects of the cold.

The amphibian's skin, on the other hand, is very different than the fish's. Amphibians have a much thicker epidermis without the underlying dermal scales seen in fish, and have well-developed mucous glands in their dermis to help keep their skin moist when out of water. And since adult amphibians have no gills, respiration in amphibians takes place through their moist, highly permeable skin and mouth, and through their well-developed lungs. Larval forms of amphibians, however, are aquatic and breathe through gills, which disappear as the larvae metamorphose into adults and move out of water onto dry land. Amphibians' dependency on both water and land also requires them to have a well-developed set of legs for locomotion. The amphibian foreleg has twenty-five distinct muscles, a humerus, radius, ulna, carpal, metacarpal and phalangeal bones, all of which are connected to a well-developed pectoral girdle, which elevates the amphibian off the ground so it can move about without scraping its belly. The amphibian hind leg, in turn, has twenty-five muscles, a femur, tibia, fibula, tarsal, metatarsal, and phalangeal bones, which are all connected to a well-developed caudal girdle, which, likewise, raises the amphibian off the ground for faster locomotion. The fins of a fish, on the other hand, have only two muscles, a dorsal and ventral muscle, and a small number of fan-shaped bones with no well-developed pelvic or shoulder girdles, making it virtually impossible for fish to walk about on dry land with any efficiency. Certain species of fish can use their fins to move short distances from one pool of water to another during the dry season, but that doesn't make them land animals anymore than flying fish are birds.

Third, a fish out of water must have the ability to quickly adapt to living on dry land, but again, its anatomy makes that impossible. Fish have no eyelids or lachrymal glands or tear ducts to keep their eyes moist like amphibians do, which causes their eyes to cloud up and dry out within minutes of being out of water. Every fisherman with a stringer of fish has seen this happen. Fish also have a flattened cornea and lack the elaborate eye muscles that amphibians have, which would make it harder for them to see on dry land. Fish only have an inner ear, with no middle ear or eardrum as amphibians do, so they would be hearing impaired out of water. They also lack the complex balance-maintaining mechanism of the land animal's semicircular canals located in the inner ear and the proprioceptor (muscle-nerve) feedback mechanism from limbs to the brainstem's medulla oblongata that land animals also have that keep them oriented upright: a definite survival disadvantage. In addition, a fish's nasal cavity is designed for smelling in water and would be of little or no use on dry land, and absolutely no use for breathing because it is not open to its pharynx, which is the passageway to the lungs in animals. Fish have no salivary glands or tongues to help them chew their food before they swallow as land animals do, so they would die from starvation. Fish have no pivoting axis-atlas complex (the first two cervical vertebrae of the neck) like amphibians have, which allows head movement independent of the body, thereby making it much harder for fish to locate a meal or a dangerous predator on dry land. Also, the fish's body musculature is in an axial arrangement from head to tail and its ribs and muscles are not connected to its vertebrae like amphibians, which makes it ideal for swimming but terrible for walking or running on land. Gravity, among other reasons, causes a fish to flop over on dry land because it can't support its own weight. Also, water resists changes in temperature a lot more than air, so a fish out of water would quickly succumb to rapid fluctuations in air temperature.

As you can see, it takes a whole lot more for fish to survive out of water than just simply growing a pair of legs and feet, as Darwinists would have you believe. It would take the simultaneous restructuring of the fish's entire anatomy, physiology, and biochemistry, with the development of new skin, eyes, ears, nose, mouth, throat, lungs, heart, guts, nerves, brain, muscles, bones, arteries, veins, lymphatics, visceral organs, and glands. This would be a statistical impossibility if it were done by random chance alone because the entire fish genome of millions and millions of precisely coded triplet codons would have to instantaneously mutate in concert with no guiding force or intelligent design to work with.

The Coelacanth Can't

Darwinian evolutionists believe that all land animals, including man, evolved from one of three species of prehistoric fish of the subclass *Sarcopterygii,* which means "fleshy fin." They are the lungfish, the Rhipidistia, and the Coelacanth. Each one of these particular species of fish, which have lobe-shaped fins instead of the usual ray-shaped fins of most other fish, at one time or another has been declared by Darwinists to have been our ancient ancestor from the sea. Following the discovery of the first Coelacanth fossil by the Swiss scientist Louis Agassiz in 1839, there was a general disagreement amongst Darwinists over which of these three species of fishes was the true ancestor of modern man. Some argued it was the lungfishes of the southern tropics, others the extinct Coelacanth, still others the extinct Rhipidistia. With the fossil records showing that the Rhipidistia supposedly died out in the early Permian Period some 300 million years ago, and no new evidence of its living past that time discovered, most Darwinists in the mid-nineteenth century leaned towards either the lungfish or the more recently extinct Coelacanth as being our ancient piscine ancestor.

For a while, Darwinists thought they had found the missing link between fish and land animals when the first living lungfishes were discovered in the mid-nineteenth century. The first lungfish was discovered in the Amazon river area in 1836, then a separate species of lungfish was discovered in West Africa the following year, and yet another in

Australia in 1869. The discovery of three distinct species of lungfish caused quite a sensation amongst Darwinian evolutionists, but there was an obvious problem with the lungfish's anatomy, which soon cast serious doubt on its being Darwin's missing link. Lungfish, which have gills and intestines like a fish, can breathe air from lungs but have very poorly developed, almost string-like fins. The lungfish can only squirm about like an eel in or out of water, which virtually ruled them out as being Darwin's legendary walking fish. Lungfish didn't meet one of the two necessary requirements laid down by Darwin: that a fish simultaneously evolved lungs and sturdy legs capable of walking about on dry land.

With the elimination of the lungfish as the missing link, majority opinion among Darwinists in the late-nineteenth and early-twentieth centuries shifted back towards the Coelacanth as the most likely predecessor of land animals. Coelacanths were an ancient species of fish having large, lobe-shaped fins that lived and thrived unchanged throughout the world for close to 300 million years, but according to paleontologists, suddenly became extinct about seventy million years ago. Fossil records of the Coelacanth clearly showed the presence of large, lobe-shaped fins, but no visible evidence of anything that could be construed as a lung. That, however, didn't dissuade Darwinists from declaring the Coelacanth to be man's ancestor from the sea. Future fossil discoveries, they said, would prove them right.

As fate would have it, or maybe as Providence would have it, a bombshell was dropped on the scientific community that would seriously damage the evolutionists' theory that man evolved from fish. A couple of days before Christmas in 1938 a five-foot-long, 127-pound fish resembling the supposedly extinct Coelacanth was found floating belly-up off the east coast of South Africa. News of this amazing discovery spread like wildfire throughout the world. The media likened it to the discovery of the Loch Ness Monster or the huge eighty-foot dinosaur *Diplodocus.* Ichthyologists offered huge rewards to any South African fishermen who could catch a live Coelacanth and keep it alive until they arrived to study its internal organs, which, unfortunately, had been removed and discarded by the time the notable South African ichthyologist, J. L. B. Smith, arrived in February of 1939 to examine the carcass of the Coelacanth for evidence of lungs. If evidence could be found in the internal organs of Coelacanths that supported Darwin's theory that land dwellers evolved from sea creatures, it would definitely shore up the case for evolution. Every effort was made to hook one, but it would take much longer than expected.

Fourteen years later, a second Coelacanth was caught off the

Comoros Islands near Mozambique, well north of South Africa where the first one was found, and this time the visceral organs were left inside, although they were somewhat damaged from the rough handling by the fisherman who had caught the huge fish and clubbed it to death. Terribly disappointing to the Darwinists, the second Coelacanth was found to have no lungs. In fact, after hundreds of millions of years of successful oceanic existence, Coelacanths hadn't changed at all, and resembled sharks more than they did land animals because they had a blood chemistry high in urea, no ribs, a brain the size of a pea, and a primitive notochord rather than a highly specialized vertebral column, which is the main requirement for being classified as a vertebrate land animal.

Within a couple years time, the Coelacanth was being caught in abundance off the Comoros Islands. None of them, however, survived very long once caught, even though they were kept in shallow ocean water, which threw even more cold water on Darwin's theory that fish just simply crawled out of the oceans one day and established themselves on dry land. According to the local fisherman who landed them, Coelacanths prefer darker, colder water at a depth of around two hundred meters, and died within hours when hauled up to the shallower, warmer waters where there is much less dissolved oxygen and too much heat from the tropical sun. Furthermore, upon examination it was plain for scientists to see that the Coelacanth's flimsy fins would never support its massive weight on dry land—some of the specimens which were being caught weighed well over 150 pounds and measured over six feet in length.

Now that it was apparent that both the lungfish and Coelacanth were not Darwin's hoped-for missing link, scientific opinion shifted back to the Rhipidistia as Darwin's legendary fish out of water, as it remains today. The Rhipidistia is a much safer choice as the missing link because it became extinct about 300 million years ago, much longer than when the Coelacanth was thought to have become extinct, and the likelihood that someone will catch a living Rhipidistia someday and disprove Darwinism all over again is very small. So Darwinists hope.

Evolutionists failed to find their long-hoped-for missing link in the Coelacanth, yet studies still needed to be done on a living Coelacanth to see how they moved about in their natural habitat, and whether or not they could walk around on the ocean floor with their lobe-shaped fins where their natural buoyancy made them weigh much less. The deep-sea diving pioneer Jacques Cousteau was summoned by the French government to bring his boat, the *Calypso,* to the Comoros Islands in 1954 to photograph a living Coelacanth in its natural habitat, but he failed to find any trace of the elusive behemoth. It seemed that Coelacanths, with their highly reflective eyes that were capable of seeing in the dark, were well-hidden during the day in deeper water or in the protective rocky outcroppings and caves beyond the reach of Cousteau's submarine cameras. Cousteau and the crew of the *Calypso* tried again in 1963 and 1968 to photograph the Coelacanth, but again failed to locate any in the open.

In 1986, a marine biologist named Jurgen Schauer of East Germany took the first photographs of a living Coelacanth in its natural environment at a depth of around two hundred meters off the west coast of the Comoros Islands, this time at around 9 P.M. at night from inside a small, custom-built, two-man, yellow submarine named *Geo.* Schauer noted that Coelacanths show no signs of being able to walk about on the ocean floor, but instead do a graceful headstand as they approached it with the powerful lights of their submersible craft, gently flapping its large fins as if it were an underwater ballerina. Schauer returned the following year with his partner and fellow marine biology enthusiast, Hans Fricke, and observed many more Coelacanths in their natural habitat, taping their every movement. "Though we observed several individuals resting with their fins braced against the sea bottom," Fricke later wrote in the *National Geographic* magazine, "We never saw any of them walk, and it appears the fish is unable to so."[46]

The Darwinian myth of "Old Fourlegs," as the Coelacanth was affectionately nicknamed, being able to walk on its fins was forever exploded by Schauer and Fricke. In response to their discovery that Coelacanths can't walk in or out of water, Fricke said he had always been skeptical of the Darwinists' claims that the Coelacanth could walk on its fins as if they were legs. In fact, on his first dive with Schauer back in 1986 to photograph a Coelacanth in its natural environment, Fricke had skeptically remarked that "Everyone thought they rested on their fins, like they were legs. I had a hunch they didn't walk underwater, but I didn't think that all those famous professors could be wrong. It was just that their fins were not like bottom-dwelling fish. I am not an expert in fish locomotion, but I have seen so many fish in my life, and it just struck me that it was rather fishy!"[47]

Hans Fricke returned to the Comoros Islands in 1989 with his new submarine, *Jag,* which was built to go beyond a depth of two hundred meters. Thinking that Coelacanths went deeper during the day, Fricke descended to a depth of around four hundred meters to observe them in daytime, but found no evidence of Coelacanths. Returning to the depths where his partner found the first Coelacanth two years before, Fricke found them hiding in caves during the day, only coming out at night when the currents slowed and the water darkened and cooled. Altogether, Fricke recorded about 180 different Coelacanths.

In 1997, American marine biologist Mark Erdmann photographed the first living Coelacanth off the coast of Indonesia, which caused another media sensation not unlike that of sixty years earlier. Apparently, the fish that was once declared to be extinct, is more common and widespread in the oceans than scientists realized. So back to the evolutionary drawing board for Darwinists.

More Phony Evidence for Evolution

For a century and a half now, Darwinists have been endlessly searching for evidence, any evidence, that would substantiate the theory of evolution. One of their most commonly used "proofs" of evolution is a widely-circulated educational film that supposedly depicts the "natural selection" of peppered moths in action during the time of the industrial revolution of nineteenth- and twentieth-century England. The British physician and biologist Bernard Kettlewell, in collaboration with the animal behaviorist Niko Tinbergen, produced this 1950s documentary that you were probably shown in your elementary biology class as a kid. In this epic documentary, Kettlewell attempted to show how the peppered moth of England, *Biston betularia,* which typically comes in various shades of gray and black and white, began to dramatically change in the number of pure black or melanic forms during the industrial revolution in England because of Darwin's theory of natural selection at work.

According to Kettlewell, during the industrial revolution in England the number of melanic moths to peppered moths increased dramatically because of the unnatural blackening of nearby tree trunks by clouds of factory coal dust, which provided much better camouflage for the melanic forms than it did for the peppered forms. The British biologist J. W. Tutt, from whom Kettlewell got this idea, was the first to suggest all the way back in 1896 that the increase in melanic-colored moths over peppered-colored forms in Manchester, England was primarily due to black soot from the factories' smoke stacks settling on the trunks of trees in nearby forests. This enabled birds to better locate and capture the peppered-colored moths, which, up until the industrial revolution, had been safely hidden from the eyes of hungry birds by their wings that resemble lichen, the naturally-occurring plant life growing on the bark of tree trunks. As a result of this "industrial melanism," by the end of the nineteenth century more than 90 percent of the peppered moths in and around Manchester were melanic, thereby proving that Darwinian evolution had occurred through the processes of natural selection. So Kettlewell claimed.

Because of its highly visual content, Kettlewell's 1950s documentary on the evolution of peppered moths through natural selection was very convincing to his student audience, aimed primarily at grade school kids not in advanced science and critical thinking. It seemed to show positive and convincing visual proof of natural selection in action, which in turn, supposedly proved Darwinian evolution. Kettlewell was so proud of his documentary that he dubbed his peppered moth research as "Darwin's missing evidence" in an article written for *Scientific American.* Other scientists and critics, however, were not so convinced, so they began to investigate Kettlewell's claims on their own in the 1970s and 1980s. What these researchers discovered was scandalizing. Kettlewell's documentary was intentionally rigged from the start with staged pictures and misinformation in a deliberate attempt to deceive the audience that evolution and natural selection had taken place. For instance, researchers discovered that peppered moths do not rest on the trunks of trees as Kettlewell insinuated in his film, but under the branches of the higher up limbs. It was also discovered that Kettlewell had artificially gathered peppered moths of both forms and placed them directly onto the trunks of trees by hand during the day when they are normally inactive, making them easy prey for hungry birds because they will just sit there and not fly away. It was also discovered that peppered moths are night-fliers and find their daytime resting spots on the underneath of higher branches right before dawn when they start to become inactive. Kettlewell's experiment, therefore, was bogus and invalid from the get go.

Other Darwinists who just assumed Kettlewell's experiments were true redid his experiment even more deceitfully by pinning or gluing dead peppered moths to the trunks of trees. Kettlewell himself though that this sort of artificiality was a bad idea, but it just goes to show the lengths that Darwinists are willing to go to deceive the public. In one particular instance, University of Massachusetts biologist and Darwinist Theodore Sargent admitted to a *Washington Times* reporter in 1999 that he glued some dead specimens on a tree trunk for a TV documentary about peppered moths.[48]

Even if they didn't know better before, Darwinists should have discontinued the practice of artificially gluing or pinning peppered moths to the trunks of trees at least by the 1980s when it became well known to scientists that they don't even rest there, yet this deceitful practice and story continues to be shown to our kids in science films and biology books all over the world. For example, the 2000 edition of Joseph Levine's *Biology* includes faked photos of peppered moths on tree trunks, and calls Kettlewell's work "a classic demonstration of natural selection in action." Similarly, Burton Guttmann's 1999 *Biology* includes the usual faked photos and calls the peppered moth experiment "a classic contemporary case of natural selection." These are just some of the many examples of Darwinists' willingness to deceive the public.

Kettlewell and Tutt's notation of industrial melanism has been found to be totally unrelated to the supposed "natural selection" of predatory birds and moth camouflage. When biologists in the 1970s looked beyond Birmingham and Dorset, where Kettlewell first did his experiments, they found some highly unusual discrepancies in Kettlewell's theory. If melanic moths enjoyed as much of an advantage over typical peppered moths in Birmingham and Dorset's polluted woodlands as Kettlewell claimed, then we should expect to find the typical peppered variety almost completely replaced in much heavier polluted areas, such as Manchester, England. But this was not what researchers found, suggesting that factors other than camouflage and hungry birds were involved in industrial melanism. In addition, in rural Wales far away from industrial pollution, the frequency of melanic moths was much higher than expected, inspiring Liverpool biologist Jim Bishop to conclude in 1972 that "as yet unknown factors" were involved in industrial melanism. Complicating matters further, in rural East Anglia, where there was again little industrial pollution and typicals seemed much better camouflaged, melanics reached a frequency of 80 percent, prompting two other biologists to conclude in 1975 "that either the predation experiments and tests of conspicuousness to humans are misleading, or some other factor or factors in addition to predation [of moths] are responsible for maintaining the high melanic frequencies."[49] Conversely, in highly industrialized South Wales where melanics would seem better camouflaged than typicals, melanics comprised only 20 percent of the population.

Something other than camouflage was definitely contributing to industrial melanism. Compiling data from 165 separate sites in England, researcher R. C. Steward discovered a correlation between melanism and the concentration of sulfur dioxide (an industrial pollutant) north, but not south, of latitude 52°N. Steward concluded that "in the south of Britain non-industrial factors may be of greater importance" than camouflage and predation.[50] Surprisingly, after passage of tough anti-pollution legislation in the 1970s, the percentage of melanics north of England decreased as expected, but inexplicably increased in the south. Whatever caused industrial melanism, it was clearly more than just camouflage and bird predation as Kettlewell had maintained.

Still, the peppered-moth myth remains as popular today as ever with teachers of evolution because it is so visual and seems so logical. The wise person, however, doesn't believe everything he or she sees is told, and remembers what President Ronald Reagan's mantra used to be: "Trust, but verify." There is a lot of fraud in science, especially when it comes to the theory of evolution, so we must always verify what science claims to be true.

Evolutionists' Problem with Species Interdependency

One of the major problems facing evolutionists is species interdependency, the commonly observed phenomenon where a certain species requires the existence of one or more different species in order to survive and reproduce, as in the case with parasites. Parasites mature in a complicated succession of morphologically-different generations using multiple hosts or vectors, for example, and defy explanation by Darwin's simple-sounding descent with modifications theory. Take, for instance, the common human parasite *Fasciolopsis buski,* the giant intestinal fluke (flat worm). This large, flat, fleshy worm lives by attaching itself to the duodenal and jejunal walls of human beings, producing devastating consequences from the combined effects of massive tissue destruction and toxic poisoning from the fluke's release of nitrogenous wastes.

In the small intestine of man, the adult fluke produces twenty-five thousand eggs per day, which are discharged into the environment via human feces. When the fluke's eggs reach quiet, fresh water of about 80 to 90 degrees Fahrenheit, it completes its incubation in about three to seven weeks. A free-swimming *miracidium* then hatches from the egg, which, in turn, finds a snail and penetrates its soft tissues. Within a few weeks *sporocysts* are produced within the snail by the *miracidium.* Sporocysts then transform within the snail into first and second-generation *rediae.* The second-generation *rediae,* in turn, produce thousands of *cercariae* within the snail, which then exit the snail and become free-swimming again. The free-swimming *cercariae* attach themselves to water chestnuts and waterlings and become encysted. Eating the raw water plants infected with the encysted *cercariae* infects humans with the worm. Once ingested, the *cercariae* then release hundreds of larvae within the human gastrointestinal system, which quickly attach themselves to the walls of the small intestines where they mature into egg-producing adults, and the fluke's complex life cycle is repeated all over again.

In order to survive and reproduce, *Fasciolopsis buski* requires calm, warm water, a snail, a water plant, and a human host who eats the water plant raw. The problem for evolutionists here is that according to Darwinian evolution, man didn't appear until hundreds of millions of years after worms appeared, so how could this complicated host-parasite relationship have happened? A leading parasitologist by the name of Asa Chandler once said that the theory of evolution is just too inadequate to explain how extremely complex host-parasite relationships came to be. "It would be difficult, if not impossible, to explain, step by step, the details of the process of evolution by which some of the highly specialized parasites reached their present condition," Asa said.[51]

There are many more examples of human parasitic diseases that require multiple hosts and vectors and life cycles in order to survive and reproduce, defying the odds of evolution through random chance. The list of the most commonly known human parasitic diseases includes malaria, West Nile virus, and Rocky Mountain Spotted fever, to name just a few. None of these parasitic diseases can be adequately explained by Darwin's simple theory of descent with modification.

The All or Nothing Principle: Another Dilemma for Evolutionists

Another problem for evolutionists is what scientists call the "all or nothing principle." Simply stated, the all or nothing principle says that Darwin's intermediate forms would never have survived, or at least would have had no advantage in their newly mutated forms relative to their original anatomy. Take, for example, the supposed evolution of birds from reptiles, which practically every biology class in America teaches. In order for birds to have survived after evolving from reptiles, they must have simultaneously evolved a very specialized anatomy that includes a lightweight beak, an internal gizzard to help crush and digest food, a pair of wings with feathers that gives lift off the ground, fused bones in the skull and pelvis, the development of hollow bones connected to a system of air sacs dispersed throughout the body for light weight, parabronchial lungs required to breathe in flight, a new mechanism for the dispersal of heat during flight since birds do not have sweat glands, and a large sternum or breast bone that bears a central ridge or keel known as the "carina," which supports the major muscles used in flying.

All of these specialized anatomical changes are necessary for a bird's survival, and no bird is without them in one form or another; and none of them are of any use until they are all fully formed and functional, like a bird's feather. And speaking of feathers, microscopic examination has revealed that feathers have a complicated system of interlocking hooks and barbules that are perfectly fitted together to trap air, which gives the bird lift during flight. Since no bird can fly until its feathers' hooks and barbules are fully formed and interlocking, any gradual evolution of reptilian scales or mammalian hair towards bird feathers, as is proposed by the evolutionists, is of absolutely no use, which is contrary to Darwin's favorable mutations theory and survival of the fittest. As the famous French naturalist and expert in comparative anatomy and paleontology, Georges Cuvier (1769-1832), once said in criticism of Darwin's theory of transitional forms:

> Nature . . . has been settled in . . . all these combinations [transitional forms] which are not coherent and it is these incompatibilities, this impossibility of the coexistence of one modification with another which establish between the diverse groups of organisms those separations, those gaps, which mark their necessary limit.[52]

Darwin's transitional forms, therefore, are unworkable, and that is why we don't see them in the fossil record.

The lack of transitional forms in animals is also seen in plants. The sudden, abrupt appearance of higher plants like the angiosperms in the Cretaceous Period was very troubling to Darwin. "Nothing is more extraordinary in the history of the Vegetable Kingdom, as it seems to me, than the apparently very sudden or abrupt development of the higher plants," Darwin admitted.[53]

Another fact of life that supports the all or nothing principle over Darwin's theory of evolution through a multitude of transitional forms is the Cambrian explosion of underwater life some 500 million years ago. The numerous fossil records of the Cambrian Period indicate that the earliest known representatives of the major phyla of plants and animals during this period were found to have suddenly appeared without going through any evolutionary transitional forms. The only reasonable way to explain this phenomenon is that the underwater plants and animals of this period were created by God as accounted in the Book of Genesis.

The Eye: An Organ Too Complex to Have Evolved by Chance

One of the main problems facing Darwinists is how extremely complex organs such as the eye could have evolved all by themselves without a working model or an intelligent designer. Anyone who has ever studied photography and is familiar with the inner workings of cameras knows just how complicated film and lenses and filters and diaphragms and apertures and shutters are, which are all designed and manufactured by man according to the physical laws of light transmission, optics, photo-emulsion, and now digital technology. No one in their right mind would believe that a camera could ever evolve all by itself over eons of time by random chance from the common elements of the earth that are used by man to manufacture them, so why should we believe that time and random chance created extremely complex organs such as the eye, which is much more complicated and sophisticated than any human camera, but works along the same physical principles? Random chance never created anything resembling a camera, let alone a highly polished lens with its exacting diameters and curvatures, which are so skillfully grinded by lens makers out of sand, one of the most abundant minerals on earth. So why believe that Darwinian evolution created our eyes, which likewise have sophisticated lenses and covers and diaphragms? Even Charles Darwin admitted having trouble believing that his theory of evolution could account for the existence of such a complicated organ as the eye:

> Although the belief that an organ so perfect as the eye could have been formed by natural selection, is enough to stagger anyone. . . . I have felt the difficulty far too keenly to be surprised at others hesitating to extend the principle of natural selection to so startling a length.[54]

> If it could be demonstrated that any complex organ existed which could not possibly have been formed by numerous, successive, slight modifications, my theory [of evolution] would absolutely break down.[55]

If Darwin himself had troubling believing his own theory of evolution because of the existence of tremendously complex organs such as the eye, why should we believe it, especially since we know so much more about eyes than Darwin did? Innumerable volumes have since been written on the complicated anatomical, physiological, and biochemical processes of the eye, yet we still don't know exactly how it works. Pick up any medical book on the eye and start reading for yourself, and when you find that it is just too complicated to comprehend, ask yourself how random chance can understand it if you can't.

Evolutionists believe that the eye evolved over eons of time by means of natural selection from some sort of primitive ancestral sea worm that somehow evolved a photosensitive spot on its skin containing some sort of pigment that was capable of responding electrically to light waves in the visible light spectrum. Evolutionists contend, without giving any evidence to support their claims, that a sea worm's photosensitive spot evolved over time into a much more complicated light-gathering organ that we call the eye. Darwinists have come up with lots of cartoon-like drawings of how this primitive photosensitive spot of sea worms has evolved into the complex eyes of higher creatures, not unlike the common drawings you see of crawling lemurs and monkeys evolving step-wise into walking men. And without a science or medical background, you may be inclined to believe that eyes really did evolve this way, that it was as simple as what evolutionists illustrate in their drawings. But don't believe everything you see—there is nothing simple about the eye. And no evolutionist can explain how it evolved all by itself by random chance. The typical evolutionary theory that some primitive sea worm evolved a photosensitive skin spot, which later evolved into a bilateral pair of working eyes, is nothing but bunk. Some of the oldest known creatures in the world, like the extinct marine trilobites[56] of the Cambrian Period some 570 million years ago, already had a pair of well-developed eyes, and didn't need to wait around hundreds of millions of years for their eyes to evolve from a primordial sea worm.

The experts in the field of comparative anatomy and developmental embryology know this and admit they are clueless on how the eye evolved. Said Leslie Brainerd Arey, Ph.D., Sc.D., LL.D., former professor of anatomy, emeritus at Northwestern University: "Comparative anatomy fails to give any clue to the evolution of the vertebrate eye, since it is highly organized even in the lowest groups."[57]

Still, the Darwinists expect the uneducated masses to blindly accept their claim that our eyes evolved over hundreds of millions of years through random trial and error, even though they can't explain how it happened and have no evidence that it did. The fossil record shows that there are absolutely no fossils of any species having anything less than fully formed eyes.

As a car or a watch can't create movement until it leaves the assembly line or workman's bench, an evolving eye is of absolutely no use to a creature because it doesn't work. The slowly evolving eye provides no advantage, slight or great. Are we to believe that random chance and natural selection waited patiently over hundreds of millions of years for the eye to be built through trial and error until the first creature was finally able to see the light of day? This implies that natural selection has intelligence, foresight, patience, and a knowledge of optics, physics, and photochemistry, which contradicts Darwin and Dawkins' belief that evolution is like a blind watchmaker, with no intelligence, no purpose, no vision, no mind, and no mind's eye. In truth, random chance has no intelligence or foresight or patience or blueprints, and could not have created such a complex organ as the eye without an intelligent designer guiding its formation. God alone created the eye at the same time he created all his creatures who needed eyes because he loved them and wanted them to see the world in which they lived and, therefore, be fully capable of surviving. Darwinists love to be skeptical of religion, but it's high time that people of faith be skeptical of Darwinists. You needn't swallow a whole cow to know what baloney tastes like, and Darwinian evolution is nothing but baloney.

Tough Questions Darwinists Can't Answer

I could write an entire book on why evolution is unsound scientifically, but that is not my intention for this book, which is mainly to cast doubt on the godless human sciences which seek to eliminate God and Christianity from our lives. Nonetheless, before I part from the controversial topic of Darwinian evolution it is important to bring up some valid questions that Darwinists can't answer.

Darwinism's whole house of cards rests on the premise that simple molecules randomly formed from eternally existing matter by random chance over long periods of time, and that these molecules survived by first replicating themselves, and then evolving by natural selection into more and more complex forms of life. According to this simple-sounding evolutionary theory, the first molecules and primitive cells in Darwin's primordial soup replicated by making simple copies of themselves (a division process called mitosis). Mitotic cell division is highly efficient at producing progeny, in that a parent cell first makes copies of its DNA, then divides in two, then these two cells divide in two, and so on and so on. By simple mitotic division, a high number of progeny can be produced in a very short amount of time from one parent cell, which guarantees the survival of the species; which is in keeping with Darwin's theory that the only purpose to life is replication and that those who reproduce the most are the fittest.

Here is the dilemma for Darwinists: if simple mitotic division was so successful at the beginning of life on earth, why did sexual reproduction (meiotic division) evolve? Sexual reproduction necessitates that there be two separate and unique individuals who join together to make one, which, evolutionary speaking, is very inefficient and energy consuming. A male sperm, which has half the number of chromosomes of the father, must locate and join with a female egg, which has half the chromosomes of the mother. Only when the mother and father find each other and mate is a new individual created with a full complement of chromosomes (forty-six in humans). In contrast, asexual reproduction only requires one individual to split in two, and, therefore, is much more efficient at replication.

Sexual reproduction is an ingenious way to create love and bonding and good social relationships between individual members of a species, but is a very inefficient way to reproduce. So why did it evolve at all? How did asexual precursor species suddenly go from being totally alike to totally opposite (i.e., male and female)? If males and females evolved from the same asexual precursor, which gender evolved first, the male or female? Since the evolution of either the male or female requires the simultaneous evolution of the opposite sex to reproduce again, how could a species know that the evolution of the opposite sex would soon follow? If it didn't, did the evolution of both sexes occur at exactly the same time? The odds for this happening are infinitesimal.

In genetic terms, how could random variations in asexually reproducing DNA produce complementary pairs of sexually reproducing male and female DNA at the very same time? That statistical probability is beyond all imagination because it requires godlike knowledge of complementary pairs of sexual opposites and the meiotic division of sperm and egg cells, where the genetic information of the mother and father is cut in half in the production of sperm and eggs. When an egg and sperm unite to form a new life, a process known as conception, the mother and father's genetic material unite and form a new person (the father supplies twenty-three chromosomes and the mother supplies twenty-three, which gives forty-six chromosomes in the offspring). In humans, chromosomes, identified as X and Y, determine our gender. An XY chromosome pair defines a human male, while an XX chromosome pair defines a human female. The male, therefore, determines the sex of the offspring by having either an X or Y chromosome sperm (after meiotic division in the germ plasm of the male testes which reduces the amount of genetic information in half). An X or Y sperm then unites with an X chromosome ovum or egg, resulting in either a baby boy or girl. Theoretically, a male had to have existed before a female because his sex chromosomes determines the sex of the next generation, but a female had to have been present before the male so as to give birth to the offspring—an impossible paradox evolutionists can't explain. The real evolutionary question is not which came first, the chicken or the egg; it is which came first, the male or the female?

If evolution is true, then why is there so much complexity and beauty in the universe, especially since beauty and complexity are not necessary for survival and reproduction? Colorless amoebas and bacteria that replicate by simple cell division are very good at surviving. Why does man see in color and why is he able to visualize the beauty of the world while other creatures can only see in black and white or can barely see at all? Why is man the only creature that has a charitable heart towards his neighbor, while all other creatures are egocentric and beastly, especially when it comes to eating, mating, and finding shelter? Why is man the only creature that worships and seeks out the one who created all the wonderful things on earth, and ponders the reason for his existence and where he goes when he dies? There are just far too many questions that Darwinism can't explain for me to believe in evolution. I just don't have enough faith to be an atheist.

Chapter Three

The Pseudoscience of Freudian and Jungian Psychology: Religion is a Universal Neurosis

> "In view of these similarities and analogies one might venture to regard obsessional neurosis as a pathological counterpart of the formation of a religion, and to describe that neurosis as an individual religiosity and religion as a universal neurosis."
>
> —Sigmund Freud, *Obsessional Actions and Religious Practices*

> "At a very early stage, therefore, the real Christ vanished behind the emotions and projections that swarmed about him from far and near; immediately and almost without trace he was absorbed into the surrounding religious systems and molded into their archetypical exponent."
>
> —Carl Jung, *Psychology and Religion*

A Short History of Atheistic Psychology

In today's hedonistic and perverted society, Christianity, the Bible, and any mention of the belief in the Judeo-Christian God have been all been eliminated from school.[1] Nowhere is the elimination of Judeo-Christian beliefs more evident than in the field of psychology, where textbooks and curriculums have been thoroughly censored of any mention of the supernatural elements of God. Just open any textbook on psychology or psychiatry and try to find where it says that there exists a supernatural reality outside of the material reality that can and does affect the mind and spirit. Nowhere will you find in any of these science books that disorders of the mind are caused by anything other than an imbalance of neurotransmitters in the brain caused by some sort of psycho-physiological malady.

Traditional Christianity's belief that evil spirits exist and often torment our minds with temptations and destructive thoughts has been relegated by modern atheistic science to a bygone era of inquisitions and witch hunts. Demons and devils don't exist, say today's Freudian psychologists, and for that matter, neither does God. Every psychological disorder of the human mind has a material basis, which can be successfully treated with drugs and psychotherapy. Prozac and the therapists couch have made the long-established practice of examining one's conscience and going to confession obsolete.

Modern psychology's disbelief that evil spirits are a contributing factor to mental illness is a fairly recent phenomenon, historically speaking. In every ancient culture around the world, doctors, priests, shaman, and medicine men were very well aware of the existence of evil spirits and the harm they could cause the human mind, long before Jesus Christ appeared with his teaching about the real existence of Satan.[2] In ancient Egypt, Africa, and the Middle East it was believed that mental illnesses were sent by God or gods as punishment for the violations of divine laws. In the Hebrew book of Deuteronomy, the fifth book of the Old Testament written around 1274-1235 B.C., it is said that Almighty God punished those who forsook or rebelled against him with "madness, blindness, and panic."[3] Similarly, in the Old Testament's first book of Samuel (1090-1004 B.C.), it is written that the first King of Israel's disobedience to God caused an evil spirit of depression and melancholy to possess him: "The spirit of the Lord had departed from Saul, and he was tormented by an evil spirit sent by the Lord."[4]

It is likewise recorded in the Book of Daniel chapter four that in the seventh century B.C., Nebuchadnezzar, the ruthless King of Babylon, was inflicted with madness by God for a period of seven years, so that he would know that it was God, not him, who is sovereign above all things:

> Let his mind be changed from the human; let him be given the sense of the beast, till seven years pass over him. By decree of the sentinels is this decided, by the order of the holy ones, this sentence; That all who live may know that the Most High rules over the kingdom of men: 'He can give it to whom he will, or set it over the most lowly of men.'[5]

Nebuchadnezzar roamed the earth eating grass like a beast for seven years until he suddenly regained his senses and gave thanks to God for freeing him from his mental torture:

> His dominion is an everlasting dominion. And his kingdom endures through all generations. All who live on the earth are counted as nothing; he does as he pleases with the powers of heaven as well as those who live on the earth. There is no one who can stay his hand or say to him, 'What have you done?'[6]

Other historical writings not associated with the Bible show that the learned men of antiquity, Jew and Gentile alike, were far better informed on the causes of madness by evil spirits than modern psychologists are today. In ancient Greece around the tenth to ninth centuries B.C., the Greek mythologist Homer wrote in the *Iliad* and the *Odyssey* that the Greek gods influenced and directed man's conscious inner thoughts. Likewise, Homer wrote that an awareness of an internalized conscience explained the punishments meted out by the gods upon man for their transgressions of moral principles.[7] Like many of his pagan philosophic contemporaries, Homer was very astute about the existence of a spiritual realm, and how it could have substantial power and control over the material realm. Homer, by the way, was the one who coined the term "psyche" as the "life breath" of man; that immortal substance which keeps the human being alive and persists after death as the eternal soul or spirit.

While not privy to a revealed knowledge of the One True God as Jews were, Homer and other philosophers in early Greek culture theorized that mental disorders were caused by two evil spirits named Mania and Lyssa who were sent by the angry gods to torment man. Thus, the origin of the word "mania" in modern psychiatric lexicon as a synonym for being out of control of one's thoughts.

Prescribed treatments for being tormented by evil spirits during ancient times included the playing of soothing songs, fasting, rigorous purification rituals, herbal medicines, and exorcisms.[8] Yet, somewhere between the sixth and fifth centuries B.C. skeptical philosophers and medical doctors began to reject the commonly-accepted belief that evil spirits adversely affect the mind, and began replacing it with today's popular modern belief that only physiological mechanisms are responsible for mental illness.

The most famous of the early Greek skeptics was Hippocrates (c.460-377), from whom the Hippocratic oath that graduating medical students recite comes from. Hippocrates, the so-called "Father of Modern Medicine," believed that all diseases, including mental illness, were caused not by evil spirits but by an imbalance in the brain's four bodily humors—blood, phlegm, black bile, and yellow bile—which, as we know, is medically untrue. Said Hippocrates:

> Men ought to know that from the brain, and the brain only, arise our pleasures, joys, laughter, and jests, as well as our sorrows, pains, grief, and tears. . . . These things that we suffer all come from the brain when it is not healthy but becomes abnormally hot, cold, moist, or dry. . . . Madness comes from its moistness. When the brain is abnormally moist, of necessity it moves, neither sight nor hearing is still, but we see or hear now one thing and now another, and the tongue speaks in accordance with the things seen and

> heard on any occasion. But when the brain is still, a man is intelligent. The corruption of the brain is caused not only by phlegm but by bile. You may distinguish them thus: those who are mad through phlegm are quiet, and neither shout nor make a disturbance; those maddened through bile are noisy, evil doing and restless. . . . The patient suffers from causeless distress and anguish when the brain is chilled and contracted contrary to custom; these effects are caused by phlegm, and it is these very effects that cause loss of memory.[9]

Thanks to Hippocrates' popularity and prestige among physicians, his erroneous humoral theory of disease spawned the dangerous practice of bloodletting, which uses surgical cuts or leeches as a means to rid the body and brain of these harmful, "excess" humors. The practice of bloodletting lasted until around the eighteenth century, when physicians finally realized that draining blood did far more harm than good.

Two other important names in early secular medicine and psychology were Cicero and Galen. Cicero (106-43 B.C.), the famous Roman orator and statesman, was the first to use the word "libido" in a psychological way to describe our passionate carnal desires. Cicero believed that diseases of the human mind were caused not by excessive humors as Hippocrates had suggested, but by excessive disorders stemming primarily from errors in judgment, especially about sex. Galen (130-201 A.D.), the Greek anatomist and personal physician to the Roman Emperor Marcus Aurelius, was the first medical doctor to try to correlate the repression of our sexual libido with psychological anxiety. Galen believed that hysterical disturbances were caused by the retention of sperm in men and the delay of puberty in women. Frequent sexual relations and/or daily masturbation was prescribed by Galen as a treatment for psychological stress and mental illness. As you will later see, the theory of the sexual etiology of mental illness suggested by Cicero and Galen was very influential on the famous nineteenth-century psychologist Sigmund Freud.[10]

During the periods of the Renaissance (1400-1600) and the Age of Enlightenment (1600-1800), scientific knowledge exploded with new discoveries by such famous men as Descartes, Newton, Halley, Fermat, and Mariotte. It was during this wonderful time of scientific discovery that many changes were made on how psychiatrists and psychologists viewed and treated the mentally ill. For instance, the Spanish humanist Juan Luis Vives (1492-1540), the "Father of Empirical Psychology," was one of the first physicians to recommend special hospitals just for the mentally ill.[11]

Because of the increase in scientific knowledge, many physicians no longer saw religion, prayer, and the ceremonial rituals of the past as valid treatments for mental diseases. An Italian physician by the name of Paolo Zacchia was one of the first secular doctors during the Age of Enlightenment to insist that only medical doctors were qualified to diagnose and treat diseases of the mind. Following Zacchia's lead on the secularized treatment of mental illness, physician Thomas Sydenham was the first to establish the "clinical" approach to the diagnosis of mental diseases. Another secular psychologist named Robert Burton, in his book *Anatomy of Melancholy,* totally rejected any supernatural or occult influences as the causation of insanity; citing the "true causes" of mental illness as pathological emotional factors, including excessive religiosity. Many psychologists were now viewing religion as the cause of mental illness rather than the cure.[12]

During the Scientific Revolution and Age of Enlightenment many ludicrous theories were bandied about on what really causes mental illness. The famous French mathematician Rene Descartes (1596-1650) proposed a new explanation of human behavior based on a mechanical-hydraulic theory of "animal spirits" filling the ventricles of the brain (religious terminology was still popular and not easily dismissed by the materialists). To Descartes, animal spirits or unseen fluids flowed out of the brain through the nerves and somehow affected psychological functioning and passions in a reflexive manner. Disruptions of these animal spirits by intense passions negated the soul's ability to control the pineal gland, where Descartes erroneously thought the body and soul joined, thereby causing actions contrary to the free-will of the soul. As science later discovered, Descartes ridiculous theory of animal spirits had no basis whatsoever.[13]

One of the first psychological materialists to treat mental illness using the scientific discoveries of the Age of Enlightenment, such as magnetism and gravity, was a Dr. Franz Anton Mesmer (1734-1815). This charlatan and quack, who at one time aspired to be a priest, wrote in his doctoral dissertation that invisible bodily fluids respond to planetary gravitational forces that can positively or negatively affect mental health. Mesmer no doubt was influenced by Descartes' theory of animal spirits and Newton's discovery of gravity in 1666. Using the advice of a Jesuit priest named Maximilian Hell,[14] Mesmer claimed to have cured a young woman of hysterical neurosis by simply touching her body with magnets. Impressed by his apparent success with magnets, Mesmer substituted his earlier gravitational forces theory of good physical and mental health for a magnetic forces theory, which he called "animal magnetism." Thus the origin of the commonly-used term, which at one time meant something totally different than it does today. Maladjusted magnetic fluids were, according to his theory, responsible for poor mental health.[15]

Going even further, Mesmer claimed that his "magnetic cures" could be attained without the use of magnets simply by touching, gesturing, or intensively staring into his patients eyes, a treatment that came to be known as "mesmerism." It didn't take long, however, before Mesmer was shown to be what he really was: a quack. In the mid-nineteenth century Dr. James Braid from Scotland discovered that the success of mesmerism was due not to magnetism flowing from mesmerists to the patient, but to the susceptibility of the patient to the power of suggestion, a process that Braid called neuro-hypnology (later called hypnosis).[16] Soon the practice of hypnosis permeated the field of psychology and every quack in town was practicing it, including Freud.

Speaking of quacks, the nineteenth century had more than its fair share of quack psychologists. A German neurophysiologist by the name of Dr. Franz Joseph Gall (1758-1828) dreamed up a supposedly scientific theory he called "phrenology," which claimed to reveal personality types, behavior, and higher mental capabilities by simply measuring the exterior contour of the cranium.[17] Based on the popular supposition that man is more intelligent than animals because of his greater development of cortical gray matter, which results in a larger braincase, Gall theorized that differences in individual intelligence and personality in man would likewise be reflected by protrusions in the contour of the skull overlying areas of the cortex where Gall believed a particular personality trait or behavior was located. Gall and his colleague, Johann Christoph Spurzheim, examined hundreds of heads and mapped out thirty-seven regions on the skull where they thought underlying cortical areas contained specific personality traits such as love, kindness, aggression, reverence, and so on. Gall's research was well received and for the next century phrenological societies, parlors, and publications were the rage throughout Europe and the United States. People everywhere were blowing good money to literally have their heads examined. But a fool and his money are soon parted and there is a fool born every minute. Believing that knots on your head can explain your personality, behavior, and intelligence is as foolish as believing that the stars can foretell your future.

Further scientific advances in medicine in the nineteenth century and man's pride in his discoveries prejudiced more and more psychologists against the belief in God. The discovery of the areas in the brain which regulate speech, vision, touch, hearing, motor control and higher mental functioning, together with a clearer understanding of nerve cells structure, electrical transmission of nerve impulses, sensory systems, motor co-ordination and reflex pathways, convinced many famous scientists such as Emil Du Bois-Reymond and Hermann Helmholtz that only physical-chemical forces of attraction and repulsion are active within the body. Supernatural forces, souls, and spirits do not exist in the human body, only mechanical forces.

Due to the widespread acceptance of atheistic materialism within the scientific community in the nineteenth century, psychologists no longer believed in the dogmatic teachings of orthodox Christianity, yet some still borrowed from its teachings in order to explain their experimental observations. For instance, the so-called "Founder of Experimental Psychology" was a German physician who lectured at the University of Leipzig and who believed that all living things—humans, animals, and plants—had souls or minds. Fechner set about trying to quantitatively measure, by mathematical formulas, the elicited responses to barely noticeable differences in external stimuli in order to explain complex psychological behavior. While all his findings were later dismissed as meaningless, the atheistic psychologists of his day nonetheless adopted his experimental methodologies in order to advance their own materialistic theories. The only truly noteworthy thing about Dr. Fechner was his fascination with the occult and the paranormal, which many psychologists continue to have.

At an early age, Fechner suffered from obsessive-compulsive thoughts, headaches, and an exhaustion that nearly killed him. During his tenure at Leipzig, a bizarre interest in the experimental psychology of afterimages led him to peer at the sun through tinted glasses for long periods of time. Peering at the sun nearly blinded him and caused him to become extremely photophobic. Confined to his room, which he had painted black, Fechner became emotionally disturbed. In the solitude of his forced darkness, he was overcome by thoughts that he was close to discovering the secret of the world through scientific means. After three years of living in darkness, Fechner's eyesight began to improve somewhat and he was again capable of seeing light without pain. While walking through his garden one day, as if he'd suddenly seen the light, an outside voice entered into Fechner's mind that told him of the existence of souls in flowers and plants. Fechner later wrote a book on the mental life of plants, which many people still believe in today, and devoted the rest of his life to promoting and proving his pantheistic theory that all living things have a conscious mind and soul, which is still very popular in reincarnation theories of New Age teachings and Eastern mysticism.[18]

Following Fechner's lead in experimental psychology were several scientists attempting to make psychology a more legitimate science by their careful experimental observations of conscious mental processes responding to varying external physiological stimuli. The most notable of these psychological positivists and materialists was a lab assistant to the eminent German physiologist Hermann Helmholtz, a medical doctor and professor by the name of Wilhelm Wundt (1832-1920). Wundt is credited by many academic scholars as being the founder of modern psychology because of his numerous rigidly-controlled experiments that attempted to define human behavior in terms of neurophysiological brain activity rather than metaphysical processes of mind and soul. Wundt, like most of his atheistic colleagues in German psychology and psychiatry, believed that consciousness was the end result of brain and nervous system activity, that an eternal mind or soul does not exist apart from the brain, and that there is no such thing as spiritual immortality.[19]

Taking exception to the strict materialism of the German psychophysiologists was the American philosopher, pragmatist, and psychologist William James (1842-1910). James, who is often called the founding father of American psychology, disagreed with the German scientific authorities that mental states were only a reflection of the physiological states of the central and peripheral nervous systems. Instead, James believed that there exists a higher conscious or "state of mind" within the brain that can reason, imagine, will, remember, feel, and be aware of things independent of the current physiological states of the brain and nervous system. James' belief in a higher state of mind did not, however, mean that he was a God-fearing dualist like Plato. On the contrary, James, a philosophical pragmatist, didn't believe in a transcendent soul capable of existing eternally outside of the body as Plato did. Rather, James believed that higher mental processes were a consequence of Darwinian natural selection and the enhanced survivability within the human race evolving over eons of time.

Like his father before him, William James suffered from recurring bouts of acute anxiety and depression, which drove him close to suicide. While seeking treatment in Europe for severe depression, James studied under Hermann Helmholtz in Germany and became very interested in the field of psychology. After obtaining his medical degree and while trying to reconcile his atheistic scientific beliefs with the orthodox Christian beliefs of his theologian father, James suffered a nervous breakdown. Only when he began reading an essay on free will written by the free-thinking French philosopher Charles Renouvier did he begin to come out of his depression, which enabled James to devote much more of his time to the study of psychology. In his studies of psychology, James showed a keen interest in the power of the occult and began conducting numerous psychological experiments on hypnosis, spiritualism, seances, paranormal experiences, and other forms of altered states of consciousness. It is important to note that while atheistic psychologists don't believe in evil spirits, they still have an uncanny attraction and fascination with the occult and the demonic forces behind them.[20]

William James founded the American Society for Psychical Work in 1884, and from 1892 to 1907 he published four major literary works based on his experimental observations: *Principles in Psychology, Talks to Teachers, Varieties of Religious Experiences,* and *Pragmatism.* In *Varieties,* James formalized his psychological explanation to the paranormal and mystical observations he made; which, thanks to the influence of Sigmund Freud, he attributed not to anything supernatural or preternatural, or demonic, but to the power of the unconscious mind.[21]

Following James, Sigmund Freud (1856-1939) was the most celebrated psychologist at the turn of the twentieth century, and his atheistic theories on the powers of the unconscious mind on mental diseases were heralded as groundbreaking revelations by a growing number of secular psychologists and medical doctors in Europe and America, which included such famous names in psychology and psychiatry as Jung, Adler, Rank, Klein, and Reich.

Shortly before the advent of atheistic Freudian psychology, a medical doctor and amateur psychologist by the name of Francis Galton (1822-1911) conceived of the idea of intelligence testing for the purpose of advancing society by selectively breeding a superior race of intellectuals. Cousin to Charles Darwin, Galton believed that intelligence and mental acuity were inherited from one's parents, and that society would be much better off if it restricted inferior people from having children—a Hitler-like social-engineering concept Galton referred to as "eugenics." To accomplish eugenics, Galton devised several anthropometrical (physical measuring) tests for the sole purpose of identifying individual physical differences that he thought correlated to higher intelligence and natural superiority. Galton's idea of selectively breeding a superior race of people struck a positive chord with the Nazis and, yes, many Americans. In fact, by 1917 it became the leading field of psychological study in America.[22]

The person most responsible for the introduction of intelligence testing into America was a French lawyer, naturalist, and quack psychologist by the name of Alfred Binet. Early in his career, Alfred Binet, a devotee of Anton Mesmer, had been accused of being a charlatan and a quack for claiming to transform his patients' emotions under hypnosis by the strategic use of magnets. Later on, when magnetic transformations were shown to result from the power of suggestion, Binet resigned from Jean Martin Charcot's Salpetriere Clinic of Neurology and Hypnosis in France. Several years later, as director of the Laboratory of Physiological Psychology at the Sorbonne, Binet switched his interests to intelligence testing and boasted that he could measure a person's intelligence by craniometry (skull measuring). Binet contended that the larger the skull, the smarter the individual, which, of course, is totally ridiculous. It wasn't until 1905 that Binet was finally taken seriously by the public when he and Theodore Simon developed a way to measure the intelligence of children by comparing individual intelligence-test results to normal children of the same age. Binet's intelligence test was given a numerical value by the psychologist William Stern in 1912, who modified the Binet-Simon test a bit by dividing a person's mental age by his chronological age to come up with a mental or intelligence quotient, commonly known as IQ.[23]

IQ testing became extremely popular in the early-twentieth century and was commonly used by psychologists in American schools, hospitals, and courts of law to identify the so-called "mentally inferior," so that eugenic sterilization could be used to eliminate their genes from society. In 1907, Indiana became the first state to formalize the practice of forced sterilization on the "confirmed criminals, idiots, rapists, and imbeciles."[24] By 1931, twenty-seven states had passed laws enforcing eugenic sterilization on social undesirables.[25] All told, at least sixty-four thousand Americans in thirty states were forcibly sterilized until the government in the early 1960s finally outlawed this spurious practice.[26]

Intelligence testing was used by the American government's immigration agencies to severely restrict the influx of Jews and Catholics immigrating from Southern and Eastern Europe. Based on the results of Binet-Simon IQ tests, which supposedly showed that up to 80 percent of the Jews, Hungarians, Italians, and Russians were imbeciles and feeble-minded, the U.S. government passed a tough anti-immigration law in 1924, reducing the total quotas for Southern and Eastern Europe to less than one-fifth of that for Western and Northern Europe.[27] It would take years before this practice was outlawed by the government as patently discriminatory.

Lewis Terman, a professor of psychology at Stanford University, later revised the Binet-Simon intelligence test and by 1937 the newer Stanford-Binet test, as it is now called, had become the standard for intelligence testing in the U.S. Other intelligence and personality tests have become vogue at one time or another, such as the Weschler intelligence scales, Cattell intelligence scale, Raven Progressive Matrices, Leiter International Performance Scale, Columbia Mental Maturity Scale, Minnesota Multiphasic Personality Inventory, Rorschach Test, and the Thematic Apperception Test, but there is still great debate over what these tests really prove.

The use of intelligence and personality testing may have its place in society, but critics claim that they too often discriminate against underprivileged and culturally disadvantaged minorities. Employers, however, are resorting more and more to the use of these tests to screen out unwanted applicants, which may or may not be legal or ethical. The problem with intelligence tests is that they only measure what people know, not what they are capable of learning or achieving, so one must be very careful not to use intelligence or personality tests as the only means to judge integrity, character, and job performance capabilities.

Around the time that race and heredity were being heralded by psychologists as the determining factor behind intelligence and personality, another absurd psychological theory known as "behaviorism" was developed by animal psychologists in an attempt to explain complex human behavior in terms of neurological responses to environmental stimuli, without regards to conscious thought and experience. Based upon the Darwinian premise that man is nothing more than a highly evolved animal without an incorporeal mind or soul, behaviorists proposed that by studying the behavior of animals in certain experimental situations, they could explain and predict why human beings behave the way they do, with the intention, of course, of changing human behavior and creating a utopian society on earth without the need for any dogmatic religion as a moral guide. Behaviorists believe that there is no such thing as evil or sin, only inappropriate or untrained behavior.

The first celebrated psychological behaviorist was a man named Edward Lee Thorndike (1874-1947). Thorndike was the son of a Methodist minister from Boston who conducted behavioral experiments with chickens in mazes and cats in box-traps. Thorndike observed that by simple trial and error, animals learned which responses best rewarded them with pleasure, either food or escape. The second and perhaps most famous of the early behaviorists was a Russian physiologist named Ivan Pavlov (1849-1936), the grandson and son of Russian Orthodox priests. Pavlov is best known for his psychological experiments on dogs. Pavlov observed that after several repetitions of the sound of a bell or buzzer followed by food, the dogs learned to salivate at the sound alone, even if it were not followed by food. Pavlov called this observation a "conditioned reflex" (salivating) to a "conditioned stimulus" (ringing bell). To Pavlov and his fellow behaviorists, stimulus-response learning was the key to the explanation of complex human behavior, in that we are conditioned to behave or respond to external stimuli in a way that gives us pleasure or avoids us pain.[28]

The most influential of the American behaviorists was a man named John B. Watson (1878-1958), who, together with James R. Angell, officially coined the term "behaviorist."[29] Watson at one time intended to become a Baptist minister, but chose the field of psychology instead. It is interesting to note how many famous psychologists at one time wanted to become ministers or priests but ended up being psychologists. In 1908, Watson became the chair of the psychology department at Johns Hopkins University, but was forced to resign in 1920 by the president of the university, Frank Goodnow, for behavior unbecoming a married professor (Watson seduced his student). As an experimental behaviorist, Watson used laboratory experiments on rats and infants to affirm what he perceived as the preeminence of environment stimuli on the influence and development of human behavior. Drawing absurd conclusions from animal experiments, Watson concluded that all human behavior was the direct result of stimulus-response conditioning, and his kind of psychological conditioning could be used to condition any kind of human behavior he wished:

> Give me a dozen healthy infants, well-formed, and my own specified world to bring them up in and I'll guarantee to take any one at random and train him to become any type of specialist I might select—doctor, lawyer, artist, merchant-chief and, yes, even beggar-man and thief, regardless of his talents, penchants, tendencies, abilities, vocations, and race of his ancestors.[30]

Perhaps the most recognized name in the field of behaviorism, in terms of being studied in psychology and biology textbooks, was the controversial animal psychologist from Harvard University, B.F. Skinner (1904-1990). Skinner demonstrated in his animal experiments that rats and pigeons learn to act or behave in certain ways that give rewards, supposedly without any use of conscious thought, reason, free-will, or moral reflection. Skinner's experiments showed that random animal actions operating on the environment, such as a rat pressing a bar down that automatically drops food pellets into a tray, provide them with immediate rewards, or positive reinforcements, thereby producing what he termed "operant conditioning." According to Skinner, operant conditioning could be used by social educators to teach good moral actions, thereby eliminating antisocial behavior.[31] Humans are like rats that can be trained to behave if just given the right amount of pellets.

As time has shown, however, behavioral modification and operant conditioning doesn't work for everyone. Some humans are just plain evil and no matter what you offer them as a reward they are going to do evil. Tyrants like Hitler, Stalin, and Saddam Hussein are just a few examples. No incentives or promises of rewards would have gotten them to act in a civilized manner. To them, evil and antisocial behavior was the reward in itself.

Freudian Psychoanalysis: The Couch Supplants the Confessional

> "To me, psychoanalysis is a hoax—the biggest ever played on humanity. By showing who [psycho]analysts are, how they work, what they believe, and what they have done, I hope to show Freud as a Fraud."
>
> —Edward R. Pinckney, M.D. and Cathy Pinckney
> *The Fallacy of Freud and Psychoanalysis*

Most Americans generally view Freudian psychoanalysis as being very helpful in diagnosing and treating afflictions of the human mind and spirit, but we Americans are too easily duped. The so-called "scientific" practice of psychoanalysis, which was invented by Freud, a hardened atheist, is more of a pseudoscience than real science because it adversely modifies and reforms a person's core beliefs, morals, values, ethics, actions, and behaviors according to what godless Freudians believe, which is not in Jesus Christ, who is Truth itself. The false science of Freudian psychology, which is really a false religion, is the largest and most influential pseudoreligion in the world because people of all faiths and religions have put their faith and trust in it over and above their own religious beliefs. Nowhere has that been more evident than right here in the United States, where Freudian psychology has been at the forefront of the assault on traditional Christian doctrines and values.

Let me give you just one example where the belief in Christianity has been overshadowed by belief in Freudianism. It has been common practice in the past for pastors and priests who've suffer from mental illness or who've exhibited socially-unacceptable behavior such as homosexuality, pedophilia, or alcoholism to be sent by their superiors to secular Freudian psychologists and psychiatrists for treatment. Nowhere is this more evident than in the recent sex-abuse crisis that has recently swept the Catholic Church in America, where offending priests and deacons have been systematically shipped off to secular Freudian institutes for treatment, only to have their long-held Catholic beliefs put in doubt by non-believing Freudians who believe in no moral absolutes.[32]

For a layperson, too, it can be very dangerous to see a Freudian psychiatrist or psychologist you don't know. The most important question all patients must ask of themselves before undergoing Freudian psychoanalysis is "Is this person trustworthy when it comes to treating my psychological disorder?" The answer to that question is "Probably not." For one thing, the fee for psychoanalysis at $100 to $150 per hour can end up costing a fortune because there is a financial incentive for unscrupulous and greedy psychologists and psychiatrists to schedule as many psychoanalysis sessions as they can before declaring the sessions over. For another thing, after receiving Freudian psychoanalysis you may be much worse off than before because the atheistic biases of Freudians clash with most people's deeply-held religious beliefs. Patients can leave psychoanalysis totally confused over what is the truth and what actually caused their psychological disorder in the first place, especially if they originally believed that the devil or demons were in some way oppressing their minds, which they often do. Try telling a Freudian psychoanalyst you think the devil is real and attacking your mind and see what his or her reaction is. You might just end up in a straight jacket and locked away in a padded room, or have your mind fried by electroshock treatments. Psychiatrists and psychologists have tremendous power and authority with the law and the courts and can have you committed to a mental institute without your consent. Numerous are the reports of abuse by mental health care professionals on helpless patients in spiritual distress.

It is an indisputable fact that most Freudian psychologists and psychiatrists don't believe in God, let alone the existence of evil spirits. Like Sigmund Freud, their idol, most Freudians think that a belief in God and demons is a form of "universal neurosis" in which the poor, the weak, the uneducated, and the religiously brainwashed seek out some sort of mythological father figure to help them cope with the difficulties of everyday life. According to Freudians, this is how the biblical story of God got started—through our wish for a strong and protective father figure.

The history of godless Freudian psychology is an interesting one to read and, yes, to psychoanalyze. Freud hated his father and his father's religion, Judaism, and thought that all religions were evil and harmful to one's psyche—especially Christianity. Trying to keep God's commandments in order to avoid the punishments of hell, especially the commandment forbidding sex outside of marriage or between two people of the same gender, Freud thought, was the underlying cause of all the neuroses in the Victorian era, especially among females. Therefore, Freud felt it absolutely necessary to re-educate his patients during psychoanalysis that hell doesn't really exist and that they needn't worry about going there for engaging in immorality. After all, Freud surmised, Darwin had just proven that God doesn't exist, so why worry about hell anymore? Sex outside of marriage between two consenting adults of whatever gender is perfectly normal and should not be avoided.

Like they did Darwin, the secular intellectuals and atheistic scientists of the late-nineteenth and early-twentieth centuries lauded Freud as a genius of great understanding and insight into the human psyche. Nearly everyone in American psychology and psychiatry at the beginning of the twentieth century worshipped Freud and wanted to become a Freudian. It wasn't until the later decades of the twentieth century, however, that the Freudian stranglehold on psychology and psychiatry began to loosen because of mounting criticism by lay people outside the field of psychology when they began to study for themselves Freud's original writings, and rightly came to see his theories about sex, especially his theory of psychosexual repression in young children and adolescent females, as the perverted work of a dirty old man.

For instance, Freud thought it completely natural and healthy for young boys to want to have sex with their mothers and, likewise, young girls with their fathers. The so-called "universal" tendency for young boys to want to have sex with their mothers and their wish that they could kill their father, whom they see as the primary rival for mom's sexual favors, was called by Freud the Oedipal complex, a perverted theory that Freud borrowed from Sophocles' Greek tragedy of a fifth-century-B.C. boy named Oedipus, the son of Laius and Jocasta who kills his father, marries his mother, and sires a beautiful daughter by mom named Antigone, who is also his sister. So, according to Freud, incest with mom and patricide of dad are completely natural desires in line with Darwinian evolution. Failure to recognize and understand the Oedipal complex (Electra complex in girls) causes mental anxiety and hysterical neuroses, which, if left untreated, precipitates into more severe forms of mental illness.

Towards the latter half of the twentieth century, when Freudianism began to fall from grace, the non-sexual psychological theories of Carl Jung, Freud's star pupil and protégé, began to gain in popularity over the Freudian theory of childhood psychosexual development. Jung was Freud's favorite pupil but later came to believe just the opposite of Freud regarding the dangers of religion and the relevance of proper childhood psychosexual development. Unlike Freud, Carl Jung believed that having the correct understanding of God and religion was the cure for mental illness rather than the cause of it. Jung's fascination with understanding and psychoanalyzing religion, however, didn't mean that he believed in religion—Jung was a dyed-in-the-wool atheist like Freud—rather, he believed that the "myths" and "symbols" of traditional religion, especially Catholicism, were extremely useful in understanding the human psyche and could be successfully used to diagnose and treat psychological disorders and the social apathy and malaise of the mid-twentieth century. Jung's psychoanalytical interpretations of the symbolic meanings of religion in diagnosing and treating mental illness, however, were just as harmful as Freud's belief that religion is a form of universal neurosis. Jung thought all religions, including Judeo-Christianity, were myths and not meant to be taken literally. The stories in the Bible are great allegories for teaching ethics and morals and understanding our "collective unconscious," Jung thought, but not historic fact. Believing otherwise, as biblical fundamentalists do, can be psychologically damaging.

The tremendously evil influences Freud and Jung have had on modern society cannot be overstated. Today's psychologizing of American language, values, religion, and ideals comes mainly from the widespread teaching of Freudian and Jungian psychobabble in our schools. Every student who takes a course in psychology gets exposed to Freudian and Jungian psychology, which has significantly contributed to the decline in moral values and the rise of antichristianity in America. As an example, the immensely popular appeal that Freudianism and Jungianism enjoys in American can be seen in the widespread use of Freudian and Jungian terminology in our everyday language. For example, common expressions you hear today about someone being "anal" or "anal retentive," or having "penis envy," derives from Freud's perverted theory of childhood psychosexual development. In the same way, when one is accused of committing a "Freudian slip." it generally means that he or she accidentally verbalized what was in their unconscious minds.

Freudian jargon has so permeated American society that one often hears and uses Freud-speak without ever realizing it or knowing what it really means. Examples of this includes such commonly used Freudian terms as id, ego, superego, libido, free-association, unconscious repression, denial, sublimation, inhibitions, psychosomatic illness, ego defense mechanisms, rationalization, fixations, inferiority and superiority complexes, narcissism, transference, catharsis, and projections—to name just a few.

Likewise, psychological terminology commonly heard in America attributable to Jung include such often-heard words as the collective unconscious, archetypes, Christ consciousness, self-actualization, synchronicity, and our feminine *anima* within males or masculine *animus* within females. We don't realize it, but a lot of what we think and say comes from Freud and Jung.

The Perverted Life of Sigmund Freud: The Godfather of Atheistic Psychoanalysis

Warning: the following section contains graphic sexual content and may not be suitable for young readers.

The six main tenets of atheistic Freudianism are: 1) the key to mental health and well-being is guilt-free sex 2) the most formative influence in young children is their universal desire to have sex with their opposite-sex parent or sibling 3) the resistance to these completely natural sexual urges or the failure to outwardly express inward hostility and aggression causes mental illness 4) religion and the belief in God is a form of universal neurosis 5) those who believe in God, Satan, and supernatural spirits are mentally deranged 6) along with man's primal erotic life instinct, the *eros,* man also has a suicidal death wish, which Freud called the *thanatos.*

Freud invented psychoanalysis in 1896, which is nothing more than science's replacement of the priestly confessional, as a totally unproven pseudoscientific technique for analyzing and interpreting the unconscious phenomena of mentally and spiritually distressed patients. During his psychoanalytic sessions, Freud encouraged patients, especially his young female patients, to talk freely (free-association) in a completely uncensored way without inhibitions, self-reproach, or fear of criticism by him or punishment by others about daytime sexual fantasies, early childhood sexual experiences, or nighttime sexual dreams. The symbolic or psychological interpretation of dreams and sexual fantasies was at the forefront of Freudian psychoanalysis and he wrote several famous works on these controversial topics.[33] Once the patient's psychological problem, which Freud always thought was sexual repression hidden away in the unconscious mind, was discovered by the patient by moving it into conscious awareness, Freud could then "talk" the patient into a cure. Of course, psychotherapeutic drugs, which are so successful today at alleviating the symptoms of mental illness, had not yet been discovered.

Freud invariably told his patients that their painful unconscious memories, such as disturbances in normal childhood sexual development, were the root cause of their anxiety neuroses. Freudians have termed this the "repressed memory syndrome." Here is how it supposedly works. Completely removed from consciousness by the unconscious ego defense mechanism, which Freud called "repression," the nervous excitation associated with the unwanted sexual memory or sexual desire later on in life attempts to force its way back into the conscious mind but, instead, is pathologically converted by ego defense mechanisms into psychosomatic illnesses and/or hysterical disturbances—such as paralysis, loss of bodily sensation, hallucinations, visual and speech disturbances, somnambulism (sleep walking), depression, amnesia, suicidal thoughts, and other mental abnormalities. By helping patients recall painful memories back into consciousness through psychoanalysis, Freud claimed he could alter or remove his patient's psychologically-debilitating inhibitions to the painful memories or morally unacceptable ideas, along with the religious guilt perpetuating the strangulating affect on the mind; thereby allowing the debilitating psychosomatic symptoms to suddenly disappear like a puff of smoke. Freud called this "abreaction."

Freud's psychological theory of the repressed memory syndrome has since become so popular and so well accepted by psychologists and psychiatrists that one of the very first things that mental health care professionals suspect when patients present to them with psychological problems is that they were sexually abused as a child and are subconsciously blocking out the painful memories. While it is true that many children have suffered through child abuse, there is absolutely no valid scientific evidence whatsoever supporting Freud's theory of the repressed memory syndrome, even though reams of books have been written on the subject.[34] In fact, many horror stories have arisen in the last decade or so where determined psychotherapists were so convinced that their patients were sexually abused as children, even though they weren't, that they planted false memories into their patients' minds through hypnosis or psychoanalysis about family members or friends abusing them in their childhood.[35] Just imagine the heartache of a father or mother or a relative wrongfully accused of abusing their child because an unscrupulous Freudian psychotherapist was convinced otherwise. It happens all the time.

Just a few years ago an overblown hysteria over satanic ritual abuse (specifically, the molestation of young children by adults during satanic rituals) swept through the rank and file of Freudian psychotherapists, who began suspecting cases of satanic ritual abuse (SRA) everywhere they looked. A lot of psychotherapists became so convinced that SRA was epidemic throughout this country that they planted suggestions of SRA into the minds of their patients undergoing hypnosis. Lo and behold, false memories of SRA began surfacing so fast that you would think that the clock had turned back to the time of the Salem witch trials. Scads of books were written on SRA and it was the subject of endless talk shows and documentaries.[36] But you don't hear much about SRA anymore because after many thorough and fruitless investigations, the police, the attorneys, and the courts have discredited the testimonies of the vast majority of the alleged SRA victims, who have since recanted their stories and admitted that their psychotherapists had planted false memories in their minds.

Freud, like Darwin, believed that man was nothing more than a highly evolved monkey whose foremost drive, due to his natural psychic energy or libido, was to have as much gratuitous and pleasurable sex as he could possibly have so as to perpetuate his genes to the next generation.[37] The id, where our libido is contained, Freud said, is the center of instinctual animal drives and operates under the "pleasure principle" of self-gratification. Its primary goal is satiating the strong sexual desires and urges of the flesh. The ego, on the other hand, retains close relationship to consciousness and reality and is the center of the human will. It mediates, evaluates, integrates, coordinates, regulates, and adapts to reality, while controlling the id by conscious suppression of what Jews and Christian's call temptation. The superego, the highest part of the human psyche, represents the internalization of parental restrictions and the moral, religious, and ethical values and rules of society. It functions to reward or punish through a system of moral attitudes and guilty feelings. It judges, guides, prohibits, renounces, commands, and resolves internal conflicts, and can be said to be analogous to the Christian concept of the conscience.

According to Freud's theory of the psychosexual etiology of mental illnesses, mental anxieties occur when completely natural desires emanating from the id, such as incest with parent or sibling, homosexuality, bisexuality, adultery, and heterosexual fornication, are repressed from consciousness into the unconscious by ego defense mechanisms, where they are then pathologically converted into psychosomatic illnesses. Religion and societal mores that condemn these sexual thoughts cause guilty feelings to arise within the superego, which, in turn, commands the ego to repress the natural desires into the unconscious, which then eventually causes psychosomatic symptoms to rise to the surface. The male psychoanalyst's task (in the late-nineteenth century men were the only ones practicing psychoanalysis) is to first form a semi-erotic attachment with the female patient, a process Freud called "transference," then uncover the repressed sexual desire by bringing it back into conscious awareness (usually after many months of time-consuming and expensive psychoanalysis), so as to finally release its strangulating effect, which usually means advising the patient to give in to the sexual temptation, thereby effecting a guilt-free, long term cure. Freud's usual method of having the female patient lie down on the couch looking away from him while he was at the head of the couch looking down on her breasts and thighs is a very sexually domineering position, which, no doubt, was Freud's intention. Many a historian has suggested that there was more than just talking about sex taking place on Freud's couch.

While Freudianism has since been exposed as the work of a sexually perverted quack, gullible psychologists, psychiatrists, scientists, intellectuals, and secular humanists in America in the early to mid-twentieth century believed wholeheartedly that Freudian psychology was the key to understanding and explaining complex human behavior and mental illness. Why so? One reason was that Freud was European and Americans at that time worshipped the prestige of European doctors and scientists. Another reason was that the antichristian philosophies of rationalism, materialism, Darwinism, moral relativism, and all the other atheistic "isms" mentioned in the first chapter were so prevalent in the academic world of America at that time that anyone with enough charm, wit, and chutzpah, which Freud definitely had, who was willing to publicly challenge traditional Judeo-Christian beliefs with so-called "scientific evidence" was immediately hailed as a wise old sage and prophet. But Freud was no sage or prophet, or even the least bit wise. He was a false prophet and a false messiah, a fraudulent wolf in sheep's clothing leading the ignorant lambs to the slaughter under the clever guise of science, preaching a new liberation from God's moral laws on sexual purity and the guilt that ensues from breaking them. But as many of his critics have since pointed out, Freud was nothing but a fraud.

Sigmund Freud was born Sigismund Schlomo Freud on May 6, 1856, to Jewish parents, Jacob and Amalia Freud, at Freiberg in Morovia (now Pribor, Czechoslovakia). Amalia, Jacob's third wife, was just twenty years old when she married forty-year-old Jacob. Jacob had two sons from his first marriage—Emmanuel and Philip—and would later have five daughters and two more sons with Amalia—Julius, Anna, Rosa, Maria, Dolfi, Paula, and Josef.

In 1860, the Freud family moved to Vienna, Austria, a predominately Roman Catholic city where anti-Semitism was prevalent and growing. At this crucial and formative period in his early life, the biggest influences on Sigmund Freud were his doting young mother, his elderly Roman Catholic nursemaid, and his abusive elderly father. Sigmund's nursemaid, his "other mother" as Freud often called her, regularly took him to Mass and taught him about the Catholic doctrines of repentance and forgiveness of moral sin, until she was abruptly whisked off to jail for petty theft one day, turned in by Sigmund's older step-brother Philip. That was a terrible psychological blow to young Sigmund because he genuinely loved the elderly Catholic nursemaid who, as some historians believe but can't substantiate, had been sexually molesting him at home.

One of the most memorable childhood sexual experiences that Freud had was with his mother. Sigmund's mother, Amalia Freud, was young, attractive, and had a comely figure. At age forty-two, having undergone self-analysis, Freud recounted the time he was sexually aroused by his mother at four years old when he caught a glimpse of her naked body as she was undressing. Freud's memories of his sexual attraction towards his mother and his nursemaid (and later on his sexual attraction for his wife's virginal young sister, Minna Bernays, who came to live with the Freud's in 1896 as a nursemaid and "other mother" number two to Sigmund Freud and his children), became Freud's impetus for the formulation of what he called the Oedipus complex.

Sigmund Freud was a very gifted boy. He graduated with honors in 1873 from the Gymnasium, despite the rise of anti-Semitism in Europe and his reciprocal hatred for the Viennese Catholics and the Church of Rome. Deeply impressed by the evolutionary theories of Charles Darwin and the writings of the poet-scientist Johann Wolfgang Von Goethe (pronounced "Gerta"), author of the famous play *Faust,* Freud decided to pursue a career in medicine. The writings of Goethe, the great-grandfather of Carl Jung, were a tremendous influence on Sigmund Freud throughout his entire life. In Goethe's historic play, *Faust,* Faust makes a pact with a devil named Mephistopheles and then drinks a rejuvenating magic potion in order to deflower a young virgin named Margaret. Like Faust, Freud had a magic elixir of his own, liquid cocaine, which he took for the first time on April 30, 1884, the satanic feast of *Walpurgisnacht,*[38] and thereafter he became a cocaine addict. And like Faust, Freud went on to deflower a young virgin named Minna Bernays, his sister-in-law.

Enrolling at the prestigious Vienna University medical facility, Freud received a very liberal, humanistic education that included the study of controversial Kabbala mystics, modernist theologians, and antichristian philosophers such as Hegel, Feuerbach, Nietzsche, Voltaire, Rousseau, Schopenhauer, as well as British empiricists and naturalists such as Darwin, Tyndall, Huxley, Lyell, Sir William Thomson, and Lockyer. Other famous notables who no doubt had an influence on Freud were Meynert, Benedikt, Janet, Herbart, von Hartmann, Carpenter, Mandsley, and Fechner.

At Vienna University, Freud became the protégé of the eminent physiologist, Ernst Brucke, a scholar about the age of Sigmund's father who had an uncanny physical resemblance to Sigmund Freud with respect to Freud's piercing and scowling eyes. Brucke's atheistic empiricism, scientism, and positivism was assimilated with enthusiasm by

Freud. Other atheistic scientists who influenced Freud during his time in college were Emil Du-Bois Reymond and Hermann Helmholtz, who maintained that the only forces active in biological organisms were the forces of attraction and repulsion inherent in all matter.

After graduating from medical school in 1881, Freud continued working in Brucke's laboratory on neurological research. In 1882, Freud left research for general practice in Vienna General Hospital, and a couple of years later he began researching the properties of cocaine, a virtually unknown chemical substance at that time. In fact, Freud published a paper called "On Coca," enthusiastically endorsing cocaine as a cure-all for pain, exhaustion, low-spirits, and morphine addiction.[39]

In 1885, Freud obtained a grant to study with the world-renowned French neurologist, Jean-Martin Charcot, in Paris at the Salpetriere mental institute. Freud became fascinated with Charcot's study of hysteria and its remarkable response to hypnosis. To Freud, the idea that hysterical phenomena were psychological in origin was first suggested to him by the fact that Charcot could precipitate hysteria through hypnotic suggestions.[40] During this time with Charcot, Freud seriously began contemplating a career in psychiatry rather than general medicine. Also during this time, Freud first postulated the link between neurotic anxiety and the repressed sexuality of Victorian Europe of the late nineteenth century.

In 1886, Freud married a young Jewish girl by the name of Martha Bernays. Although he and Martha were Jewish, Freud strictly forbade any observance of religious ceremony in their home, much to the consternation of his new wife. It is interesting to note that the apartment house in which the young Freud and his wife lived in was built upon the ruins of Vienna's Ring-Theater where four hundred people were burned to death in a horrible fire during a performance of Offenbach's *The Tales of Hoffman* in 1881, the year Freud graduated from medical school.[41]

Shortly after his marriage to Martha, Freud befriended a Jewish ear, nose, and throat specialist by the name of Wilhelm Fliess—at that time a reputable doctor but now known to be a "crank and a pathological numerologist."[42] Fliess was Freud's most adamant supporter at a time when Freud's promotion of psychoanalysis was being met with skepticism and criticism by the prominent scientists and doctors of the late 1880s and early 1890s because of his XXX-rated theories. Even Martha viewed her husband's ideas on psychoanalysis as "a form of pornography."[43] As was later discovered by historians, it was Wilhelm Fliess, as Freud's homosexual lover, who inspired Freud with many of his weird pornographic ideas on psychosexual development, infantile sexuality, bisexuality, and homosexuality.[44]

Freud believed that homosexuality "was not a crime, not a sin, not a disease, not a form of madness or a symptom of decadence,"[45] and spent a great amount of time psychoanalyzing historic figures whom he thought were homosexual or bisexual like himself. Freud published a long psycho-biographical essay in 1910 entitled "Leonardo da Vinci and a Memory of His Childhood," which psychoanalyzed Leonardo's supposed homosexual affairs with his younger male students. Freud viewed his own homosexual attraction towards his younger colleagues Adler, Ferenczi, and Jung, and his friend and lover Wilhelm Fliess, as similar to Leonardo's, even though it has never been proven that da Vinci was gay.

Freud's growing confidence in his psychological theory of the sexual etiology of neuroses became even stronger in the early 1890s. In a letter written to his friend Wilhelm Fliess in 1893, Freud stated that he believed that nervous disorders and mental illnesses had only one primary cause—sexual repression—rather than many: "It may be taken as well known that neurasthenia [nervous physical symptoms] is a frequent consequence of an abnormal sexual life. The assertion, however, which I should like to make and test with observations is that neurasthenia can in fact *only* be a sexual neurosis."[46]

To fend off these nervous disorders Freud advocated "free sexual intercourse," and "the use of condoms" as a method of birth control in heterosexual couples and a preventative against sexually transmitted diseases for both heterosexual and homosexual couples.[47]

In *Studies on Hysteria,* a now-famous collection of theoretical papers and case histories published by Joseph Breuer and Sigmund Freud in 1895, one particular case of Breuer's stands out as the primary impetus to the development of Freud's psychoanalysis. It involved a young, sexually naive Jewish women by the name of Anna O. who developed a variety of very bizarre symptoms characteristic of demonic possession, including paralysis of the limbs, memory lapses, sleepwalking, rapid mood swings, speaking in foreign tongues not know to her, hallucinations of black snakes, skeletons, and skulls, feelings of being tormented by demons, suicidal impulses, and the presence of multiple personalities. Breuer noted that by making Anna remember her past traumatic events and the time when her pathological symptoms first occurred, he could talk her through her debilitating symptoms, causing them to magically "disappear," but only temporarily, as they always came back. This new, groundbreaking form of psychotherapeutic treatment was nicknamed the "talking cure" by Anna O.,[48] but was referred to as "psychoanalysis" by Freud. An actual case of demonic possession, therefore, became the foundation of Freudian psychology.

Before Freud formulated his universal theory of the Oedipus complex, he had originally claimed that psychoneurotic anxiety in adults stems primarily from their sexual molestation in childhood by fathers or father figures. Freud called this original theory of his the "sexual seduction theory." Freud, as it is generally believed, formulated his sexual seduction theory in an attempt to explain the origin of his own psychological neuroses, which evidence suggests resulted from his own father's sexual molestation of him and his brothers and sisters. Just how many of Freud's brothers and sisters were molested is unknown. In a letter written to his homosexual friend Wilhelm Fliess on February 8, 1897, a year after his father died, Freud likened the symptoms of his patients' hysterical headaches to the oral sodomy of children by their fathers, his included, which are "characteristic of the scenes where the head is held still for the purpose of actions in the mouth. . . . Unfortunately, my own father was one of these perverts and is responsible for the hysteria of my brothers (all of whose symptoms are identification) and those of several younger sisters."[49]

Historian have thus theorized that Freud's hatred of his father stems from Jacob Freud's molestation of young Sigmund and his brothers and sisters, and was the primary reason why Freud hatred God the Father so much. As with many famous historical figures who were either atheistic or tyrannical dictators—such as Hitler, Stalin, Nietzsche, Feuerbach, or Marx—the father was either absent or abusive. Hatred for one's own father has been consistently shown to be one of the biggest contributing psychological factors for the rejection of God.

After abandoning his original sexual seduction theory in 1905, Freud published *Three Essays on the Theory of Sexuality,* outlining his new belief in infantile and childhood sexuality, and the universal stages of normal childhood psychosexual development, which included the resolution of the Oedipal complex. The inspiration behind Freud's publication of his *Essays* came from four different sources: 1) the psychoanalysis of his own childhood fantasies towards his mother and nursemaid 2) the fear and hatred he had for his abusive father 3) from his interpretation of the sexually symbolic content of his adulthood dreams 4) case studies of his hysterical young female patients Anna O., Katharina, Elisabeth Von R., Lucy R., and Dora.

Freud divided childhood development into several stages. Beginning at about age one, the nursing infant primarily receives sexual stimulation and erotic gratification through the suckling of the mother's erect nipples. This is the first stage of Freud's theory of psychosexual development, which he called the "oral stage." During the anal stage, from one to three years of age, the toddler becomes sexually aroused and stimulated by the maturation of his or her anal sphincters and the control over the expulsion or retention of the feces. At three to five years of age, the child enters the phallic stage, and the child's sexual attention becomes focused on the pleasurable sensations and excitation of the genital area. This stage is characterized by the fascination of the male penis by both little boys and little girls, and the beginning of, or the increased frequency of, masturbation and sexual fantasies of having sex with the opposite-sexed parent or sibling. The Oedipal desires little boys have for their mother elicits fear in them that the father will cut off their penis, which Freud called "castration anxiety." Little girls, on the other hand, go through an Electra complex of wanting their fathers to have sex with them, and instead of having castration anxiety like little boys, they are jealous of boys because they don't have a penis, which Freud called "penis envy." Freud believed that the establishment of the Oedipal complex, and its subsequent resolution by the boy's realization that his penis was safe from his father's knife, was crucial for normal psychosexual advancement into the next childhood sexual phase, the "latency stage."

The latency stage in children, Freud said, is a quiescent period between ages five and thirteen, where sexual and aggressive energies are turned towards learning, same-sex play, and development of one's superego. This is a period of same-sex bonding and completely normal homosexual experimentation for both little boys and little girls with persons the same age or older. Many gays and lesbians have used Freud's theory of the latency stage as justification for their early homosexual attractions and hyperactive proclivities. Finally, according to Freud, the homosexually-oriented latency stage is followed by the "genital stage of older adolescence and young adulthood." This last stage of psychosexual development is a time of separation from the parents, intensification of libidinal drives focusing on non-incestual, non-homosexual relationships, which culminates in marriage and the production of offspring to perpetuate the species, according to Darwinian principles.

Pathological personality traits or "complexes," according to Freud, can be traced back to disruption of any of these five stages of childhood psychosexual development. Narcissism, excessive optimism or pessimism, demandingness, dependency, envy, and jealousy all result from disruption of the oral stage. Orderliness, obstinacy, stubbornness, frugality, paranoia, ambivalence, messiness, defiance, rage, and sadomasochism result from non-resolution of the anal stage. If you ever heard anyone say someone was "anal retentive," this is what it means. All kinds of mental disorders, according to Freud, derive from the unresolved Oedipal conflict of the phallic stage. Adult homosexuality stems from an unresolved childhood Oedipal involvement characterized by an unfulfilled longing for a strong father-son relationship, together with a strong maternal identification, plus a "fixation" at the anal stage and/or latency stage. Agoraphobia and stage fright results from a repressed desire to be an exhibitionist, and schizophrenia and psychosis result from the sexual libido turned inward (narcissism).

With the growing acceptance of Freud's sexual theory, in December of 1908 Freud received an invitation from Stanley Hall, the president of Clark University in Worcester, Massachusetts, to present a series of lectures on his new psychosexual psychiatry. Invited along with him were Sandor Forenzi and Freud's protégé, Carl Jung, the young Swiss genius from Zurich whose father and uncles were prominent and respected Protestant ministers. Looking towards the future, Freud saw in Jung, his eventual successor, an excellent opportunity to win over many Gentile converts to his atheistic beliefs on psychoanalysis, which, up until now, was considered primarily a Jewish theory and practice.

The satanic parody of God the Father and God the Son was obvious in the duo of Freud and Jung. As the founder of atheistic psychoanalysis, Freud was the Jewish "godfather" of psychoanalysis, while Jung, a former Christian, was the Gentile "godson."

Elated to have Jung as his adopted godson and heir apparent, Freud was ever mindful of the story of young Oedipus killing his father, and grew paranoid over time of having the same patricidal rebellion against him by Jung. On two occasions the fearful Freud fainted in Jung's presence when conversations drifted to talk of death. The first time was during Freud's "peat bog corpses conversation" with Jung on August 20, 1909, and the second was three years later at the Psychoanalytic Congress in Munich when conversation turned to the Egyptian Pharaoh Amenophis IV, who had removed his father's name from all Egyptian monuments.

And as Freud had so feared, the split between him and Jung over his sexual theory of neuroses would soon come about. During their trans-Atlantic voyage to America in 1908, Freud told Jung of a dream that he'd had that involved Freud's wife and his sister-in-law Minna Bernays. When Jung pressed Freud for more details so that he could interpret his dream, as Freud had taught him to do, Freud suddenly became suspicious of Jung and refused to go any farther; afraid of revealing his relationship with his sister-in-law. This would have revealed to Jung the real motive behind Freud's theory of the universal Oedipus complex: Freud's own perverted incestual desires. "I could tell you more, but I cannot risk my authority," Freud said to Jung, who immediately became suspicious of some kind of hanky-panky going on between Freud and Minna. Jung recalled fifty years later in his autobiography that it was at that moment when he realized that Freud was nothing but a fraud. "At that moment he [Freud] lost it altogether. The sentence burned itself in my memory; and in it the end of our relationship was already foreshadowed. Freud was placing personal authority above truth."[50]

As Jung realized, Freud's psychosexual theory of the Oedipus complex was not the key that unlocked the hidden secrets of the human psyche, but was only a justification for Freud's own sexual desires for his mother and sister-in-law. Before the establishment of the universal Oedipus complex, Freud had postulated the sexual seduction theory because of what his perverted father had done to him and his brothers and sisters. But then along came Freud's comely young sister-in-law, Minna Bernays. Since childhood Freud had incestual desires and dreams for his mother and nursemaid, and now he had them for his sister-in-law. Naturally Freud reasoned that incestual desires and hatred towards one's father had to be universally experienced by all males, or else he was sick. Trying to rationalize his own deviancy, Freud said, "It is the fate of all of us, perhaps, to direct our first sexual impulse towards our mother and our first hatred and our first murderous wish against our father. Our dreams convince us that this is so."[51]

In a 1912 treatise entitled "On the Universal Tendency to Debasement in the Sphere of Love," Freud wrote that true peace of mind only comes when every man acknowledges his incestual desires for mother and sister. "It has an ugly sound, and it is also paradoxical, but nevertheless it must be said that whoever is to be free and happy in love must have overcome his deference for women and come to terms with the idea of incest with mother or sister."[52]

Freud's Assault on God and Moses

After Freud finished with his theory of the Oedipus Complex and had nothing new to say about the subject, beginning in 1907 until the time of his death in 1939, the rapidly-aging Jewish psychoanalyst shifted his attention away from the psychoanalysis of incestual child/adult sex to the psychoanalysis of the Jewish and Christian religions. As a self-avowed atheist, Freud was convinced that to understand religion one must stand completely outside of it as an unbeliever like himself. In the psychoanalysis of religion, Freud attributed the idea of gods and demons to the "projections" of the unconscious mind, which are nothing more than illusionary personifications of unconscious complexes in the struggle between our instinctual Darwinian drives and our environment. In his first major work brutally attacking religion, entitled "Obsessional Actions and Religious Practices" (1907), Freud compared the sacred acts and ritual practices of believers to the obsessive/compulsive actions of neurotics: "People who carry out obsessive actions or [religious] ceremonials belong to the same class as those who suffer from obsessive thinking, obsessive ideas, obsessive impulses and the like. Taken together, these form a particular clinical entity, to which the name of 'obsessional neurosis' is customarily applied."[53]

According to Freud, all religions were a form of universal neurosis: "In view of these similarities and analogies one might venture to regard obsessional neurosis as a pathological counterpart of the formation of a religion, and to describe that neurosis as an individual religiosity and religion as a universal neurosis."[54]

In four essays published in the periodical *Imago* between 1912 and 1913, collectively titled *Totem and Taboo,* Freud attempted to trace the anthropological origin of religion, particularly that of Christianity, to the pagan practice of totemism—a thesis proven by others since Freud to be totally without basis. According to Freud, totemism is a ritual of primitive Native American, Australian, and African tribes whereby certain sacred animals (symbolic of God) cannot be killed or eaten except when the animal is made a victim of ritual sacrifice (symbolic of communion). The totem animal protects and gives warnings to the tribes, prophesies the future, and, if a dangerous animal, spares the tribal members from harm. To Freud, this totem animal represents the father of the clan who was killed and eaten by his jealous sons because the powerful father wanted all the women to himself. After killing him in their jealousy, the sons mourned their father's absence, so they instituted the totem ritual in the father's honor. Freud believed that this "exaltation of the father" by the murderous sons is at the root of all religious worship, principally Christianity:

> The psychoanalysis of individual human beings, however, teaches us with quite special insistence that the god of each of them is formed in the likeness of his father, that his personal relation to God depends on his relation to his father in the flesh and oscillates and changes along with that relation,and that at bottom God is nothing other than an exalted father.[55]

According to Freud, original sin in Christian "mythology" represents the Oedipal elimination or "murder" of the primal father and the subsequent establishment of the worship of the son over the father in his place, with the celebration of the murder of the father by the son in the ritual of the Holy Eucharist:

> In the Christian doctrine, therefore, men were acknowledging in the most undisguised manner the guilty primaeval deed, since they found fullest atonement for it in the sacrifice of this one son [Jesus]. Atonement with the father was all the more complete since the sacrifice was accompanied by a total renunciation of the women on whose account the rebellion against the father was started. But at that point the inexorable psychological law of ambivalence stepped in. The very deed in which the son offered the greatest atonement to the father brought him at the same time to the attainment of his wishes *against* the father. He himself [Christ] became God, beside, or, more correctly, in place of, the father. A son-religion displaced the father-religion. As a sign of this substitution the ancient totem meal was revived in the form of communion, in which the company of brothers consumed the flesh and blood of the son—no longer the father—obtained sanctity thereby and identified themselves with him. Thus we can trace through the ages the identity of the totem meal with animal sacrifice, with the anthropic human sacrifice and with the Christian Eucharist, and we can recognize in all these rituals the effect of the crime by which men were so deeply weighed down but of which they none the less feel so proud. The Christian communion, however, is essentially a fresh elimination of the father, a repetition of the guilty deed.[56]

Freud's most controversial psychoanalytical assault on religion came in 1927 with the release of his best-selling book *The Future of an Illusion*. In it, Freud argued that religion is an "illusion" or fanciful wish in response to the harsh realities of this world. "I have tried to show that religious ideas have arisen from the same need as have all the other achievements of civilization: from the crushingly superior forces of nature."[57] He also said that:

> These, [psychological origins of religious ideas], which are given out as teachings, are not precipitates of experience or end results of thinking: they are illusions, fulfillments of the oldest, strongest and most urgent wishes of mankind. The secret of their strength lies in the strength of those wishes. As we already know, the terrifying impression of helplessness in childhood arouses the need for protection—for protection through love—which was provided by the father; and the recognition that this helplessness lasts throughout life made it necessary to cling to the existence of a father, but this time a more powerful one."[58]

Always the consummate empiricist and science worshiper, Freud went on to say that there are no absolute truths outside of what can be proven experimentally by science, especially in regards to the "delusions" of religious doctrines:

> Having thus taken our bearings, let us return once more to the question of religious doctrines. We can now repeat that all of them are illusions and insusceptible of proof. No one can be compelled to think them true, to believe in them. Some of them are so improbable, so incompatible with everything we [scientists] laboriously discovered about the reality of the world, that we compare them—if we pay proper regard to the psychological differences—to delusions . . . But scientific work is the only road that can lead us to a knowledge of reality outside ourselves.[59]

In his 1932 paper entitled "The Question of Weltanschauung" (Ideology), Freud boasted of atheistic science's inevitable triumph someday over religion:

> The scientific spirit, strengthened by the observation of natural processes, has begun, in the course of time, to treat religion as a human affair and to submit it to a critical examination. Religion was not able to stand up to this. What first gave rise to suspicion and scepticism were its tales of miracles, for they contradicted everything that had been taught by sober observation and betrayed too clearly the influence of the activity of the human imagination. . . . It seems not the case that there is a Power in the universe which watches over the well-being of individuals with parental care and brings all their affairs to a happy ending. On the contrary, the destinies of mankind can be brought into harmony neither with the hypothesis of a Universal Benevolence nor with the partly contradictory one of a Universal Justice.[60]

Warning of the inherent "dangers" of religion, Freud predicted that the salvation of mankind would ultimately come through "science," not God:

> Our best hope for the future is that intellect—the scientific spirit, reason—may in process of time establish a dictatorship in the mental life of man. . . . Whatever, like religion's prohibition against thought, opposes such a development, is a danger for the future of mankind.[61]

In 1938, the same year that Nazi Germany marched into Freud's homeland of Austria, and a year before he died in England after fleeing Nazi persecution, Freud, full of Judas-like remorse for having abandoned his orthodox Jewish brethren, launched a scathing, hateful counterattack on his increasingly vocal critics by excommunicating and murdering Moses, the symbolic "Oedipal father" and moral law-giver of the Jews. In his scandalous book entitled *Moses and Monotheism* (1939), Freud contended that the Moses of the Old Testament was not really a Jew but an Egyptian who had attempted to convert the ancient Hebrews to monotheism. Unwilling to renounce their polytheism, Freud said, the Jews murdered the Egyptian Moses and didn't adopt his monotheism until several centuries later when another person, also named Moses, forced them to worship a strict single deity.

Needless to say, this insulting, revisionist-history blasphemy of orthodox Judaism created a tremendous uproar in the European Jewish community already under siege by Adolph Hitler and the Nazis, who murdered three of Freud's sisters in Nazi concentration camps. Unfortunately for Freud, he never lived long enough to see the defeat of the Nazis who had murdered so many of his friends and family because he died the same year his *Moses and Monotheism* was published. His book undoubtedly was an arcane attempt by Freud to discredit the one man that reminded him most of his guilty conscience, Moses, and to justify his incestual desire for his mother and sister-in-law because the Egyptian pharaohs practiced incest and they considered themselves gods.

Sigmund Freud, a.k.a. the "Jewish Scourge of Science," died on September 23, 1939, from oral cancer, the result of his life-long cigar addiction, or should I say his "fixation."[62] It is just and right that Freud should die of oral cancer because, like the Second Beast of the Apocalypse, so many malignant lies against God came out of his mouth. Mimicking the three unclean spirits that resembled frogs coming out of the mouths of the dragon, the beast, and the false prophet of Revelation 16:13, out of Freud's cancerous mouth came forth a cynical and deceitful demonic spirit bent on discrediting Christianity, ridiculing Judaism, confusing moral values, destroying natural inhibitions, legitimizing sexual depravity, and replacing God's moral laws with the false religion of a lewd and lascivious theory he had the audacity to call science.

Regrettably for society, exaggerated veneration, aggrandizement, and overblown reverence of Freud's name has made Freudian psychoanalysis the dominant secular religion within the profession of mental health today. It is estimated by the American Psychiatric Association and American Psychological Association that over half of psychiatrists and psychologists in America are atheists and don't believe in the literal existence of supernatural spirits. Psychotherapists have become so enamored with the name of Freud that they've made him into an icon—Freud the Magnificent, the New Moses, the New Messiah, High Priest, Jewish Hero, and Scientific Savant. But in reality, Freud's perverted gospel of atheism and deceit attacks clean social values, promotes moral relativism, licenses illicit behavior and drug abuse like cocaine, distorts Judeo-Christian ethics, proclaims sex to be the key to happiness, undermines religious faith, equates a guilty conscience with mental neurosis, promotes occultism (hypnosis and dream interpretation), secularizes medicine, deifies science, encourages infidelity and promiscuity, sexually exploits women and children, legitimizes homosexuality and bisexuality, destroys natural inhibitions, advocates sexual permissiveness in youngsters, teaches hatred towards parents, dwells on painful memories, plants false memories, deludes patients into thinking that they're getting better, fosters over-introspection, makes people suicidal, breaks up families and marriages, and reduces the totality of human behavior and existence to the level of brute animals.

Unfortunately for America, blind faith in Freud's pansexual theory of childhood psychological development has made die-hard Freudians impervious to any substantive charges of malfeasance and moral relativism. Despite their history of failures (the advent of powerful psychotherapeutic drugs has rendered Freudian psychoanalysis obsolete), and their obvious charlatanism, militant Freudians still control most of the key positions of power and importance in the tremendously influential profession of mental health, and have caused immense psychological damage to the fragile minds of mental patients in need of God's forgiveness and spiritual healing. Frustrated non-Freudian, Christian psychotherapists who believe in God and who have tried to accuse Freudians of iatrogenic misdiagnosis, misdirection of treatment, and medical malpractice have been savagely counterattacked with denigration, intimidation, harassment, persecution, and censorship by their militant Freudian enemies who, still holding the reigns of power in psychotherapy, are able to spread the infectious pestilence of atheistic Freudianism to an unsuspecting and gullible public, especially in America.[63]

Freudianism in America

"Americans are Jung and easily Freudened." —James Joyce

Originating in Vienna, Austria, in the early 1880s, Freudianism first came to America around 1906 when Freud's followers from Vienna and other parts of Europe immigrated to North America. The first Freudian to come over to North America was A. A. Brill of New York, then Ernest Jones of Toronto, followed by J. Putnam of Harvard, Smith Ely Jellife of

New York, W. A. White of Washington, Franz Alexander of Chicago, Herman Nunberg of Philadelphia, Sandor Rado of New York, Hans Sachs of Boston, and Karen Horney of Chicago. By the middle of the twentieth century, the number of practicing Freudians in Europe and North America was substantial, and included such famous names in psychology as Erik Erikson, Alfred Adler, Harry Stack Sullivan, Eric Berne, Otto Rank, Melanie Klein, Wilhelm Reich, Erich Fromm, Abraham Maslow, Gordon Allport, Henry Murray, Kurt Goldstein, Andreas Angyal, Gardner Murphy, Kurt Lewin, David Rapaport, and Karl and William Menninger.

Freudianism was enthusiastically welcomed into the so-called "new psychiatry" centers in America (such as Menninger's, Chestnut Lodge, Austin Riggs, Boston Psychopathic Hospital, Sheppard-Pratt, Phipp's Clinic, and St. Elizabeth's Hospital) by Jewish psychoanalysts from the Freudian school fleeing Nazi persecution during the early- to mid-twentieth century. As their numbers steadily increased in the 1940s, 1950s, and 1960s, Freudians held a virtual monopoly over American psychiatry and clinical psychology, usurping the majority of the highest administrative seats and teaching posts in the most prestigious hospitals, medical schools, universities, and psychiatric institutes throughout America.

Freudians gained a success in the United States they never got in Europe because the vast majority of them had a medical degree behind their names, which gave them the credibility that ordinary laymen don't otherwise have. As Americans, we tend to worship the prestige of anybody with a doctor in front of their name (such as Dr. Ruth, Dr. Phil, and Dr. Laura) and foolishly believe whatever they say because of the years of schooling they've completed and the amount of money they make. Doctors, however, are people just like anyone else and can be brainwashed into believing whatever their instructors and administrators say because of the power and authority their peers lord over them, and because of the tremendous career risk student doctors take if they are perceived as being nonconformists. Thankfully, though, the stranglehold that Freudians once held over American clinical psychology and medical psychiatry is not as strong as it was in the past. This is primarily because of all the negative publicity it's received in the media in the last thirty years, particularly by conservatives groups and religious organizations who have managed to expose Freudianism for what it truly is: atheism. Nevertheless, it is estimated that there are currently about twenty thousand trained Freudian psychoanalysts practicing in the U.S.

The major national umbrella organizations for psychoanalysts includes the American Psychoanalytic Association (with its 3,200 members), the American Academy of Psychoanalysis, the American Psychological Association's Division 39 on Psychoanalysis, and the National Association for the Advancement of Psychoanalysis. The largest, oldest, most prestigious, and historically the most important national psychoanalytic organization is the American Psychoanalytic Association (APsaA), which is affiliated with the worldwide accrediting organization founded by Sigmund Freud himself, the International Psychoanalytic Association.

The APsaA was founded in 1911 and currently has twenty-nine accredited training institutes, forty-one affiliate societies, and four affiliate study groups throughout the U.S. The APsaA provides a forum for the exchange of psychoanalytic ideas between the various Freudian associations and publishes the major American psychoanalytic journal *The Journal of the American Psychoanalytic Association.* Its principal function, however, has been the accreditation and maintenance of professional standards within the psychoanalytic institutes, which are reviewed and evaluated periodically by the APsaA through on-site visits, and individual analysts are evaluated for certification as analysts belonging to the APsaA. The powerful and prestigious APsaA, however, plays little part in the regular educational, clinical, or intellectual activities of the individual psychoanalytical institutes.

To become tops in the field of psychoanalysis it is necessary for one to first earn a four-year medical degree, followed by several years of internship and residency in an accredited psychiatric teaching hospital, followed by several more years training in a psychoanalytic institute as a licensed psychiatrist or clinical psychologist, and last but not least, to undergo an extensive period of self-analysis by a senior staff psychoanalyst at one of the top four psychoanalytic institutes in America located in Boston, New York, Chicago, and Los Angeles. The New York Psychoanalytic Institute was the first and, for decades, the most prestigious Freudian institute, which set the model for the others. Then and only then is one considered to be a true member of the "Freudian club."

According to *U.S. News and World Report's* annual survey in 2001-2002 of "America's Best Hospitals," the top ten psychiatric hospitals in America where Freudianism is still being practiced are, in order of ranking: Massachusetts General Hospital in Boston; New York Presbyterian Hospital; McLean Hospital in Belmont, Massachussettes; C. F. Menninger Memorial Hospital in Topeka, Kansas; Johns Hopkins Hospital in Baltimore; University of California, Los Angeles, Neuropsychiatric Hospital; Yale-New Haven Hospital in New Haven, Connecticut; Mayo Clinic in Rochester, Minnesota; Duke University

Medical Center in Durham, North Carolina; and the University of Pittsburgh Medical Center. Although people may think that they are getting the best of care in these prestigious medical institutes, they don't realize that many of their mental health care providers don't believe in God and view those who do with suspicion and condescension.

Godless Freudianism remains the majority ideology among the rank and file of mental health care professionals in America, but atheistic Jungianism is closing in fast.

Carl Jung: The Godson of Atheistic Psychoanalysis

Dr. Carl Jung was a legendary medical doctor and psychoanalyst from Zurich, Switzerland, who from 1907 to 1913 was the most influential member of the Freudian school. Today he is universally regarded by liberal scholars as one of the foremost psychoanalysts of the twentieth century, having authored a massive amount of critically-acclaimed publications on the psychogenesis of mental illness, the psychology of the unconscious, the psychological interpretation of dreams, the classification of personality types such as introversion and extroversion, the structure and dynamics of the psyche, the psychoanalysis of alchemy[64] and the occult and paranormal phenomena, the psychology behind word association, the non-coincidental occurrence of mysterious phenomena (which Jung called "synchronicity"), the psychological symbolism of numbers, the naturalness of femininity in males and masculinity in females, the history of civilization, self-realization (individuation), archetypes, the inherited (collective) unconscious, mythology, the psychology of flying saucer sightings, gnosticism, the psychology of Buddhism, I Ching, T'ai chih, and, most important, the psychology of Western religion. The publication of Jung's voluminous writings, lectures, and teachings have inspired countless numbers of post-modern scholars, theologians, and liberal intellectuals to wholeheartedly embrace and advance Jung's ideology that "Christianity is nothing but myth and symbolism."

Sigmund Freud was so impressed with this young man from Zurich that he proudly referred to Carl Jung as his heir apparent, his "eldest son . . . anointed . . . as my successor and crown prince."[65] Ironically, Jung means "young" and Carl means "man" in German. So smitten was Freud with Jung's potential as his successor that in 1910 Freud insisted that he, Jung, the heir apparent godson of psychoanalysis, be appointed permanent president of Freud's newly founded International Psychoanalytical Association.

In spite of Freud's deep respect and admiration for Jung, these two monoliths of psychoanalytic psychology would part ways in 1913 over their disagreements on the definition of libido, Jung's growing skepticism of Freud's sexual theory of neuroses, and Freud's equally incredulous attitude towards Jung's interest in the investigation and explanation of traditional religion, paranormal, and occult phenomena. Remarking about Freud's attempt to steer Jung away from the psychological study of religion and occultism, Jung recalled:

> I can still recall vividly how Freud said to me, 'My dear Jung, promise me to never abandon the sexual theory. That is the most essential thing of all. You see, we must make a dogma of it, an unshakable bulwark.' . . . In some astonishment I asked him, 'A bulwark—against what?' To which he replied, 'Against the black tide of mud'—and here he hesitated for a moment, then added 'of occultism.'[66]

Contrary to Freud's dogmatic materialism and rejection of anything metaphysical or spiritual, Jung believed that man has both a physical and a spiritual side. Jung, however, didn't believe in the existence of supernatural spirits per se. Jung denied the existence of the God of the Bible and the true divinity of Jesus Christ and believed, instead, in an inherent cosmic divinity dwelling within the mind of every individual (what Jung called our "Christ consciousness").

Similar to Freud's psychological theory of the id, ego, and superego, Jung theorized that the human mind likewise has three distinct parts: 1) the conscious ego, 2) the personal unconscious, and 3) the collective unconscious. The collective unconscious, Jung said, consists of genetically inheritable psychic contents outside the realm of our awareness, common to all but to no one in particular. These common psychological contents are ubiquitous ideas and instincts that Jung, borrowing from Plato, called "archetypes." Archetypes transcend cultural and ethnic differences and explain the similarities between the world's myths, folktales, and religions. For instance, the archetypical idea of God the Father commonly surfaces as symbolic images in our psyches, and is often misconstrued as reality rather than psychic symbolism. Like Freud, Jung believed that the archetypical idea of a supernatural male deity first originated in the unconscious minds of primitive men and women due to a child-like need of protection from the dangers of the world by a strong, yet benevolent father figure. Man created the image of God the Father in his primitive unconscious, so when we think of God or pray to him we are actually communicating with our inner selves: our "divinity within." To realize this is tantamount to a religious epiphany, which Jung called "individuation." We are God!

Although Jung denied the traditional teachings of Christianity and the divinity of Jesus Christ, he did attribute the evils of the twentieth-century to a decline in Christian values and a corresponding rise in scientific rationalism and secular humanism. And although he believed that all religions were just myths, Jung felt that Christianity's ancient mythic symbols were still necessary to insure good social order by giving the ignorant commoner a sense of meaning and purpose to his dull and wearisome life. It was becoming more and more apparent to Jung that with the rise of godless Nazi fascism and Marxist communism, Christianity had lost its credibility and was in desperate need of a spiritual rejuvenation. What was needed, Jung felt, was a neo-Christian revival coming from Freud's new science of psychoanalysis.

Having delusions of grandeur about his messianic importance to the world, Jung was convinced that he was that much-needed spiritual rejuvenator who alone was capable of reviving and rehabilitating Christianity's ancient doctrines and archetypical symbols, by assimilating them into a society-rejuvenating, self-centered New Age psychology (it's important to note that the name "Jung," in German, also means "new" or "rejuvenated").[67] What is strange about Jung, however, is that he, the son of a Protestant minister from a strong Protestant family, would choose Roman Catholicism as the foundation for his rejuvenated, "we-are-gods" Christian psychotheology. Jung's reason for doing so had to have been an insult to his Protestant relatives:

> It [Protestantism] proclaims doctrines that nobody understands and demands a faith that nobody can manufacture. . . . the Catholic Church possesses a richly developed and undamaged world of dogmatic ideas, which provide a worthy receptacle for the plethora of figures in the unconscious.[68]

Convinced of his mission to enlighten the world about the mythology and symbolism of the Catholic Church, Jung devoted the last twenty years of his life to dissecting and psychoanalyzing Christianity in general and Roman Catholicism in particular. Jung's major writings and lectures on these subjects are included in the *Collective Works of C. G. Jung* by Princeton University Press (1967-), particularly volume 9, part I, *The Archetypes and the Collective Unconscious;* volume 9, part II, *Aion: Researches Into the Phenomenology of the Self;* volume 11, *Psychology and Religion: West and East;* volume 12, *Psychology and Alchemy;* and volume 14, *Mysterium Coniunctionis: An Inquiry Into the Separation and Synthesis of Psychic Opposites in Alchemy.*

Aion: Researches Into the Phenomenology of the Self (1951) includes Jung's defense and glorification of evil in one of his most popular works, "The Shadow," and Jung's denial of the divinity of Christ in his works "Christ, a Symbol of the Self," "The Sign of the Fishes," "The Historical Significance of the Fish," "The Ambivalence of the Fish Symbol," "The Fish in Alchemy," "The Alchemical Interpretation of the Fish," "Background to the Psychology of Christian Alchemical Symbolism," and "Gnostic Symbols of the Self." Jung's *Psychology and Religion: West and East* includes his controversial Catholic reductionist works "A Psychological Approach to the Dogma of the Trinity" (1942/1948), "Transformation Symbolism in the Mass" (1942/1954), and "Answer to Job" (1952) which blasts God for putting Job through living hell just to win a bet with Satan over Job's faithfulness in the midst of terrible adversity. *Psychology and Alchemy* (1944) contains Jung's major works on alchemy, and includes "Introduction to the Religious and Psychological Problems of Alchemy," "Religious Ideas in Alchemy," and "Individual Dream Symbolism in Relation to Alchemy." *Mysterium Coniunctionis: An Inquiry Into the Separation and Synthesis of Psychic Opposites in Alchemy* contains Jung's psychological works on his theory of opposites, which includes "The Paradoxa," "The Personification of the Opposites." "Rex and Regina," "Adam and Eve," and "The Conjunction."

In Jung's controversial essay "A Psychological Approach to the Dogma of the Trinity," contained within Jung's book *Psychology and Western Religion,* Jung wrote that the archetypical concept of the Father, Son, and Holy Spirit "sprung forth from the unconscious [mind] of the human race" not in Palestine during the time of Jesus Christ, but in ancient Egypt, Babylon, and Greece, and could "rearise anywhere at any time," as it later did in the Holy Land.[69] For example, Jung wrote that in ancient Egyptian mythology, the trinity of Osiris (god of the underworld), Horus (god of light), and Isis (goddess of nature, wife and sister to Osiris), was preceded by an earlier trinitarian archetype of god-king-ka; where god is the father, pharaoh is the king and son of god, and ka is the procreative power of god in the human mother of pharaoh.[70] Furthermore, Jung said, in ancient Babylon the first known trinity of Anu the sky god, Bel the earth god, and Ea the god of knowledge, preceded the later Babylonian trinity of Sin the moon god, Shamash the sun god, and Adad the storm god.[71] Likewise, in ancient Greece the Pythagorean number symbolism of threes and Plato's cosmology predated the pre-Christian archetype of the Trinity of Father, Son, and Holy Spirit, whom traditional Christians worship as one God.[72] After having citing all these ancient pagan myths as examples of pre-Christian archetypes of the Trinity, Jung stated his true belief that it was ancient

Greek gnosticism, not Judaism, that "ultimately gave birth to Christianity."[73]

In Jung's *Psychology and Western Religion,* Jung wrote that Christianity's archetype of the redeemer-god and Original Man is age-old, and is traceable to as far back as the Purusha (primeval male) of Hindu mythology and Gayomart (primeval man and culture hero) of Persian mythology.[74] Thus, Jewish expectations of the coming messiah, the redeeming Savior-God, were transferred upon the real Jesus, who was merely human and not divine, by the anticipations of his followers in the Middle East: "At a very early stage, therefore, the real Christ vanished behind the emotions and projections that swarmed about him from far and near; immediately and almost without trace he was absorbed into the surrounding religious systems and molded into their archetypical exponent."[75]

Erroneously, messianic Jews awaiting the arrival of the Savior unconsciously elected Jesus as their symbolic-Christ, the archetypical representation of the incarnate God-Man arising out of our collective unconscious since the first man evolved from a monkey. Jesus the God-Man, therefore, is not really God but a symbolic representation of our individual selves: our Christ consciousness within, "since the image of the God-Man lives in everybody and has been incarnated (i.e., projected) in the man Jesus, to make itself visible, so that people could realize him as their own interior *homo,* their self."[76]

The archetypical representation of Jesus Christ as our inner-self, however, was too unrealistic for Jung, who felt that striving for Christ-like perfection is psychologically damaging because our inferior dark side, which Jung referred to as the "shadow," contains some "positive" factors we should not exclude, such as natural instincts and sexual desires, reactions, insights, and creative impulses that, according to Jung, should never be repressed.[77] Therefore, to maintain good psychological health or "wholeness," as Jung put it, our shadowy self must be accepted as a good and positive force rather than an evil force. This was one of Jung's main criticisms of traditional Christianity: the rejection of the shadow within us, which, of course, Christians equate with evil and Satan. To make our shadow more acceptable to Christians, Jung rewrote the Christian doctrine of the Trinity to include, rather than exclude, evil and imperfection. Jung called this new doctrine of his the "Quaternity."

According to Jung, the Christian Trinity of Father, Son, and Holy Ghost is hopelessly incomplete because it omits the "fourth" and perhaps most misunderstood member of the Godhead, who is Lucifer. As the two "Sons of God," Christ and Lucifer personified equal and opposite aspects of the Father's nature (creative and destructive, good and evil) missing in the traditional Christian doctrine of the Trinity. For us to continue to omit Lucifer, Christ's "eldest brother," from the Godhead, Jung said, was a big mistake because the natural order in the cosmos always includes complementary pairs of mutually dependent, inseparable opposites, such as light and dark, summer and winter, good and evil, Christ and Antichrist (the *mysterium Coniunctionis*). "How can one speak of 'good' at all if there is no 'evil'? Or of 'light' if there is no 'darkness,' or of 'above' if there is no 'below'?" Jung said in his defense of Lucifer.[78]

The devil, therefore, should be included as a fourth member of the Godhead, Jung believed, because it was absolutely necessary for there to be an adversary opposed to the will of God the Father, so that man could be redeemed with the Father by the death of the Son on the cross. Lucifer knew this better than anyone else, Jung said. Therefore, he was the most faithful and obedient Son of God; and, hence, the better of the two[79]: "Therefore Lucifer was perhaps the one who best understood the divine will struggling to create a world and who carried out that will most faithfully. For, by rebelling against God, he became the active principle of a creation which opposed to God a counter-will of its own."[80]

Jung defended his new doctrine of the satanic Quaternity against his critics by postulating that modern man can no longer rely on the Christian doctrine of the Trinity for psychological health or wholeness because it's just too out of touch with reality concerning the beneficial existence of our dark side. Said Jung: "This relationship to reality is completely lacking in the idea of the [Christian] Trinity, so much that people nowadays tend to lose sight of it altogether, without even noticing the loss . . . As part of the doctrine of salvation it [the fourth person, Satan] must, indeed, have a saving, healing, wholesome effect."[81]

Jung's main goal as an unbelieving psychoanalyst, of course, was to cast as much doubt and throw as much cold water on traditional Christian beliefs as he could, so as to a create a whole new generation of unbelievers and skeptics like himself. For example, in his essay "Transformation Symbolism in the Mass," first published as a lecture in 1940-41, Jung stated that the archetype of the Roman Catholic Mass, with its celebration of the Eucharist as the body and blood of Jesus Christ, was already "rooted in the human psyche" and practiced long before the advent of Christianity in Near Eastern religions, and in geographically isolated peoples of Central America.[82] For example, Jung said that in the ancient Near Eastern pagan religion of Mithraism, which was popular in Persia, India, Greece, and the Roman Empire from approximately 1400 B.C. until 400 A.D., the pagan god Mithras (god of light, heat, fertility, and mediator between the gods and men) carried the bull-sacrifice to a meal at which bread marked with crosses were laid on the table and eaten.[83] In another example of the Mass being appropriated by Christians from ancient pagan practices, Jung cited the ancient Aztec rite of *teoqualo* (god-eating), in which poppy bread is molded into the form of the god Huitzilopochtli (the war god associated with the sun) and eaten in his commemoration. Yet another pre-Christian archetype of the Mass, Jung said, comes from the supposed visions of the philosopher and Gnostic alchemist of the third century A.D. named Zosimos. Zosimos supposedly had visions of a priest who voluntarily sacrifices himself to be pierced with a sword, scalped, dismembered, and burned until transformed into spirit, and then consumes his own flesh. Psychologically, Jung said, the similarities of the Zosimos visions and the Catholic ritual of the Mass symbolize "the archetype of sacrifice,"[84] in which our mature ego comes into realization of its transcendent self, unites the material world of the flesh with that of the spirit, and then sacrifices its own selfish claims.[85] All these pre-Christian archetypes, Jung said, are what really gave rise to the religion of Christianity, not Jesus Christ himself.

In *Psychology and Alchemy,* first published in 1944, Jung wrote that the occult practices of alchemy and astrology, with their emphasis on the redemption of matter, human nature, and the shadowy "fourth," have been able to achieve psychological wholeness for their practitioners, whereas Christianity has failed: "Whereas in the [Catholic] Church the increasing differentiation of ritual and dogma alienated consciousness from its natural roots in the unconscious, alchemy and astrology were unceasingly engaged in preserving the bridge to nature, i.e., to the unconscious psyche, from decay."[86]

In 1951, Jung published a more detailed psychoanalytical interpretation on the supposed origins and history of Christianity entitled *Aion: Researches Into the Phenomenology of the Self,* with its main theme, of course, being the need for a total revamping of traditional Christianity. A year later in "Answer to Job" [1952], which is included in *Psychology and Religion,* volume 11 of the *Collected Works of C. G. Jung,* Jung plays the part of a criticizing and confrontational psychotherapist to the God of the Old Testament, Yahweh, by answering what he refers to as the "amoral" shortcomings and "contradictoriness" of God as the "sum of all good," exemplified in the Book of Job by God's dark side, which accepts a satanic challenge to take away everything of value from Job, the righteous servant of God, just to test his faithfulness and obedience. Jung roundly criticizes God for doing so, which typifies all those throughout the ages who demand to know why, if God is all good, does he allow evil to exist in the world.

In "Answer to Job," Jung also says that God has a feminine spirit of wisdom named Sophia, who as the "matriarchal bride" and "co-creator of the world" with him at the beginning of time, was abandoned by God when he chose as his new bride Israel. But like the ever-faithful-but-forsaken Lucifer, the abandoned Sophia, who of course to Jung represents godless human science, needs to be brought back into the original Godhead of the Quaternity.

In his 1955 work *Mysterium Coniunctionis,* Jung feared that the unresolved conflict between human science and orthodox religion might never be resolved. Only when the harmful split in the psyche of modern day Christians between the concept of good and evil is resolved, Jung said, which is primarily due to Christianity's unnatural repression of the flesh and the rejection of Satan, the "missing fourth," will psychological healing be obtained. But for Christians to achieve resolution of their psychological schizophrenia, Jung said, they must reintegrate the "alchemical" symbolism of the "conjunction of opposites," such as darkness/light, good/evil, body/soul, nature/spirit, masculine/feminine, and Christ/Antichrist back into their psyches where it belongs.

The Demon-Possessed Life of Carl Jung

> "Woe to those who call evil good, and good evil; who substitute darkness for light and light for darkness; who substitute bitter for sweet, and sweet for bitter! Woe to those who are wise in their own eyes, and clever in their own sight!" —Isaiah 6:20-21

So why was Dr. Carl Jung, a Freudian-trained psychoanalyst from Zurich, Switzerland, so obsessed with reinventing Christianity? Why did he so vehemently insist that Christianity was some sort of mythological fairytale concocted out of the prototypical imaginations of our collective unconscious minds, and that believing in it literally, rather than symbolically, was unwholesome and dangerous to one's mental health? Why did Jung believe that evil was good but that good was evil, that darkness was light but that light was darkness, and why was he so insistent on recognizing Lucifer as the fourth person of the Godhead and elevating mere mortal men to the level of God? The startling answers to these questions become rather clear when one closely examines Carl Jung's bizarre life, a biography you'll never read in any college psychology or medical school psychiatry textbook.

Born on July 26, 1875, Carl Jung was only six months old when his parents moved to Laufen castle and vicarage above the falls of the Rhine in Switzerland, where bodies of drowning victims were swept over the rocks during a disastrous flood, and the cemetery sexton was kept busy throughout the night burying the dead. As an adult, Carl Jung often recalled the fearful nights at the castle listening to frightful sounds of things walking about the house as he lay in bed praying to God for sleep to come. But nighttime prayers to the Lord Jesus, taught to him by his mother, were of no help because of his close association of the name of Jesus with the burials of dead people and the dark, terrifying figures of Jesuit priests in "women's clothes" coming down the hills of the cemetery.

At the tender and impressionable young age of three, Jung had his first dream of the devil that haunted him for the rest of his life. In that dream, young Carl found a stone-lined hole in the meadows near his home that had a rock stairway leading far down into the earth. As any curious young boy would do, he descended the rocky stairs and at the bottom found a doorway with a round arch, closed off by a sumptuous green curtain. Pushing the curtain aside, Carl entered a long, dimly lit stone chamber with red carpeting leading up to a low platform where a golden throne sat. On the golden throne was a fifteen-foot phallus: a huge, erect penis the width of a massive oak tree, the likes of which he had never seen before. At that very instant his mother's voice rang out from above in his dream, "Yes, just look at him. That is the man-eater!"[87] Jung's dream, needless to say, scared the living daylights out of him, and memories of it would reappear in his childhood whenever the name of Jesus was mentioned.[88]

At six years old, Carl's parents took him on trip to Arlesheim. Passing a Catholic church for the first time, the curious youngster slipped away from his mother and peaked inside the open door at the congregation celebrating Easter services. As fate would have it, suddenly and inexplicably, Carl stumbled on a step and busted his chin wide-open on a piece of iron fence. For years afterwards, Jung was unable to enter another Catholic church or come close to a priest without having panic attacks of falling, blood, or scary Jesuits in black robes.[89]

Not long after his fatefull fall at church, strange and frightful nighttime happenings, as Jung described them, "had begun to thicken":

> From the door to my mother's room came frightening influences. At night Mother was strange and mysterious. One night I saw coming from her door a faintly luminous, indefinite figure whose head detached itself from the neck and floated along in front of it, in the air, like a little moon. Immediately another head was produced and again detached itself. This process was repeated six or seven times . . . At that time I was sick with pseudo-croup, accompanied

> by choking fits. One night during an attack I stood at the foot of the bed, my head bent back over the bed rail, while my father held me under the arms. Above me I saw a glowing blue circle about the size of the full moon, and inside it moved golden figures which I thought were angels. This vision was repeated, and each time it allayed my fear of suffocation. But the suffocation returned in the anxiety dreams. I see in this a psychogenic factor: the atmosphere of the house was beginning to be unbreathable.[90]

Jung's father and uncles were Protestant ministers, so it was taken for granted that someday Carl would become one too. But as the months went by, Jung's interest in divinity classes and going to Protestant worship services steadily declined, but not the study of the "exotic" religions of the East, such as Buddhism and Hinduism, which he discovered in his father's private library, and which he said were "an inexhaustible source of interest." Other interests such as a love for nature and make-believe caves, and a fascination with fire began to grow in young Carl's soul.

When Jung turned eleven, the recurring thought of a God totally different from his father's gloomy and frightening "Lord Jesus" began to interest him enough that he began praying to it. As it happened, not long afterward, while on his way to the cathedral square after school one day, a dreadful thought tried to enter Carl's mind while he was contemplating the beauty of the sky and the sunlight reflecting off the cathedral roof:

> The world is beautiful and the church is beautiful, and God made all this and sits above it far away in the blue sky on a golden throne and . . . Here came a great hole in my thoughts, and a choking sensation. I felt numbed, and knew only: 'Don't go on thinking now! Something terrible is coming, something I do not want to think, something I dare not even approach. Why not? Because I would be committing the most frightful of sins. . . . The most terrible sin is the sin against the Holy Ghost, which cannot be forgiven. Anyone who commits that sin is damned to hell for all eternity.'[91]

Desperately trying to block the sacrilegious thought trying to force its way into his mind, the torment became so unbearable that after three nights of hardly sleeping a wink, Jung's resistance was finally broken by total exhaustion:

> Sweating with fear, I sat up in bed to shake off sleep. 'Now it is coming, now it's serious! I *must think.* It must be thought out beforehand. *Why* should I think something that I do not know? I don't want to, by God, that's sure. But *who* wants me to? Who wants to

> force me to think something I don't know and don't want to know? Where does this terrible will come from? And why should I be the one to be subjected to it? I was thinking praises of the Creator of this beautiful world, I was grateful to him for this immeasurable gift, so why should *I* have to think something inconceivably wicked? . . . Oddly enough, I did not think for a moment that the devil might be playing a trick on me. The devil played little part in my mental world at that time, and in any case I regarded him as powerless compared with God. Hence there was no question in my mind but that God Himself was arranging a decisive test for me, and that everything depended on my understanding Him correctly. I knew, beyond a doubt, that I would ultimately be compelled to break down, to give way, but I did not want it to happen without my understanding it, since the salvation of my eternal soul was at stake.' God knows that I cannot resist much longer, and He does not help me, although I am on the point of committing the unforgivable sin. In His omnipotence He could easily lift this compulsion from me, but evidently He is not going to. Can it be that He wishes to test my obedience by imposing on me the unusual task of doing something against my own moral judgment and against the teachings of my religion, and even against His own commandment, something I am resisting with all my strength because I fear eternal damnation? Is it possible that God wishes to see whether I am capable of obeying His will even though my faith and my reason raise before me the specters of death and hell? . . . I dare not trust my own reasoning as far as that. I must think it all through once more.[92]

Reasoning to himself that the irresistible thought trying to force its way into his mind was God's will, that God would give him "grace and illumination," Jung's resistance was overcome and he willingly gave in to these overwhelming thoughts:

> I gathered all my courage, as though I were about to leap forthwith into hell-fire, and let the thought come. I saw before me the cathedral, the blue sky. God sits on His golden throne, high above the world—and from under the throne an enormous turd falls upon the sparkling new roof, shatters it, and breaks the walls of the cathedral asunder.[93]

At that very moment a tremendous burden was lifted from Jung's mind and he was at peace, a sign he mistook as God's grace. What puzzled Jung, however, was why God defecated on his own cathedral:

> That, for me, was a terrible thought. But then came the dim understanding that God could be something terrible. I had experienced a dark and terrible secret. It overshadowed my whole life, and I became deeply pensive.[94]

Reflecting on this irresistible and sacrilegious thought, Jung reasoned that God crapped on his own church because the traditions of men, the Bible, religious institutions, and dogmas or doctrines do not bind him. "God Himself had disavowed theology and the Church founded upon it."[95] God was free to do as he pleased, even if it meant contradicting himself on his Ten Commandments of morality. What was most important to God was that you have the courage to obey whatever he asks you to do, even if it bothers your conscience. You see, Jung reasoned, God can't be confined to churches or religions, but is found in the "mountains, the rivers, lakes, trees, flowers and animals." God is nature and is in everything and is everything! Trying to limit God to the traditional ideas of Christianity denied the true nature of God, which his father and his eight uncles (all of whom were Protestant parsons) knew nothing about. The fabulous Gnostic secret that Carl now had, that only he was elected to know, was that God had a dark side, a shadow.

Thereafter, Jung began to doubt everything that his father and his form of Christianity stood for, and church for him became "a place of torment." Carl's first communion was a terrible disappointment: the bread was "flat" and the wine was "sour." There was no life in going to church and receiving communion, only death:

> Slowly I came to understand that this communion had been a fatal experience for me. It had proved hollow; more than that, it had proved to be a total loss. I knew that I would never again be able to participate in this ceremony. 'Why, that is not religion at all,' I thought. 'It is an absence of God; the church is a place I should not go to. It is not life which is there, but death.'[96]

Not long after the renunciation of his father's Christian faith, Jung's grades in school began showing remarkable improvement, and he succeeded in reaching the top of his class by age fifteen, despite frequent bouts of depression and an increasingly violent temper. It was during this time of outstanding grades and inward spiritual battles that Jung began to sense that he had two persons living inside him. That second person he called "personality no. 2":

> Somewhere deep in the background I always knew that I was two persons. One was the son of my parents, who went to school and was less intelligent, attentive, hard-working, decent, and clean than many other boys. The other was grown up—old, in fact—skeptical, mistrustful remote from the world of men, but close to nature, the earth, the sun, the moon, the weather, all living creatures, and above all close to the night, to dreams, and whatever 'God' worked directly in him.[97]

Jung's mother also possessed a second personality that often manifested itself in the night. Jung spoke affectionately of her personality no. 2 as the "natural mind," capable of seeing things as they truly are, having psychic "insight" into unseeable or unknowable things:

> In the course of my life it has often happened to me that I suddenly knew something which I really could not know at all. The knowledge came to me as though it were my own idea. It was the same with my mother. She did not know what she was saying; it was like a voice wielding absolute authority, which said exactly what fitted the situation.[98]

Unaware that he was falling deeper and deeper into demonic possession, Jung's personality no. 2 was luring him into a curiously sympathetic preoccupation with the devil, who up until this time had no special appeal to him. Out of the blue one day, Carl's mother suggested that he read Goethe's *Faust.* a famous story and play about a man named Faust who sold his soul to a devil named Mephistopheles so that he could have material success in life and a young maiden's virginity. It is very important to note that Jung's paternal grandfather, also named Carl Gustav Jung, was, according to family records, the illegitimate son of Goethe's mistress, Sophie Ziegler Jung. Sophie was the wife of Jung's great-grandfather, Franz Ignaz Jung, when she supposedly "met" many writers, Goethe included, at the Mannheim Theater in Germany.[99] This bloodline relationship to Goethe provides an excellent explanation for why Jung was chosen by Satan to be one of his minions.

As the years progressed and Carl's knowledge increased, he became extremely well versed in the teachings of liberal philosophers. While in college, Jung studied the philosophies of the ancient Greeks, Hegel, Leibniz, Carus, Hartmann, Schopenhauer, Kant, and Nietzsche. And because of his love for nature which he equated with God, Jung's academic interests outside of philosophy included the study of zoology, paleontology, geology, prehistoric archeology, evolution, and anatomy, which eventually led him into the field of medicine.

When Jung turned twenty-three, strange things began to happen again at home while on summer break from school. From the dining room a loud explosion, not unlike a gunshot, revealed that the solid walnut table from the dowry of his paternal grandmother (the cheating wife of Carl Gustav Jung mentioned earlier), had amazingly split against the grain from edge to center for no apparent reason.[100] Two weeks later, the steel blade of a bread knife inexplicably shattered into pieces and caused another ear-shattering report, coming this time from the cupboard.[101]

Intrigued by the cause of these strange events, Jung began attending the seances of his mother's relatives, the Preiswerks (German for "Praise Works"), who were deeply involved in the occult practice of necromancy, conjuring up spirits of the dead.[102] The medium at the Preiswerks' seances was Carl's fifteen-year-old cousin, Helly Preiswerk, who frequently went into cataleptic trances and conjured up the spirits of the dead (including her grandfather spirit-guide), spoke in altered voices, diction, and tongues (sometimes in the third person), and experienced visions of demons, premonitions, out-of-body experiences, levitation, table turning, and automatic writing episodes (sometimes in reverse order or transposed sentences).[103]

Fascinated by these occult events, yet scientifically skeptical of their origin, Jung kept detailed records of Helly's seances for two years and used them in his M. D. dissertation entitled "The Psychology and Pathology of So-Called Occult Phenomena," published in 1902. Jung used the pseudonym "Miss S. W." for Helly to protect her identity and his family's reputation.[104] Irreversibly prejudiced by his rationalistic medical training and immersion in Freudian psychology and Darwinian evolution, Jung concluded that none of these paranormal phenomena were caused by anything metaphysical, supernatural, or demonic, but were all a result of a type of mental anxiety where Helly's unconscious psyche autonomously split off psychic fragments of her sexually repressed memory-images, disguising them as other personalities capable of taking control of her motor functions and consciousness. This is called anamnesis by modern secular psychiatrists. This coming into the unconscious of sexually-repressed memory-images, which Jung termed "cryptomnesia," was common to horny young girls undergoing puberty. Jung also noted that Helly's trances were consistent with other so-called medically-related symptoms such as somnambulism (sleep-walking), hypnagogic hallucinations (the half-awake, half-asleep drowsiness preceding full sleep), epilepsy, hysteria, auto-hypnosis, nervous exhaustion, and various other physical diseases that had nothing to do with the devil or demonic possession. Everything could be explained by science.[105]

According to Dr. Jung, always the Freudian-trained skeptic, visions, apparitions, and communication with spirits are projections of unconscious autonomous complexes that have no direct association with the ego.[106] In other words, evil spirits and ghosts are not real, but are psychogenic mental images emanating from the conflicts within one's unconscious. As Jung frequently explained it, "They are delusions of the senses."[107]

In 1909, while on a trip to Vienna to discuss the psychogenic origins of precognition and parapsychology, an interesting thing happened as

Freud was needling Jung for his dabbling in the occult. This strange occurrence was a typical example of what Jung called "synchronicity," a non-coincidental compensatory event resulting from intrapsychological happenings:

> While Freud was going on this way, I had a curious sensation. It was if my diaphragm were made of iron and were becoming red-hot—a glowing vault. And at that moment there was such a loud report in the bookcase, which stood right next to us, that we both started up in alarm, fearing the thing was going to topple over on us. I said to Freud: 'There, that is an example of a so-called catalytic exteriorization phenomenon.' 'Oh come,' he exclaimed. 'That is sheer bosh.' 'It is not,' I replied. 'You are mistaken, Herr Professor. And to prove my point I now predict that in a moment there will be another such loud report!' Sure enough, no sooner had I said the words than the same detonation went off in the bookcase.[108]

After the second bookcase bang, Freud truly began fearing Jung. When Jung finally broke away form Freud in 1913, as Freud feared would happen someday, Jung began having visions of floods and fantasies of blood that nearly drove him insane. Jung felt a constant inner pressure, a "demonic strength" as he put it, pressing down on his mind. Determined to find the reason for these attacks, lest he commit suicide, Jung totally surrendered to these inner pressures:

> There was a demonic strength in me, and from the beginning there was no doubt in my mind that I must find the meaning of what I was experiencing in these fantasies. When I endured these assaults of the unconscious I had an unswerving conviction that I was obeying a higher will.[109]

On another occasion, Jung had a vision in which he descended into the abyss of the dead and met an old man claiming to be Elijah, the prophet of the Old Testament, and a beautiful young girl named Salome, the wanton little temptress who asked King Herod for the head of St. John the Baptist on a plate. Both claimed that they had a black serpent living with them that had "an unmistakable fondness" for Jung.[110]

Another vision involved a figure of an old man with horns, a lame foot, and wings who claimed to be Philomen, the well-to-do Christian of Colossae who was a personal friend and follower of St. Paul the Apostle. Jung was told by the spirit of Philomen that there are things or forces in the psyche that produce themselves and "have their own life."[111] Philomen, who was really a familiar spirit from the underworld and not the real Philomen of the New Testament, became Jung's personal mentor and represented to Jung his "superior insight." Later on, the Egyptian spirit named ka of the Egyptian trinity of god-king-ka, which, as you remember, Jung said was stolen by Christians in their formulation of the Trinity, began appearing to Jung. Ka, thought Jung, was the helpful "earth demon" representing the benevolent "spirit of nature" that Jung worshipped so much as God.[112]

Taking a moment here to analyze what was happening to Jung, all these so-called "helpful" spirits of Jung were actually demons in disguise sent from the underworld to deceive mankind into believing that the devil and his demons don't exist, or are at least belevolent and helpful in nature. The Blessed Virgin Mary warned the world in the Secret of La Salette in 1846, which is contained in its entirety in my book *The Woman and the Dragon: Apparitions of Mary,* to be on guard against such deceiving spirits who masquerade as righteous souls to lead men astray from Jesus Christ and his true Gospel. Evidently, Jung never read Mary's message from La Salette:

> Evil books will be abundant on earth, and the spirit of darkness will spread everywhere a universal slackening in all that concerns the service of God. They will have great power over Nature; there will be churches built to serve these spirits. People will be transported [spiritually] form one place to another by these evil spirits, even priests, for they will not have been guided by the good spirit of the gospel which is the spirit of humility, charity and zeal for the glory of God. On occasion the dead and the righteous will be brought back to life. "That is to say that these dead will take on the form of righteous souls which have lived on earth, in order to lead men further astray; these so-called resurrected dead, who will be nothing but the devil in this form, will preach another gospel contrary to that of the true Jesus Christ, denying the existence of Heaven; that is to say, the souls of the damned."[113]

Jung's helpful spirits were really demons in disguise. They were using him to deny the reality of Jesus Christ to the world, helping him concoct nonsensical psychological theories that sounded scientific and true in order to fool the general public into accepting demonic lies about Christianity. The above passage of Mary in La Salette warning the world about the spread of darkness and the masquerading of evil spirits as righteous souls actually happened to both Jung and his niece Helly (whose name, by the way, sounds a lot like a pet name for Hell). Helly frequently went into trances while acting as a medium for the family, and was transported to far away places while her body remained behind in a cataleptic or comatose state, exhibiting the usual symptoms of possessed people such as low blood pressure, extreme pallor, shallow breathing, and frigid temperature. While Helly's spirit was away from her body, other spirits with their own voice, language, name, experiences, and personality, who pretended to be righteous souls who once lived on earth, would speak through Helly's mouth, informing Jung and his relatives during her seances that they were looking after her body while Helly was away so that she would not die. Often times these spirits would recommend to the family to put cold compresses on Helly's forehead while she was away so that when she returned her headaches from the experience would not be so severe. Helly would often complain to her spirit guides that leaving her body always exhausted her physically, and she protested having to go on many occasions.

Jung's data-collecting experience with Helly from 1899 to 1900 would later on assist Jung in his own detailed record keeping of the demons that were in possession of him. While making written record in his little "Black Book" of what his own "helpful" spirits were saying to him, Jung began questioning his purpose for doing so, whereupon he heard a woman's voice, recognized to be a former mental patient of his who was now dead, tell him sternly that "It is art." After angrily telling her it was not art but "nature," Jung got no answer from the woman, so he voluntarily let her take control of his speech centers, whereupon she began to use his voice and speak through him. Jung called this woman-within himself *anima;* the primitive feminine soul of man (a Jungian concept that many psychologists have since used to justify homosexuality in men[114]). Whenever Jung's *anima* appeared and took control over his speech centers, Jung said "It was like an invisible presence in the room." Often times while writing notes in his little Black Book, *anima* would interrupt him with humiliating vulgarities and obscenities, which Jung tried his best to ignore because of his dependence on her as the "mouthpiece of his unconscious." Said Jung in praise of his vulgar *anima* within: "It is she who communicates the images of the unconscious to the conscious mind, and that is what I chiefly valued her for."[115]

In 1916, shortly after a vision Jung had in which his soul "flew away" from his body, the house he and his family were living in at 228 Seestrasse in Küsnacht suddenly became infested with evil spirits. Said Jung:

> It began with a restlessness, but I did not know what it meant or what 'they' wanted of me. There was an ominous atmosphere all around me. I had the strange feeling that the air was filled with ghostly entities. Then it was if my house began to be haunted. My eldest daughter saw a white figure passing through the room. My second daughter, independently of her elder sister, related that twice in the night her blanket had been snatched away; and that same night my nine-year-old son had an anxiety dream.[116]

On Sunday, a couple of days later, Jung's doorbell began ringing on its own while the whole family looked at each other with astonishment and palpable fear:

> The atmosphere was thick, believe me! Then I knew that something had to happen. The whole house was filled as if there were a crowd present, crammed full of spirits. They were packed deep right up to the door, and the air was so thick it was scarcely possible to breathe. . . . Then they cried out in chorus, 'We have come back from Jerusalem where we found not what we sought.'[117]

As soon as Jung took up his pen and began to let *anima* write freely, the doorbell hauntings ceased. Jung then completed in just three nights time a Gnostic work that he titled *Septum Sermones ad Mortuos (Seven Sermons to the Dead),* which Jung said was a "prelude to what I had to communicate to the world about the unconscious."[118]

Jung's haunting episodes at home had the profoundest influence on the evolution of his attacks on Christianity and his formulation of the theory that the unconscious, rather than God, is the primogenitor of all religious ideas. And like Freud, the remainder of his life was spent trying to destroy Judeo-Christianity.

Hauntings of Jung by evil spirits continued into the summer of 1920. While giving lectures in London, the small cottage that Jung was staying at began to have manifestations of being haunted by mischievous ghosts. While Jung was lying semi-paralyzed in bed, the air suddenly became stuffy, filled with noxious odors that smelled to him like an "open carcinoma." On subsequent nights, the hauntings progressively worsened. Sounds of dripping water, furniture creaking, footsteps, rustling noises, animals running about loose in the bedroom, bangings on the walls, all began to rattle Jung who couldn't rest. On the fifth weekend of his stay, Jung awoke again in the middle of the night to the same noises, but this time beside his pillow was the severed head of an old hag, glaring at him with her right eye and the left half of her face missing. Dr. Jung, always the religious skeptic, explained away the ghost-like phenomena as consistent with hypnagogic hallucinations, noises in his ears, sounds of his own heartbeat, subliminal intuitions, and projections of his unconscious.[119]

Another bizarre haunting episode followed not long afterwards. In 1922, Jung bought some land in Bollingen, formerly owned by the Catholic monastery of St. Gall, where he began constructing a house that he nicknamed "The Tower." During the winter of 1924, while alone in the house, Jung began having a dream of hundreds of dark-clad figures, like Wotan's army of the dead, pouring down from the mountainside around his Tower. Jung was awakened from this dream not once but twice by the sounds of footsteps, music, talking, and laughter coming from somewhere outside the house. When Jung got up and peered out the window a second time, no one was there. Now convinced that it was no dream, Jung said to himself, "Why this is a simple case of haunting!"[120] Four years later Jung found a possible explanation for this haunting episode. While excavating an annex to his house, Jung discovered that he had built his house on a graveyard where numerous French soldiers had been buried in 1799.

Several months later, the night before his mother's sudden death in January 1923, Jung had a terrifying dream about a gigantic wolfhound with a gaping maw tearing past him in a dark forest, which he immediately understood from his mythological readings as the "Wild Huntsman's hound" which had been commanded to carry away a human soul to the underworld. The very next day, Jung received the shocking and painful news that his mother, who was possessed just like her son with demons, had just died. Said Jung about this disquieting dream the night before his mother died: "Seldom has a dream so shaken me, for upon superficial consideration it seemed to say the devil had fetched her."[121]

Many other occult phenomena and demonic dreams were yet to come. In 1927, Jung had what he called his "Liverpool dream." The Liverpool dream confirmed to Jung his previous assumption that the self, not God, is the center of man's life. In this dream Jung saw an island in the center of a pool that was surrounded by darkness. The island had a single magnolia tree radiating with light. To Jung, this symbolized that our central self, "is the principle and archetype of orientation and meaning."[122] God is not central in our lives; the "self" is central.

While traveling in India in 1938, a year after he challenged a Yale audience to break the bonds of established religion and accept the challenge of "immediate religious experience," Jung had what he termed his "Holy Grail dream," in which he alone was chosen to swim across an island channel to bring back the Holy Grail. This dream symbolized to Jung the definite need for a spiritual revitalization within Christianity by him returning the Holy Grail, the symbolic "healing vessel" of the world.[123] So much more was Jung convinced of his mission in life as the spiritual rejuvenator of Christianity because of this dream.

While working on his writings of necessary opposites in his *Mysterium Coniunctionis* in the 1940s, Jung had another dream in which his father was an ichthyologist and his mother was an innkeeper for visiting dead spirits. Jung interpreted this dream as an affirmation of his destiny to be the "cure of souls," symbolized in his dream by his father's interest in fish, which to Jung meant the care taking and reeducation of naive and ignorant Christians, and his mother's guardianship over departed souls.[124]

In the 1950s, during his struggle on his work "Answer to Job," Jung had another dream in which his earthly father lead him to the "highest presence," but Jung refused to touch the floor with his forehead while kneeling before the presence, which to him was an important sign of defiance to God the Father. This defiant act symbolized to Jung "the idea of the creature that surpasses its creator by a small but decisive factor."[125]

In addition to all these strange dreams that were directing his destiny as the greatest mythicizer of religion the world has ever known, Jung had a very disquieting near-death experience following a sudden heart attack in 1944. On the precipice of eternity, Jung had a vision of himself leaving his body and the earth and entering a black Hindu temple orbiting the planet. But before he could enter it, the image of a doctor trying to resuscitate him rose up from the earth with a message that there was a "protest" against Jung's premature leaving, so he must return at once. Jung immediately regained consciousness and survived the heart attack, but soon afterward a horrible thought came to him that this doctor would have to die in his stead. Sure enough, the doctor that saved his life came down with a fever and died several weeks later from septicemia.[126]

The Symbolic Death of Carl Jung: A False Prophet of Human Science

Because of the overwhelming evidence of occult and paranormal activity that I've highlighted in the bizarre life of Carl Jung, there can be no doubt that he was chosen by Satan to be a key instrument in the destruction of traditional Christianity. Even Jung himself admitted that he was possessed by a demon that overpowered him. Writing in his memoirs, Jung spoke of having feelings at times that his life was not his own, and that he was being possessed and manipulated by a powerful and evil spirit: "I have much trouble getting along with my ideas. There was a daimon [demon] in me, and in the end its presence proved decisive. It overpowered me, and if I was at times ruthless it was because I was in the grip of the daimon."[127]

For once, Jung was right: the powerful and evil spirit that the Bible defines as the Second Beast of the Apocalypse possessed him. Through

Carl Jung, his false prophet of human science, the Second Beast of the Apocalypse attacked one by one the long-held doctrines and teachings of Christianity. By reinterpreting divine revelation as mere symbolic mythology, Carl Jung destroyed the faith of millions of Christians in the literal truth of the holy sacraments, sacred scriptures, the divinity of Jesus Christ, and the real existence of heaven and hell. Jung's contribution to the development and propagation of the New Age movement, Christian modernism, the homosexual crisis now taking place within the Catholic Church, creation spirituality, feminist theology, goddess worship, inclusive language, neo-paganism, religious mythology, witchcraft, nature worship, secular humanism, pantheism, shamanism, Eastern mysticism, parapsychology, satanism, and the occult cannot be denied.

In his book *Catholicism and Jungian Psychology,* author Marvin Spiegelman gives witness to the tremendous modernistic influences that Jungianism has had on Catholic priests and the religious in the last forty years. In his collection of essays written by liberal Jungian authors, which includes Jungian priests and religious who have undergone psychological training from the C. G. Jung Institutes, of Zurich, Los Angeles, San Francisco, Boston and Chicago, Spiegelman's book proves that many clerics have forsaken their perpetual vows of poverty, chastity, and obedience to Jesus Christ to follow the ways of Mammon and Asmodeus. They are becoming Jungian promoters, counselors, consultants, instructors, psychologists, psychoanalysts, psychotherapists, and philosophical doctors rather than priests and nuns.

The recent sex scandal and cover-up crisis within the American Catholic Church, which is closely associated with the acceptance of liberal Freudian and Jungian values, was prophesied to happen by the Blessed Virgin Mary as far back as 1846 in her Secret of La Salette. The following is an excerpt from Mary's prophesies about the current crisis within the Catholic Church that is extremely pertinent for today:

> The priests, the ministers of my Son, the priests by their wicked lives, by their irreverence and their impurity in the celebration of the holy mysteries, by their love of money, their love of honors and pleasures, the priests have become cesspools of impurity. Yes, the priests are asking for vengeance, and vengeance is hanging over their heads. Woe to the priests and those dedicated to God who by their infidelity and their wicked lives are crucifying my Son again! The sins of those consecrated to God cry out for vengeance, and now vengeance is at the door, for there is no one left to beg mercy and forgiveness for the people. . . .
>
> Woe to the Princes of the Church whose only occupation will be

> to heap wealth upon wealth, and to preserve their authority and their proud domination! . . .
>
> The vicar of my Son [the pope] will have much to suffer, as, for a time, the Church will be the victim of a great persecution: this will be a time of darkness. The Church will suffer a terrible crisis. . . . The Holy Father will suffer greatly, I will be by his side to the end in order to receive his sacrifice. The wicked will make several attempts on his life, but they cannot harm him. . . .
>
> In convents, the flower of the Church will decompose and the devil will make himself like the King of all hearts. May those in religious communities be on guard on the people they must receive, for the devil will resort to all his evil tricks to induce sinners into religious orders, for disorder and the love of carnal pleasures will spread all over the earth. . . . Many convents are no longer houses of God, but the grazing ground of Asmodeus and his like.[128]

The homosexual sex abuse scandal within the American Catholic Church has been ongoing for decades and was predicted to happen many years ago by the Blessed Virgin Mary. But the true scandal is the American Catholic Church's attempt by its heirachy to cover up the abuses, denying their cover-ups, paying out hush money to keep victims quiet, transferring serial abusers to far-away parishes where they could prey again on innocent and unsuspecting parishioners, paying out hundreds of millions of dollars in lawsuit settlements both in and out of court, money which was collected from the faithful laity for charitable causes, neglecting to formulate effective policy in accordance with civil law on members of the American Church who commit crimes, refusing to step down as leaders of the Church when their credibility and authority is no longer accepted by the laity and the non-Catholic American citizens, as in the case of Cardinal Law of the archdiocese of Boston who thought he was "above the law," and refused to resign to the very end when it became obvious that he was complicit in the abuse of minors. The American Church hierarchy's usual response in the past was to put priests on administrative leave and send them to some restful place where they would receive Jungian and Freudian counseling. The American Catholic Church has fallen into the satanic trap of trusting secular human science instead of divine science, which condemns homosexuality and child abuse as abominations and extremely grave sins.

> You shall not lie with a male as with a woman: such a thing is an abomination.[129]

> Whoever causes one of these little ones who believe in me to sin, it would be better for him to have a great millstone hung around his neck and to be drowned in the depths of the sea.[130]

> They exchanged the truth of God for a lie and revered the creature rather than the creator, who is blessed forever. Amen. Therefore, God handed them over to degrading passions. Their females exchanged natural relations for unnatural, and the males likewise gave up natural relations with females and burned with lust for one another. Males did shameful things with males and thus received in their own persons the due penalty for their perversity.[131]

Now the chickens have come home to roost and the secular public and Catholic laity in America is demanding justice and swift action for clerical cover-ups. Accusations of clerical cover-ups have even reached the highest members of the Vatican, including Cardinal Ratzinger and Pope John Paul II, as lawsuits have been brought against the Vatican by American victims and their attorneys. But in defense of Pope John Paul II, who is a living saint, and the good priests and nuns within the Roman Catholic Church who are still in the majority, the Catholic Church isn't the only religious or secular organization to be hit by sex abuse scandals, but it is the one receiving the most attention. Child sex abuse is most common not in the church but in the home, with an estimated one in four children being abused by their father or close family member.

The liberal media often cites priestly celibacy as the main cause of pedophilia, but as anyone who treats cases of child abuse will tell you, the vast numbers of reported cases involve fathers and their own children. Abstinence from sex is not the cause of pedophilia, or else all celibate singles, male and female, secular and religious, would be pedophiles, which is surely not the case. Homosexuals and child abusers who have no children of their own tend to go where the chances of finding victims or meeting people like themselves are the greatest, such as in seminaries, religious orders, church summer camps, day-care centers, youth sports leagues, the Boy Scouts and Girl Scouts, Broadway, Hollywood, the entertainment industry, art schools, ballet, the theater, hair salons, restaurants, gay bars, in health careers and the like. So don't think that priestly celibacy or following Christ's commandment to be pure and abstinent until marriage is not abnormal and does not cause psychological illnesses and sexual deviancy, as Freud and Jung thought. Sexual deviancy and immorality is a personal choice of giving in to the temptations of Satan, not a result of abstinence. Abstinence until marriage is a good and moral practice. If more people practiced it the world would be a lot better off and AIDS would go away.

Through his teachings and the teaching of his many followers, Jung successfully promoted the satanic deception that evil is good and that good is evil, that psychological "wholeness" is healthier than spiritual holiness, and that since we are all gods, self-actualization rather than self-denial and obedience to God our Father is our main goal in life. The Apostle Paul warned us of this, that in the latter days people would become lovers of themselves rather than lovers of God, as is taught and practiced in Jungianism and Freudianism: "But understand this: there will be terrifying times in the last days. People will be self-centered and lovers of money, proud, haughty, abusive, disobedient to their parents, ungrateful, irreligious."[132]

Our fixation on self-actualization, self-esteem, self-centeredness, and self-worship is directly opposed to the commandment of Jesus for self-denial:

> If a man wishes to come after me, he must deny his very self, take up his cross, and follow in my steps. Whoever would preserve his life will lose it, but whoever loses his life for my sake and the gospel's will preserve it. What profit does a man show who gains the world and destroys himself in the process? What can a man offer in exchange for his life? If anyone in this faithless and corrupt age is ashamed of me and my doctrine, the Son of Man will be ashamed of him when he comes with the holy angels in his Father's glory.[133]

To illustrate the tremendous importance we put nowadays on self-centeredness, go to any bookstore and look at the many shelves on self-help and self-esteem, and you will find a plethora of best-selling titles. Thanks to Jung and people like him, we, not God, have become the central focus of our lives.

Freudian and Jungian psychology is the forbidden fruit of modern times, and many are those who eat of it. But Jung's ear-tickling dogma that "we are gods" is the same old clever lie that Satan used to seduce our first parents, Adam and Eve, in the Garden of Eden: "You shall not surely die. For God knows that when you eat of it your eyes will be opened, and you will be like God, knowing good from evil."[134]

As I have proven with an abundant amount of documented evidence, the satanic involvement in the life of Carl Jung is undeniable. But wait, there is still more proof for those of you who don't believe in a literal devil. True to his proud nature and love for himself, the devil left us a hidden reminder that he owned a certain man named Carl Gustav Jung, and that he'd come to claim his minion on a very special day. Carl Gustav Jung, the devil's advocate and denier of Christianity, died at his home in Küsnacht on the sixth day of the sixth month in the year 1961 (6-6-61). It doesn't require a theologian or biblical exegete to see the satanic symbolism of the number of the Second Beast (666) hidden in the date of Carl Jung's death, which, if I may say so without sounding too cynical, is "synchronicity" at its finest. Sorry, Carl. You said you believed in the significance of numbers. Well, the Bible has a special number for antichrists like you in chapter thirteen of the Book of Revelation, and it is 666: "A certain wisdom is needed here; with a little ingenuity anyone can calculate the number of the beast, for it is a number that stands for a certain man. The man's number is six hundred sixty-six."[135]

Joseph Campbell: America's Premier Jungian

> "For the time will come when people will not tolerate sound doctrine, but, following their own desires, will surround themselves with teachers who tickle their ears. They will stop listening to the truth and will wander into myths." —2 Timothy 4:3-4

The pre-eminent comparative mythologist of the twentieth century in America was a liberal secular humanist, mythology scholar, writer, and teacher at Sarah Lawrence College in New York named Joseph Campbell. Acclaimed by many as the world's foremost authority on comparative world mythology, folklore, and legend, Campbell earned his many praises and accolades by being one of the first American mythologists to publicly teach that Christianity is nothing but a big lie, a myth created in the imaginations and dreams of men not unlike the popular fairytales of the Brothers Grimm and Aesop's fables. Unlike Campbell, most American teachers of world mythology during the nineteenth century were far too reluctant to publicly say that traditional Judeo-Christianity was based on a lie, for fear of retribution by the religious community. But by the mid-twentieth century, the insidious specters of Jungianism, Freudianism, scientific rationalism, Eastern mysticism, naturalism, Darwinism, and religious modernism had so inculcated into the core curriculum of Christian seminaries, universities, and religious institutes, that nary a whimper of outrage was to be heard from the mesmerized theologians and faculty members of the renowned Christian institutes. Because of this inexcusable tolerance and outright acceptance of mythicizers of religion like Joseph Campbell by many in the Christian community, millions upon millions of people have lost their faith in Jesus Christ.

Joseph Campbell was the author of many books and commentaries on comparative world mythology and religion, a close friend of the Kennedy's, and a well-known lecturer and radio talk show guest. He appeared in 1985 and 1986 on TV in a series of videotaped interviews on the Public Broadcasting Station with journalist Bill Moyers, an outspoken liberal. PBS taped the interviews and entitled them *The Power of Myth. The Power of Myth* was six hours of conversations between Moyers and Campbell and was shot on location at the George Lucas Skywalker Ranch and the Museum of Natural History in New York. In 1988, *The Power of Myth* interviews were made available to the public on six videocassettes and in a book format entitled *Joseph Campbell: The Power of Myth with Bill Moyers.* Jacqueline Kennedy Onassis was the book's editor, and Doubleday, a division of Bantam Doubleday Dell Publishing Group, Inc., was the publisher, which, coincidently, was then located at 666 Fifth Avenue in New York City.[136]

Joseph Campbell wrote a surfeit of mythological books that claim that all religions, including Christianity, are founded on myths with similarly recurring themes that are formed, like Jung had previously said, out of man's collective unconscious. Campbell's many writings dealing with that theme include: *A Skeleton Key to Finnegan's Wake* (1944), *The Hero with a Thousand Faces* (1949), *Tarot Revelations* with Richard Roberts (1979, 1982), *Transformations of Myth Through Time* (1990), *Mythic Worlds, Modern Worlds* (edited by Edmund L. Epstein, 1993), *Thou Art That: Transforming Religious Metaphor* (Joseph Campbell and Eugene C. Kennedy 2001), and many others.

In addition to the many books written by Joseph Campbell, there are also a number of conversations with Campbell on his belief in the myth of Christianity that have been made into books or published as articles in magazines. They include "An Open Life: Joseph Campbell in Conversation with Michael Toms" (1988), "The Professor with a Thousand Faces" (*Esquire,* September 1977), "This Business with the Gods: Joseph Campbell in Conversation with Frazer Boa" (Ontario, Windrow Films, Ltd., 1989), and several others.

And in addition to these, Joseph Campbell edited a number of mythological works by other authors who likewise claimed that Christianity is just a myth, such as *Myths and Symbols in Indian Art and Civilization* by Heinrich Zimmer (1946), *Myth and Mythmaking,* Braziller (1960), and *Myths,* Mcgraw (1974).

Many famous artists, scholars, musicians, and filmmakers have credited Joseph Campbell's mythological writings and lectures and interviews as inspiration for their creative ideas. Included in their ranks are such notable celebrities as Steven Spielberg, George Lucas, John Boorman, George Miller, the Grateful Dead, Isamu Noguchi, Martha Graham, Richard Adams, Robert Bly, Barbara Myerhoff, and Mirija Gimbutas.

The extremely popular *Star Wars* series produced by George Lucas were based in large part, according to Lucas, on the mythological works of Joseph Campbell. The George Lucas Skywalker Ranch in Nicasio, California, contains a vast library with thousands of volumes of books on fantasy and mythology, including a complete selection of Campbell's works, which Lucas often references for new ideas.

As you probably know, the *Star Wars* movies were based upon the battle between the forces of good and evil, which as Jung and Campbell believed, are opposite poles of the same force, like the Yin and Yang of Taoism. In every *Star Wars* movies some kind of nondescript, impersonal force is the ultimate power in the universe rather than God, whose name is never mentioned. *Star Wars*, with its elimination of a transcendent almighty God, has been a perfect vehicle for the destruction of many young kids belief in traditional Judeo-Christianity.

The "force" in *Star Wars* is based on Eastern mysticism's mistaken belief that we are all part of a universal, semi-divine vast force or energy. This widespread antichristian belief, however, is just bad science. Force and energy are not ultimate reality, as George Lucas, who is a practitioner of Eastern mysticism, would have us believe; they are only the mechanical results of the interactions of matter in a physical universe. Force and energy are not intelligent, or personal, or eternal, or loving and caring, or almighty, or all knowing, or always present as God is. They are merely physical laws of the material universe that God created. Anyone who has ever taken an introductory course in physics knows that force is simply mass times its acceleration ($F=ma$), and that the energy contained within every atom, according to Einstein's theory of relativity, is equal to its mass times the speed of light squared ($E=mc^2$). Force and energy are the physical results of the movement of matter in space during time. They are dependent on the material universe for their attributes and are not the creators of anything.

As the eternal Creator and Prime Mover of the universe, God, who exists in three persons—the Father, Son, and Holy Spirit—created the material universe from nothing and put it in motion, from the largest galaxies to the tiniest atoms. It is God's putting matter into motion that generates force and energy. Despite what George Lucas would have you believe, force and energy have no power and existence outside of God. Unfortunately, scores of unbelieving scientists and educators have dedicated their lives to proving that something other than the God of the Bible, like some kind of non-judgmental cosmic power, represents ultimate reality. Joseph Campbell was one of those persons, as was documented in *A Fire in the Mind* by Stephen and Robin Larsen.[137]

From Catholic to Cynic: How Joseph Campbell Lost His Faith and Wandered into Myths

Joseph Campbell was born in New York City on March 26, 1904, the grandson of poor Irish Catholic immigrants from the County Mayo and Dublin area who, because of the 1845-47 Irish Potato Famine, immigrated to the United States. Encouraged by his dad at an early age, Joseph, who was an altar boy attending a day school convent in New York City run by Catholic nuns, developed quite a fascination with Buffalo Bill, Indian folklore, the totem poles at the local museum of natural history, astronomy, nature, music, and athletics. The Indian legends and myths that Joseph read about, which he thought resembled the virgin birth and resurrection of Jesus Christ, began to instill doubt within Campbell about his Catholic upbringing.

Joseph was a very gifted and talented boy and his grades in school were exceptional, even though he contracted bronchial pneumonia on the day after Christmas in 1917 and was forced to miss half of his freshman year at Iona high school.[138] While convalescing at his family's Pocono mountain vacation bungalow in Pike County, Pennsylvania, Joseph met an old outdoorsman named Elmer Gregor, who told him many wonderful stories about nature and American Indian folklore. Gregor's fabulous camp-fire stories had such a profound influence on Joseph's young mind that he decided then and there to someday pursue an in-depth study of comparative world mythology.[139]

In 1918, Joe enrolled at the private Canterbury Catholic prep school in New Milford, Connecticut, where he was immediately taken under the tutelage of Canterbury's headmaster, Dr. Nelson Hume, who introduced him to world literature and the art of writing. On Friday, February 21, 1919, in his second year of high school at Canterbury, the first ominous warning in Joseph's life that he was pursuing the wrong path took place at his boyhood home in New Rochelle, New York. A fire broke out in the laundry chute of his home, completely destroying his family's house, killing his grandmother, severely burning his father, and completely destroying his Indian books, diaries, and all his keepsakes.[140] Rebounding quickly form this terrible tragedy, Joe graduated first in his class in 1921, and in the fall entered Dartmouth College. Introduced to the secular humanism, naturalism, and Darwinian evolution of college life, it wasn't long before Campbell began to seriously doubt his Christian faith.

It was during this time that Joe began thinking of God in pantheistic terms as being one with nature, revealed to us through science, not religion. Displeased with the easiness of his courses and the lax discipline at Dartmouth, Joe transferred to Columbia College. There he met an English professor named Raymond Weaver, who advised him to take courses in English and comparative literature, in addition to the supplemental anthropology lectures he was attending at Barnard by anthropologist Franz Boas.[141]

Beginning in the summer of 1924, timely "coincidences" started happening to Campbell, drawing him further and further away from his boyhood faith of Roman Catholicism. On a vacation voyage with his family to Europe, Joseph met a handsome young Hindu man in his early thirties by the name of Jiddu Krishnamurti. The Theosophical Society[142] was about to reveal him to the world as the true messiah and the living vehicle of the logos for this age. Joseph was awestruck by Krishnamurti's soft-spoken mannerisms and supposed grasp of spiritual wisdom and understanding, not to mention his purported claims of mystical powers and prophetic visions. The two of them became good friends and would rendezvous again in the future several times. That summer they traveled about England discussing Hinduism, Buddhism, reincarnation, dharmic principles, atman, and the concept of the supreme universal self.[143]

Several books Campbell picked up that summer and in his fall semester at Columbia, which he proudly referred to as his "bibles," strengthened his growing belief that Christianity was just a lie. One of them was a book on Buddhism and Eastern religions given to him by one of Krishnamurti's traveling companions entitled *The Light of Asia* by Edwin Arnold. Another was Sigmund Freud's *Totem and Taboo.* Two more books, James Frazer's *Golden Bough* and Gilbert Murray's *The Rise of the Greek Epic,* both picked up in his second year at Columbia, helped convince Campbell that Christianity's doctrines were borrowed from ancient pagan folktales and myths. Yet another book Campbell read in his second year at Columbia, William Graham Sumner's social psychology book *Folkways,* which suggests that morality is relative to "the convenience of the group," convinced Campbell that all world religions evolved from local myths as a result of ethnic needs to explain and validate unique and individual requirements for maintaining good social order.[144]

In the fall of 1925, Campbell met a psychology instructor named Fenwick Holmes, who became another great influence in his life. Holmes introduced Campbell to Freudian psychology, meditation, hypnosis, altered states of consciousness, and the power of unconscious suggestion. As a result, Campbell began investigating the occult practices of astrology, fortune telling, and seances.[145]

Campbell completed his graduate degree in English at Columbia in the spring of 1927. Studies in Arthurian legend and medieval mythology, which he supplemented with courses in evolution, primitive religion, and

Chinese philosophy, culminated in Campbell's master's thesis, "A Study of the Dolorous Stroke." His thesis was well received by his peers and he was awarded an opportunity to supplement his previous course work in French, Latin, and Spanish by being granted the Columbian Proudfit Traveling Fellowship for the foreign study of Romance languages, German, Medieval Latin, Old French, and Provençal at the University of Paris under Joseph Bedier in the fall of 1927.[146]

Herbert King Stone, an American teacher in Paris, took a liking to Campbell and advised him on what classes to enroll in and offered to help him through some of his most difficult reading assignments, like St. Jerome's *Latin Vulgate.* Another new acquaintance, bookstore owner Sylvia Beach, introduced Joseph to the symbolic writings of the controversial author from Ireland, James Joyce. Joyce's *Ulysses,* which was censored and burned in America and England because of its "blasphemy," was another big influence on Campbell's future career decision.[147]

While in Paris, Campbell met Angela Gregory, an American he'd previously met in New Orleans who was now studying sculpting under the famous French artist Antoine Bourdelle. Campbell was amazed to learn that both of them were good friends with Krishnamurti and that Krishna, whom he hadn't seen in a couple of years, was now lecturing in Paris while having his bust sculpted by Bourdelle. For the first time, Campbell got to hear Krishna speak publicly on Eastern mysticism. Campbell was still continuing to attend Mass in Paris only out of habit, but was thoroughly impressed by Krishna's ability to speak convincingly on what Campbell later called the "truths" of Eastern mysticism.[148]

Campbell's fellowship in Europe was extended another year so that he could study German, Sanskrit, and Indo-European linguistics in Munich. Delighted by his extended stay, Campbell made arrangements to visit Krishnamurti in his castle in Eerde, Holland, in late summer and again in early October before proceeding on to Germany for the fall semester of 1928. In August that year, Campbell witnessed the gathering in Camp Ommen, two miles away from Krishna's castle in Holland, of three thousand people from around the world who had come to see the so-called "messiah," who was declared by the Theosophical Society to be "perfect" in every way.[149]

Campbell and other fans of Krishnamurti should have heeded Jesus' warning on the signs of the end times because one of the omens of the end time would be the rise of false messiahs and false prophets. Jesus warned his disciples to be on guard against such false messiahs as Krishna: "If anyone says to you then, 'Look, here is the Messiah!' or, 'There he is!'do not believe it. False messiahs and false prophets will arise, and they will perform signs and wonders so great as to deceive, if that were possible, even the elect."[150]

Campbell spent the last part of the summer touring Europe with his mother Josephine and his sister Alice. Shortly after visiting the Cathedral of Chartres in September, he accompanied Krishnamurti on his final trip back to Holland from Paris. After leaving Holland and arriving in Munich in early November, he began his classes, which included Sanskrit, the Hindu Vedas, Oriental culture, English literature, and selected books by such famous authors as Goethe, Mann,[151] Spranger, Nietzsche, Freud, and Jung. Of all these works, Jung's writings on the psycho-mythological origins of Christianity made the biggest impression on Campbell's impressionable young mind, especially Jung's *Symbols and Transformations,* which totally convinced Joseph Campbell, now twenty-five, that Christianity was a just as big a myth as all the other pagan religions and beliefs. After reading Jung and the other Aryan atheists, Campbell never again believed in Christianity or attended a Catholic Mass.[152]

Following a brief vacation in Turkey and Greece in the summer of 1929, Joseph returned home to New York. Eager for more books and dreaming of someday becoming a writer of short stories for magazines, the unemployed Campbell spent the next four years enjoying nudism in Woodstock, New York, and voraciously devouring the contemporary literary and educational works of authors such as Sir James Jean, Bertrand Russell, Sinclair Lewis, Eugene O' Neill, Thomas Mann, H. G. Wells, Ernest Hemingway, Theodore Dreiser, Louis Broomfield, Julian Huxley, William Somerset Maugham, Thorton Wilder, H. L. Mencken, Upton Sinclair, Thomas Wolfe, Hamlin Garland, Owen Wister, Brad Whitlock, Henrik Ibsen, Leo Tolstoi, George Bernard Shaw, and Mary Austin.[153]

Campbell had a hard time getting his short stories published because they reflected too much of his enthusiastic embrace of Russian communism and liberal socialist ideology, which didn't sit well with the conservative American publishers of that era. So, in 1932, Campbell experimented with teaching languages and history at his old college preparatory school, but quickly grew disenchanted with the idea of wet-nursing a bunch of post-pubescent teenage boys. Then a golden opportunity came up in 1934 for Campbell to teach at the liberal Sarah Lawrence College for women in Bronxville, New York. Campbell was an exceedingly handsome young bachelor with azure blue eyes and dark wavy hair, so he naturally didn't have to think twice about this opportunity of a lifetime. He accepted the teaching position at Sarah Lawrence College and remained there for the next thirty-eight years.[154]

Having finally discovered his destiny as a teacher of comparative world mythology, Campbell constructed his course curriculum utilizing his encyclopedic knowledge gained from his extensive liberal education and personal travel experiences. Imbuing into his lectures on world mythology his antireligion ideology, Campbell admitted that he intentionally tried to turn his young female students into atheists like him: "When I was teaching I would say I had almost a 50/50 split between Christians and Jews, and now and then Buddhists, Zoroastrians, and people of other kinds. And my first thought was that I am going too relieve them of their religion."[155]

The first thing he taught his students was that all religious beliefs were cyclic and persistent, sort of like the Hindu belief in the reincarnation of the soul,[156] reappearing from time to time in later generations and cultures as modified forms of the same general theme or archetype. To prove his point, Campbell pointed out to his students the many historical examples of reoccurring archetypes of the hero's triumph, death, and resurrection that existed in myths predating the arrival of Jesus Christ. Since these religious archetypes existed everywhere in the world before the time of Jesus, Christianity had to be a myth just like all the other religions. As Campbell often said with a defiantly smirky grin on his face and that nettlesome lispy voice of his: "My favorite definition of religion is 'a misinterpretation of mythology.'"[157]

During his early teaching days at Sarah Lawrence College, two more important mentors entered into Joseph Campbell's life. The first was an Indian guru named Swami Nikhilananda, who rekindled Campbell's earlier interest in Vedic philosophy by introducing him to the Hindu books *Mandukya Upanishad, The Gospel of Sri Ramakrishna, Vedanstrata, Drg Drysa Viveka, Viveka Chudamani* and the *Bhagavad Gita.* Many of Campbell's later ideas on comparative world mythology were drawn from these books on Eastern mysticism.

The second mentor, Heinrich Zimmer, whom Joseph met through Swami Nikhilananda, was to become Joseph's single most influential teacher on mythology—an eminent German scholar of Celtic and Hindu mythology and a personal friend of Thomas Mann and Carl Jung.[158] Extremely impressed with Campbell's breadth of knowledge and scholarly erudition, Zimmer recommended Joe to the megamillionaires Mary and Paul Mellon,[159] who subsequently published Campbell's first literary work in their Bollingen Series of books—a scholarly introduction to Maude Oake's book on American Indian ritual entitled *Where the Two Came to Their Father: A Navajo War Ceremonial.* It is important to note that the Mellons had been in psychoanalysis with Carl Jung in Switzerland in the 1940s, but had to leave because of the war. But before leaving they asked Jung what they could do for him to express their appreciation, and he said that it would be nice to have a foundation for the interpretation of myths and symbols. Subsequently, the Mellons established the Bollingen Foundation, which was named after Jung's home in Bollingen, Switzerland.[160]

Just a year after Campbell met Zimmer, Heinrich died suddenly of pneumonia. Zimmer's widow asked Campbell to edit his lectures in America, which he was delighted to do. Over the course of the next twelve years, the American posthuma of Heinrich Zimmer was published, which included: *Myths and Symbols in Indian Art and Civilization* (1946), *The King and the Corpse: Tales of the Soul's Conquest of Evil* by Heinrich Zimmer (1948), *Philosophies of India* (1951), and *The Art of Indian Asia: Its Mythology and Transformations* (1955). While Campbell edited the American posthuma of Zimmer, Carl Jung edited the German posthuma.

During 1944, Joseph Campbell's own writings were published twice. The first was a thirty-page commentary for a re-edited version of *The Complete Grimm's Fairy Tales,* and the second was Campbell's critically acclaimed *Skeleton Key to Finnegan's Wake* with Henry Morton Robinson, which was an in-depth interpretation of James Joyce's hard-to-decipher *Finnegan's Wake.* These two books were followed five years later by Bollingen's publication of Campbell's tremendously popular best-selling book *The Hero with a Thousand Faces,* the book that skyrocketed Campbell to fame. In his *Hero,* Campbell wrote that all legendary hero-figures, such as Jesus Christ, followed in some way or another the same romanticized, mytho-cosmogenic cycle of the hero's call to adventure, his period of trials, his atonement with the father, and his deification, glorification, and final return to the father. Campbell characterized this supposed reoccurring myth of the hero, which he called the "monomyth," as representative of the psychological battle over the dark side in all of us.

Included in *The Hero's* list of mythological champions were such legendary and historical figures as Jesus Christ, Buddha, Bodhisattva, Mohammad, Moses, Huang Ti, Tezcatlipoca, and the myriad heroes and gods and goddesses of classical Greek and Egyptian mythology. Campbell maintained that the mythological stories behind all these famous characters of antiquity have been scientifically demonstrated by Freud and Jung to have originated not from reality but in the realm of mankind's unconscious dreams and imaginations: "The symbolism of mythology has a psychological significance. Particularly after the works of the psychoanalysts [Freud and Jung], there can be little doubt, either that myths are of the nature of dream."[161]

According to Campbell, Freudian and Jungian psychoanalysis has confirmed that all the world's religions are simply reoccurring myths, which is the key to explaining why religious beliefs continue to survive in modern times: "The bold and truly epoch-making writings of the psychoanalysts are indispensable to the student of mythology . . . Freud, Jung, and their followers have demonstrated irrefutably that the logic, the heroes, and the deeds of myth survive into modern times."[162]

In his *Hero with a Thousand Faces,* Campbell claimed that "mythology is psychology misread by us as biography; history, and cosmology."[163] For example, the character and role of Jesus Christ, the redeeming God-Man, originated in man's unconscious imagination as an archetype of the ubiquitous spirit hero. The supposed existence of the divine nature in Jesus Christ, who Campbell said was just "a harmless country wise man of the semi-oriental past,"[164] is not literally true, but is psychologically representative of the "divine nature within all of us." "We are all Christs!" was one of Campbell's favorite sayings.[165] In the same way, the mythological story of the Blessed Virgin Mary as the Mother of God fulfills the archetypal role commonly seen in European fairy tales of the "helpful crone and fairy godmother."[166]

Over the years, Freud and Jung would play back and forth in Campbell's mind, but it was Carl Jung who had the greatest influence on Campbell's writings. Campbell complained that Freud never had anything new to say, but that reading Jung always gave him new ideas, and that he was the better of the two psychoanalysts when it came to interpreting the myths of religion: "As far as interpreting myths, Jung gives me the best clues I've got."[167]

]n his honeymoon in 1938. Campbell proposed to one of his young students, a minister's daughter by the name of Jean Eardman. Believing wholeheartedly in the symbolic significance of numbers, as did Jung, Campbell married Jean on the fifth hour of the fifth day of the fifth month (Thursday at 5 P.M., May 5) in honor of Thor, the pagan god of thunder in Norse mythology, whose celebrated feast day is the fifth day of the week or "Thor's day." As the Campbell's were hurriedly driving to their honeymoon cottage in the village of Woodstock, and as they were about to turn onto Maverick Road, a slow-moving funeral procession led by a hearse bedecked with the very same flowers that the Campbells had on their marriage altar suddenly cut them off. Ironically, at the same time, Joseph's sister dreamed that he had just died. But it wasn't his time to go yet, for he had much more work to do for the Second Beast.[170]

Following his near-catastrophic sojourn to the Jungian "Holy Land," Campbell in 1954 went on a one-year sabbatical to study the mythology of Near-Eastern and Asian culture and religion. His sight-seeing odyssey, which included some lecturing, took him first to the Holy Land, where he visited Jericho, the Dead Sea, Jerusalem, the Mount of Olives and the Church of the Holy Sepulcher; and then on to the Hindu and Buddhist shrines of India, Ceylon, Cambodia, Thailand, Burma, Hong Kong, Taiwan, and Japan. The most notable of his trip's experiences was his fire walking in Japan with Yamabushi ascetics and his excursion to the Bay of Bombay's cave sanctuary of Elephanta, where he viewed with reverent awe the gigantic statue of the Hindu god, Shiva *Mahesvara.*[171]

In 1959, Campbell's book *Primitive Mythology,* the first of a four-volume series that was entitled *The Masks of God,* was published. Campbell's motive for writing *The Masks of God* series was to be the first to scientifically explain the historical roots of all world religions, from the advent of Stone Age cavemen to modern times, utilizing the knowledge of recent scientific advances in comparative symbolism and mythology, linguistics, comparative world religion, philosophy, psychology, sociology, ethnology, art history, archeology, anthropology, theology, and world geography. *The Masks of God* was to be the beginning of what Campbell called "a natural science of the gods," his attempt to explain the origin of religious ideas based on the prehistoric hunter-gatherers and their experiences in the field and on the hunt, and their observances of astronomy and the cycles of nature.

According to Campbell in his *Primitive Mythology,* the universal similarities in the themes of the various world religions, such as the great deluge, the land of the dead, the virgin birth, and the resurrection, are all accounted for by the continental migration of primitive cultures and their myths, plus the spontaneous development in man's genetically-predetermined collective unconscious of Jungian archetypical motifs.

In 1962, the second volume of *The Masks of God* series, *Oriental Mythology,* was published. *Oriental Mythology* was a continuation of the main theme of *Primitive Mythology,* which focused on the origin of the mythologies in ancient Egypt, Mesopotamia, India, and the Far East; such as reincarnation, transcendence of the ego, and pantheism. Campbell didn't actually believe in the teachings of these religions but liked them a whole lot more than Christianity, which he viewed as way too intolerant: "Clearly Christianity is opposed fundamentally and intrinsically to everything that I am working and living for: and for the modern world, I believe, with all its faiths and traditions, Krishna is a *much* better teacher and model than Christ."[174]

In 1964, Campbell published his third and most critical volume of *The Masks Of God* series, *Occidental Mythology,* in which he attempted to belittle the origins of Judaism and Christianity by claiming that the cruel and vindictive patriarchal monotheism of the Old Testament was first "adopted" and then sexually "inverted," in a gender sense, from the earlier, more charitably-sublime Sumerian mother-goddess religion of the Bronze Age (4000 to 3000 B.C.) in Mesopotamia.[175] In addition, Campbell claimed that the later New Testament accounts of the resurrection of Jesus Christ, his virgin birth, the slaughter of the innocents, the baptism of Jesus in the Jordan, his temptation in the desert, his role as world teacher, his giving the power of the keys to St. Peter, and his prophecies of the last days were not novel historical events unique to first-century Palestinian Jews, but were already pre-existent in similar forms in the myths of ancient Egypt, Greece, Mesopotamia, Persia, and India. This explains, according to Campbell, why the "legend" of the "illusionary Christ" spread so rapidly throughout the region:

> It is clear that, whether accurate or not as to biographical detail, the moving legend of the Crucified and Risen Christ was fit to bring a new warmth, immediacy, and humanity, to the old motifs of the beloved Tammuz, Adonis, and Osiris cycles. Indeed, it was those early myths, filling the atmosphere of the whole Mediterranean, that had furnished the ambience of readiness within which the Christian legend so rapidly grew and spread.[176]

In 1968, volume IV of *The Masks of God* series, *Creative Mythology,* was published, in which Campbell wrote a jabbering and lengthy revisionist history of the period of time between the fall of the Roman Empire and the twentieth century. Campbell's *Creative Mythology* was a not-so-subtle, monumentally erroneous, and arrogant attempt by Campbell to discredit the Christian faith:

> Who accepts today the idea of inherited sin? And since there was no garden of Eden, no Adam and Eve, no Fall, then what is this talk about Redemption, unless by 'Fall' and 'Redemption' the same psychological states of ignorance and illumination are meant that the Hindus and the Buddhists also are talking about? In which case, what happens to the doctrine of the unique importance of the Incarnation and Crucifixion? The whole myth [of Christianity], to make any sense, must be totally reread—with honest eyes.[177]

What the world needs, Campbell argued, is not the old-fashioned religion of orthodox Christianity, but a brand-new, rejuvenated spirituality that speaks to the entire human race in modern language; one that is created by the individual from his own life experiences from the god within each of us. Like Jung, Campbell was in total agreement about his own divinity within and the need for a spiritual rejuvenation within each of us.

In 1972, Viking Press published Campbell's *Myths to Live By,* which was condensed from a series of twenty-five Joseph Campbell lectures at the Great Hall of the Cooper Union Forum in New York City between 1958 and 1971, a famous place of convocation for radical free-thinkers and puffy intellectuals. One of the primary themes of *Myths to Live By* was that, although human science had definitively proven Judaism and Christianity to be myths, no longer to be taken seriously by anyone with more than a "kindergarten education," the moral teachings of these "illusions" were still necessary for maintaining good social order, like Jung had previously said.[178]

In 1974, Princeton University Press published *The Mythic Image,* Campbell's first fully-illustrated book. It contained over four hundred pictorial illustrations from around the world. Campbell attempted to visually prove that the artistic similarities and relationships between the various world mythologies arise not from reality but from man's dreaming unconscious, a rehash of what Jung had previously said. Campbell cleverly arranged hundreds of pictographs, from ancient to modern times, of pagan and Far Eastern iconography that resembles, if one has an active enough imagination, the artistic composition and form of sacred Christian iconography. By using visual comparisons between ancient and modern religions, Campbell hoped, of course, to destroy the credibility of Christianity.[179]

M. J. Abadie was the picture researcher that was assigned by the Bollingen Foundation for *The Mythic Image.* He was later named as Campbell's seminar coordinator for his lecture circuit called the "Open Eye." Abadie was raised in a convent and was a devout Catholic before she renounced her faith at age nineteen after reading Campbell's *The Hero with a Thousand Faces.* Said Abadie about Campbell's *Hero:* "It was exactly what I needed at that point to enable me to break away [from Catholicism]. When I learned that these symbol systems were universal, not specific to Christianity, my eyes were opened wide."[180]

Campbell's last book to be published while he was still alive was *The Inner Reaches of Outer Space: Metaphor as Myth and Religion* by Van der Marck (1986), published one year before he died. In this book, Campbell defines all religions and myths as "metaphor," or figurative language used to connote or imply an imaginative idea or dream rather than denote a literal historical truth.[181] Dreams, Campbell said, produce myths, which produce religions, which can be psychoanalyzed and interpreted by their metaphors to represent truths symbolically and psychologically, but not literally.

In 1987, a one-hour film version of Stuart Brown's *The Hero's Journey: The World of Joseph Campbell,* a compilation of a series of filmed lectures, seminars, and question-and-answer sessions addressed to Campbell by his interviewers and audiences in the early 1980s, premiered in New York at the New Director's New Film's Festival out on the West Coast. Campbell proudly reminisced in one of these filmed interviews how he used to believe in the Catholic faith as a kid, but how he was now able to so easily destroy people's beliefs in Christianity, especially those in religious life:

> I remember when I was a kid in school, when I was reciting catechism. 'Why did God make you?' 'God made me to love him, to serve him, and to honor him in this world and be happy with him

> forever in heaven.' That has to do with *relationship* to God. God did not make me to realize *My godhood* There was a lovely young woman who was a nun and is now no longer a nun because she heard me lecture once. After one of my talks she came up to me and said, 'Do you believe that Jesus Christ was the Son of God?' And I said, 'Not unless all of us are.'[182]

The following year, *The Power of Myth,* videotaped interviews filmed in 1985-86 between liberal journalist Bill Moyers and Joseph Campbell, was made into a book called *Joseph Campbell: The Power of Myth* with Bill Moyers. The following quotes from *The Power of Myth,* published by Doubleday in 1988, shows just how deep Joseph Campbell had sunk into Jung's New Age, antichristian, "I am god, you are god, god is nature" spirituality that is so widespread in Western society today:

> What is a god? A god is a personification of a motivating power or a value system that functions in human life and in the universe—the powers of your own body and of nature.[183]

> Heaven and Hell are within us, and all the gods are within us.[184]

> We are all manifestations of Buddha consciousness, or Christ consciousness, only we don't know it.[185]

> You've got to find the Force inside you. This is why Oriental gurus [like Yoda in *Star Wars*] are so convincing to young people today. They say, 'It is in you. Go and find it.'[186]

> The God is within you. You yourself are your creator.[187]

In his comments on the subject of demons, devils, and the existence of hell, Joseph Campbell roundly criticized Christianity for teaching their literal existence. When asked about the concept of hell and where it came from, Campbell said, "Saint Thomas Aquinas, I guess. I don't know who came up with that horrible thought! Christianity is the only religion with that concept. I know of no other religion with a hell for all eternity."[188]

In making this incredibly outlandish statement, Campbell revealed his extreme bias and hatred towards Christianity by intentionally distorting historical facts. As well read as Campbell was, he surely knew that St. Thomas Aquinas did not invent the concept of hell. The concept of hell is as ancient as man. Every member of every culture has God-given, inborn knowledge of good and evil, of God and the devil, of heaven and hell, of damnation and salvation. That is why when we stand before the judgment seat of God no one will be able to plead ignorance of sin and its consequences—God's commandments and divine truths are written on our hearts. Take the ancient Greeks. Although not privileged with divine revelations like the Jews, they were very astute on the existence of an afterlife reward or punishment. Hell is a reality that is universally known by all peoples. Jesus spoke more about the literal existence of hell and the risks of going there than any other topic. In fact, all three of the major monotheistic religions (Judaism, Christianity, and Islam) teach the literal existence of an eternal hell. It was foolish for Campbell to blame Thomas Aquinas for inventing the concept of hell to scare people into submission to Church authority.

As far as the literal existence of evil spirits, whenever queried Campbell gave psychological explanations for the persistent ideas of devils and demons within are collective unconscious, which was more in line with Freudian and Jungian psychology, the New Age movement, Eastern mysticism, and *Star Wars* than with the truth. Said Campbell: "One of the obsessions, I think, in Christianity, is the Devil."[189] To Campbell, "A Devil is a god who has not been recognized. That is to say, it is a power in you which you have not given expression, and you push it back. And then, like all repressed energy, it builds up and becomes completely dangerous to the position you're trying to hold."[190] Campbell considered a demon or devil to be "a power in you to which you have not given expression, an unrecognized or suppressed god. Anyone who is unable to understand a god sees it as a devil." [191] Campbell also said that "Our demons are our own limitations."[192]

On the subject of an afterlife, where we go when we die, and the meaning of life, Joseph Campbell was a strict Darwinist, in that he believed life has no purpose or meaning except what we give it. To Campbell, "Life is a lot of protoplasm with an urge to reproduce."[193] That's why Campbell would always tell people to do in life whatever pleases them, to "Follow your bliss" as he would often say: "My general formula for my students is 'Follow your bliss.' Find where it is, and don't be afraid to follow it."[194]

What Profit a Man to Gain the Whole World but Lose His Soul?

Joseph Campbell's never-ending attacks on Jesus Christ and his Christian church were greeted with enthusiasm by atheists, agnostics, liberals, modernists, leftist intellectuals, socialists, communists, and New Agers of every ilk. As a result, over the course of his lifetime Campbell was awarded many grants, memberships, and high honors, such as the Proudfit fellowship (1927-28, 1928-29); grants-in-aid for editing Zimmer volumes (1946-55); the National Institute of Arts and Letters grant in literature (1949) for *The Hero with a Thousand Faces;* the Distinguished Scholar Award, Hofstra University (1973); honorary doctorate, Pratt Institute (1976); the Melcher Award for contribution to religious liberalism (1976) for *The Mythic Image;* the National Arts Club medal of honor for literature (1985); election to the American Academy of Arts and Letters (1987); membership to the American Folklore Society, the American Oriental Society, and the American Society for the Study of Religion (pres., 1972-75); and an honorary membership to the American Academy of Psychotherapists from whom he borrowed so heavily.

Campbell seemed to be very happy with his life and had it all: money, fame, and fortune. His ebullient antichristianity made him a celebrity and his gift for gab and the attractiveness of his endless stories won him many loyal fans and converts to atheism. Yes, indeed, Campbell was a very talented man, but the power within him that made him rich and famous was not from above; it was from below, from the dark side of reality. In spite of Campbell's unbelief, the devil and hell are literal realities and there is ample proof of their existence.

In this day and age it is politically incorrect to label anyone as being in league with the devil or being demon-possessed like Carl Jung was. Hardly anyone in science or upper-level academia believes in a literal devil or eternal hell anymore. Such is the effect that atheistic human science has had on our worldview. But in spite of what unbelievers may teach, the devil is a reality and so is hell. As any good jury would do in deciding a difficult case, one must weigh the evidence of the arguments for and against an allegation before reaching a verdict.

In this case, the extra-biblical evidence, for those of you who do not believe in the Bible, for the existence of a literal devil is substantial. The first proof of the existence of the devil lies in the appreciable number of documented cases of demonic possession and Church-sanctiones ritual exorcisms (yes, they still take place in the modern world). The Catholic Church maintains in each diocese a priest who is delegated the diocesan exorcist. When investigating possible cases of demonic possession, the exorcist, in consultation with trained and licensed medical doctors, psychiatrists, and psychologists who have knowledge and experience in these matters, first rules out any possibility that the afflicted person is suffering from a physical or mental disease before making a definitive diagnosis of possession. The most common indicators for demonic possession include, but are not limited to: having multiple personalities with their own name and intelligence capable of controlling the thoughts and voice of the possessed; the frequent use of blasphemous and vulgar language by these same multiple personalities; speaking in foreign languages and tongues not known by the possessed, such as Latin, Hebrew, and Aramaic; knowing past and secret things about those present in the room; having superior intelligence; angrily and powerfully striking or assaulting those nearby in superhuman ways; having extreme hatred and fear of blessed sacramentals (holy water, crucifixes, medals of saints, etc.) and communion hosts which cause great pain and burning sensations to the possessed; the presence of rapid and extreme environmental changes such as frigid coldness, or the violent movement of objects in the room; levitation of the possessed; foul odors in the presence of the possessed; bilocation (being in more than one place at a time); rigidity of the torso which becomes extremely heavy upon lifting; tremendous body strength which is beyond the realms of possibility; contortions of the body beyond normal ranges of motion; the sudden appearance of epidermal writings or claw marks on the body; extremely accurate and lengthy projectile vomiting and spitting; elongation of the tongue; distortion of facial features beyond human recognition; severe changes in vital signs such as a rapid drop in blood pressure and pulse rate; instantaneously blacking out or coming to; blanching of the skin and sharp drops in body temperature. All these medical characteristics are indicative of true demonic possession and involve someone who is possessed, but who wants to be rid of the evil spirit or spirits doing the possessing. Far more common than these kinds of malevolent demonic possession are cases of "perfect possession," in which the demonically-possessed person fully consents to being possessed by evil spirits and doesn't desire being rid of them and, thereby, shows little or no outward signs of violent possession, as was the case with Carl Jung. For those who need proof of the reality of demonic possession, three excellent references are *Hostage to the Devil* by Malachi Martin, *An Exorcist Tells His Story* by Gabriele Amorth, a former Vatican exorcist, and *Possessed: The True Story of an Exorcism* by Thomas B. Allen—the story behind the making of the movie *The Exorcist.*

Further proof of the literal existence of the devil is the tremendous number of spiritually symbolic "coincidences" that defy statistical probabilities if the world were truly created by random chance. I have already told you in the introduction section of this book that Satan, being the "Ape of God," loves to use names and dates that have symbolic meanings to autograph his work. The perfect example of this is the recent bombing of the New York World Trade Center buildings on 9-11.

Everyone in America knows that 911 is an emergency phone number, so the symbolic date of the bombings hasn't gone unnoticed by the vast majority of people in America who see this as an emblematic warning of the existence of evil in the world that is threatening to destroy civilization through radical Islamofascism.

In my first book, *The Woman and the Dragon: Apparitions of Mary,* I mentioned other examples of satanic symbolism, such as the invention of the first workable personal computer, named the "Apple," which had an original price of $666.66. The apple, of course, symbolizes the god-like power of the knowledge of good and evil that Satan offered to Adam and Eve and is now offering to us though the personal computer. We all know the great harm PCs can do, especially via the uncensored smut and evil on the internet, but did you know that the World Wide Web address is 666? W or wau is the sixth consonant of the Hebrew language. There are many more examples of satanic symbolism in the lives of Hitler, Lenin, Stalin, Isaac Newton, Charles Manson (Man's Son), the singer Madonna, and others that I mentioned in *The Woman and the Dragon,*[195] so pick up a copy and read it for yourself. Satanic symbolism is everywhere; one only has to look at the world with open eyes to see it.

Still more proof of the existence of the devil comes from the many prophecies and warnings of the Blessed Virgin Mary about the literal existence of the devil. Mary's Secret of La Salette, France (1846), and her Fatima, Portugal, prophecies (1917) have been 100 percent accurate in their prediction of the evils that would befall the world during the twentieth century and beyond. The most specific of Mary's warnings were the coming of Adolf Hitler and the start of World War II; the rapid rise, spread, and fall of Russian communism; the martyrdom of tens of thousands of priests and religious by Lenin, Stalin, and Hitler; the "Miracle of the Sun" (announced several months ahead of time) in front of seventy thousand eye witnesses in Fatima, Portugal, on October 13, 1917; the coming of a red aurora borealis that appeared in 1938 all over the northern half of the world and heralded the start of WWII; the assassination attempts on Pope John Paul II's life; and the current sexual abuse crisis within the Catholic Church.

Getting back to Campbell, without a doubt he was chosen by the powers of the dark side to help destroy man's belief in Jesus Christ. Satan delights in turning former believers like Campbell into modern-day Judases by using them to serve him instead of God. Just examine the lives of the greatest antichrists in history (Darwin, Nietzsche, Hitler, Stalin, Feuerbach, Russell, Schopenhauer, Dewey, Spinoza, Hume, and all the rest) and you will see how they were all believers in God in their youth, and many had planned on becoming ministers and priests, but the human sciences they learned in school turned them into haters of God rather than servants.

Carl Jung and Joseph Campbell's "we are god" mythology continues to have tremendous appeal to skeptics and unbelievers who doubt the truth of the Bible and who don't want to obey God's commandments. Yet, as I have already said, unbeknownst to Campbell's fans, his life's work was empowered by the dark side. I can make this bold, and I know it will seem to some, incredibly mean-spirited proclamation without apology because of the many telltale demonic signs in the life of Joseph Campbell. The three ominous warnings of death to Joseph Campbell have already been mentioned and need no further discussion. What I will now discuss are the many other signs in Joseph Campbell's life that suggest Satan's active participation in Campbell's quest to discredit Jesus Christ and make a god out of man.

Often times when Campbell gave lectures, unusual and strange things would happen. One time when he was lecturing at the Mann Ranch just north of San Francisco on the power of the Kundalini rising (the serpent power in Eastern religions), a rattlesnake came up the steps of the porch Campbell was lecturing from, curled up, and stayed the entire duration of his lecture. Immediately after Campbell's lecture was over, the snake uncoiled and left on its own. Needless to say, the audience was quite amazed at this highly symbolic event.[196] Recall Jung's familiar spirits Elijah and Salome who said they had a black serpent with them that was very found of Jung. Another time Campbell was in the dark French forest of Bretagne on a foggy day, standing beside a tree he thought was the actual tree in the Arthurian story of Merlin the Magician, and telling a story to a group of friends. In this story Campbell was describing the baying of hounds, when all of a sudden, at that very instant, out of the distant mist came the sound of baying hounds. When the surprised Campbell finished his story, the sun came out and the fog lifted.[197]

In his endless search for the symbolic meaning of everything, Joseph Campbell had an irresistible attraction to the occult in such things as tarot cards, fortune-telling, black and white magic, astrology, and Haitian Voodoo. About the same time as his *Hero with a Thousand Faces* was being published in the late 1940s, Campbell offered to help his wife's friend, Maya Deren (Eleanora Derenkowsky), publish a book on Voodoo worship entitled *Divine Horseman: The Living Gods of Haiti.*[198]

The daughter of a Russian psychiatrist, Maya, whose name means "illusion," became very interested in filming and documenting the strange gyrations and ritualistic dances of Haitian Voodoo worshipers, so she asked Joseph Campbell, known to be fascinated by what he called the "living mythology" of Haiti, to help her write *Divine Horseman.* Assisted by Campbell's expertise on the mythology of primitive cultures, the book's main theme was the psychoanalytical and mythological interpretation of the ceremonial voodoo rituals performed by Haitians in conjuring up the spirits of the dead, whom Haitians consider divine, then allowing them to completely possess or control their bodies (thus the meaning of the term "Divine Horseman"). Campbell explained to Maya, even though she, too, claimed that she had become possessed more than once during the Voodoo ceremonies, that the physical and spiritual manifestations of Voodoo possessions were due to the unconscious psyche reacting to the power of suggestion, and were not really cases of demonic possessions. How blind and ignorant the belief in atheistic human science can make some people.

The Symbolic Death of Joseph Campbell: Another False Prophet of Human Science

Like his hero Carl Jung, Joseph Campbell believed that "symbolism is everywhere," and once you understood the nature of symbolism you, too, would begin to see symbols in all things.[199] For once in his life, Campbell was right. There was plenty of symbolism in Joseph Campbell's life indicating that he, like Jung, was possessed by evil instead of good, especially with the reoccurring coincidences of numbers and dates, especially the date on which he died.[200]

Shortly before Campbell died of a sudden heart attack in the fall of 1987, Jan Lovett Keen, Eugene Kennedy, and Campbell's sister, Alice Campbell Lennining, all had dreadful premonitions of Joe dying. In a dream, about a month before her brother's sudden death, Alice recalled:

> It seemed I went to church and the Holy Supper was being served [Alice had not attended church for many years]. And I thought, 'I'll see how I react to this thing.' And sure enough, I felt very lovely somehow. But after I got back from the pew, I thought, 'I'd better get out of here. I really don't belong, something is happening.' Then I entered an apartment where Jean and Bargy would be, but there was no Joe. The dream bothered me for quite a while after that.[201]

As his sister and friends had feared, Death would soon come to claim Campbell. In August of 1987, Joseph Campbell was hospitalized at St. Francis Hospital in Honolulu with cancer of the esophagus. He was offered the holy sacrament of reconciliation with God by the resident chaplain, Father Kieran Murray, but stubbornly refused, holding on to his belief in atheism even at the hour of his death. Since he had no regrets or remorse about his life's work, and didn't believe in Jesus Christ or a literal heaven or hell, all he asked was that he die in full awareness with his wits about him—a wish that God was about to grant.

Joseph Campbell was discharged from the hospital and, when he got home, he tried going upstairs to work on the second volume of his *Historical Atlas of World Mythology,* but died suddenly of a massive heart attack. This alone was nothing out of the ordinary for a man his age (eighty-three), but the day he died was extraordinary. To the extreme embarrassment of his family and friends, Joseph Campbell, the consummate storyteller and staunch believer in the Jungian symbolism of numbers and dates, died in 1987 on the satanic feast day of the "Eve of Samhaim" (pronounced "Saw een")—better known in American as Halloween—the time when evil celebrates the lifting of the veil separating the underworld damned from the land of the living.

Just thirteen years earlier, the timing of Campbell's death on Halloween had been foreshadowed when Campbell hosted a New Age conference in 1974 at Carter Caves in Kentucky, which, ironically, concluded on Halloween with a ceremonial "descent to the underworld."[202] Campbell's family and friends were very well aware of the irony of the timing of Joseph's death and it caused them much grief and worry about his eternal fate, as well it should have.

Man's Greatest Mistake: Ignoring the Evidence for Jesus Christ's Divinity

The story of Joseph Campbell's lifelong attempt to discredit Christianity and belittle Jesus Christ and his followers is a tragic one full of erroneous assumptions, misinterpretations, intentional disregard of the evidence and facts, and outright fabrications and lies. Campbell's biggest blunder was thinking that since all religions have similar beliefs and reoccurring themes, that none of them are true; and, therefore, all religions are myths. Campbell wrongly assumed that since the Christian doctrines of the virgin birth, the sacrifice of the Son of God on the cross in atonement for our sins, Christ's resurrection from the dead, and his ascension into heaven appear in a variety of similar ways in ancient myths and religions that pre-date Christianity, that the accounts of the life of Jesus Christ in the four Gospels were borrowed from pagan beliefs, and, therefore, must likewise be myths. Here is where Campbell erred. The existence of pre-Christian beliefs throughout the world which seem similar to the historical events that happened in the four Gospels of Matthew, Mark, Luke, and John do not discredit, but rather validate the truth of Christianity. This is the reason why.

Since we are all made in the image and likeness of God and have spirits that come directly from God, it makes perfect sense that our souls remember where they came from and, therefore, retain a spark of divine memory within them of God's eternal truths. That explains our ancient ancestors' foreknowledge and understanding of God's future incarnation, his virgin birth, sacrifice on the cross, death, resurrection, and ascension into heaven, as related in the myths and folktales that pre-date Christianity.

That is why Christianity was so well accepted by people of all races and creeds around the world once Jesus Christ finally arrived. God enlightened our souls to know the messiah before he appeared so that we would have no excuse to deny him when he came. God also enlightened the minds of the prophets of the Old Testament with an uncanny foreknowledge of the virgin birth, life, death, and resurrection of our Savior. In the biblical accounts and stories within the Old Testament, which pre-date Christianity by centuries, there are literally hundreds of very detailed and specific references to the coming of the messiah, Jesus Christ, which were 100 percent fulfilled by his arrival. To all the skeptics who once argued that these messianic prophecies had to have been written by Christ's disciples years after Jesus died because they are so accurate, the discovery of the ancient Dead Sea Scrolls in the caves of Qumran in 1947 have proven that the prophesies of the Old Testament, which foretell the life and death of Christ, especially in the book of the prophet Isaiah, were in existence long before the birth of Jesus.

Like disbelieving in the many hundreds of messianic prophecies and references within the Old Testament, it is a fatal mistake to ignore or disbelieve in the many eyewitness testimonies and historical accounts of the events that truly happened in the New Testament, which accurately document the life of Jesus Christ. The historical evidence and eyewitness testimonies are abundant, but many are those who refuse to believe it, and don't even care to research it. It never ceases to amaze me how people will totally believe in the Greek philosopher Socrates but refuse to believe in Jesus Christ, even though we have much more evidence on the life, teachings, and works of Jesus than we do of Socrates. All we know about Socrates comes from the writings of Plato, Xenophon, Aristotle, and a contemporary critic of Socrates named Aristophanes. Socrates, like Jesus, left nothing in writing. What we know for sure about

Jesus comes from the many, many testimonies from eyewitnesses and contemporaries of the twelve apostles, within the Bible and from without (over thirty-nine extra-biblical sources testify to the life and death of Jesus Christ). In fact, the historical evidence for Christ's existence is so overwhelming that no scholar can reject it without rejecting every other recorded event in history.

Chapter Four

Evidence that Demands a Verdict: Jesus Christ Is God

> "For those who believe, no proof is necessary. For those who doubt, no proof is sufficient." —Unknown

Skeptics and unbelievers often say that all religions are alike, in that none of them are true. Their unfounded allegations and denunciations against Christianity, however, are based solely upon personal biases and prejudices because there is more than enough evidence, historic and scientific, to support the claim of many Christians that Jesus Christ is God. In fact, the evidence for the divinity of Jesus Christ, which comes to us from many sources, is overwhelming. Such evidence includes ancient documents, archaeological discoveries, fulfilled bible prophecies, scientifically documented miracles, ancient relics, and apparitions of saints. Yet in spite of all the evidence, which no other religion has, there are far too many people who refuse to consider any of the facts for Christ's divinity, let alone investigate them. They prefer to remain in the darkness of their unbelief, because to believe in Jesus Christ would mean they'd have to submit to God's authority and commandments. Unbelievers and deniers of the faith would rather be free to do as they please, to be their own god, which is not really freedom but enslavement to the devil so they buy into the satanic lies of Freud, Jung, Campbell, Darwin, and the like.

Historical Documents and Biblical Archaeology

The usual charge that skeptics and unbelievers make to discredit Christianity is that the four Gospels of Jesus Christ were written in the second century by Christians who were not eyewitnesses or contemporaries of Jesus Christ. They argue that Christians made up the stories of Jesus' divinity, miracles, and resurrection to empower themselves and their fledgling religious sect. Recent archeological discoveries, however, have disproved this commonly-heard accusation as totally erroneous. In 1994, Dr. Carsten Peter Thiede, director of the Institute of Basic Epistemological Research in Paderborn, Germany, concluded after years of research that the famous Magdalen papyrus fragment of the Gospel of Matthew, discovered in Upper Egypt in the late nineteenth century, was written not in the second century as skeptics originally claimed, but in the first century by St. Matthew himself around the year 50 A.D., making this the earliest known extant Gospel fragment written by an eyewitness apostle of Jesus Christ.[1]

In addition to the Magdalen fragment of the Gospel of St. Matthew, there are numerous accounts of the life of Jesus written by eyewitnesses and contemporaries who were not disciples of Christ, but who, nonetheless, recorded the miracles of Jesus. The secular historian Thallus, an eyewitness to the crucifixion of Jesus Christ, wrote in 52 A.D. on the darkness and earthquakes that occurred at the time of Jesus' death, exactly as recounted in the Bible. Although the original writings of Thallus have been lost, Julius Africanus, another early historian, recorded what Thallus wrote about the crucifixion, and his writings have survived down through the centuries:

> Upon the whole world came a most fearful darkness. Many rocks were split in two by an earthquake, and many places in Judea and other districts were thrown down. It seems very unreasonable to me that Thallus, in the third book of his histories, would try to explain away this darkness as an eclipse of the sun. For the Jews celebrate their Passover on the 14th day according to the moon, and the death of our Savior falls on the day before the Passover. But an eclipse of the sun can only take place when the moon comes under the sun, how then could an eclipse have occurred when the moon is directly opposite the sun?

Another first century contemporary historian who recorded the period of darkness during the crucifixion was Phlegon. Julius Africanus mentions a quote taken from Phlegon's *Olympiads* about an eclipse of the sun which occurred for three hours at the time of Jesus' crucifixion: "Phlegon records that in the time of Tiberius Caesar, at full moon, there was a full eclipse of the sun from the sixth hour to the ninth hour. . . . It is evident that we do not know of any such events in previous years."

There is plenty of other historical evidence on the life of Jesus Christ that comes not from early Christians, but from non-Christian Jews and Gentiles who lived during the time of the apostles. The highly-respected Jewish historian, scholar, and priest Flavius Josephus (37-100 A.D.), a personal friend of Caesar's and a contemporary of Jesus' apostles, wrote about the life and death of Jesus Christ, verifying the accuracy of the historical accounts within the four Gospels of Matthew, Mark, Luke, and John:

> About this time lived Jesus, a man full of wisdom, if indeed one may call Him a man. For he was a doer of incredible things, and a teacher of such as gladly received the truth. He thus attracted to himself many Jews and many of the Gentiles. He was the Christ. On the accusation of the leading men of our people, Pilate condemned him to death on the cross; nevertheless those that had previously loved Him, still remained faithful to Him. For on the third day He again appeared to them living, just as, in addition to a thousand other marvelous things, prophets sent by God had foretold. And to the present day [around 60-70 A.D.] the race of those who call themselves Christians after Him has not ceased.[2]

Josephus' "Jewish Antiquities" corroborate the historical accounts of the New Testament. One account was on the events surrounding the stoning of the Apostle James, son of Alphaeus and cousin to Jesus on Joseph's side (cousins were called brothers and sisters in Jesus' time):

> Upon the death of Festus [62 A.D.], emperor Nero sent Albinus to be procurator of Judea. But before he arrived, King [Herod] Agrippa appointed Annas to be High Priest. He was the son of the elder Annas. The elder Annas had been High Priest himself for a time. He had five sons who all secured the priesthood. Annas the younger, however, was a brute who observed the ways of the Sadducees who are known as being cold-hearted when they sit in judgment. With Festus dead and Albinus still traveling, Annas thought he could have his own way. Calling forth the members of the Sanhedrin, he brought before them a man named James, the brother of Jesus who was called the Christ, and others with him. He accused them of violating the law, and ordered that they be put to death by stoning.[3]

Another account of Josephus' was on the death of John the Baptist:

> But many of the Jews felt that Herod's disaster was God's judgment on him for his treatment of John, known as the Baptist. Although John was a good man and taught the Jews to worship God, and to lead righteous lives and practice justice towards others, Herod had him executed. John preached that baptism should not be used as a pardon for sins committed, but as a consecration of the body. . . . Great crowds began to gather around John because of his preaching and Herod feared that John's great persuasive power over men would lead to rebellion. Therefore, he decided it would be better to kill him before he caused an uprising. John was brought in

> chains to the fortress of Machaerus and was put to death in that prison. The Jews believed that God avenged John by the destruction of Herod's army.[4]

Many other important writings on the life of Jesus Christ and his disciples have survived from early, non-Christian historians, including those written by Pontius Pilate, Mara Bar-Serapian, Pliny the Younger, Tacitus, Suetonius, Epectitus, Lucian, Aristedes, Galenus, Lampridius, DioCassius, Hinnerius, Libanius, Ammianus, Marcellinus, Eunapius, Zosimus, Celsus, Porphyry, Hierocles, and Julian the Apostate. In fact, so numerous are the historical accounts of the life of Jesus and his disciples that he is the surest of all recorded persons in history—much more than Socrates, Alexander the Great, or Julius Caesar.

Perhaps the most personally involved non-Christian historian was Pontius Pilate, the Roman governor of Judea who condemned Jesus Christ to death. He wrote of the extraordinary events of Christ's crucifixion in an official report to Tiberius Caesar—a report that was often referred to by contemporary historians of Pilate's day. In this long and detailed narrative, after describing the miracles of Christ, Pilate wrote:

> And him Herod and Archelaus and Philip, Annas and Caiaphas, with all the people, delivered to me, making a great uproar against me that I should try him [Christ]. I therefore ordered him to be crucified, having first scourged him, and having found no cause of evil accusations or deeds. And at the time he was crucified there was darkness over all the world, the sun being darkened at mid-day, and the stars appearing, but in them there appeared no luster; and the moon as if turned to blood, failed in her light.[5]

Skeptics have tried in the past to discount Pilate's report of the crucifixion of Jesus Christ by saying that there is no archaeological evidence of Pilate's existence. But that objection was forever put to rest in 1961 when archaeologists in Caesarea Maritama found a stone plaque that was part of the steps leading up to the old Roman Caesarea Theater. The plaque bore the inscription: "Pontius Pilatus, Prefect of Judea, has dedicated to the people of Caesarea a temple in honor of Tiberius."

The existence of Pontius Pilate, verified by archaeology, is also supported by the historical records of Tacitus (the Governor of Asia in 112 A.D.), Justin Martyr (150 A.D.), and Flavius Josephus, all of whom mention Pontius Pilate in their documents.

Like the existence of Pontius Pilate, skeptics once doubted the existence of the Jewish high priest Caiaphas, who turned Jesus over to Pilate to be condemned. But recent archaeological discoveries have likewise proven that Caiaphas was a real person. In 1990, construction workers south of Jerusalem discovered the ossuary box of Caiaphas, the high priest of the Temple, which contained the actual remains of Caiaphas. The inscription on the ossuary box read "Joseph son of Caiaphas," which was the full legal name of Caiaphas the High Priest, according to the records of Flavius Josephus.[6]

Another extremely fortunate discovery of an ancient reliquary box related to the Bible was recently announced in the United Sates in October of 2002 by *Biblical Archaeology Review*. A first-century limestone bone box with the inscription "James, the son of Joseph, brother of Jesus" was recently discovered in ancient Israel.[7] The authenticity of James' ossuary box, which was displayed at the Royal Ontario Museum in Canada the same year, is supported by the unusual addition of the name of Jesus to the burial inscription. Jesus must have been someone of great importance that needed to be added, since in those days it generally sufficed to inscribe on the ossuary box only the decedent's name and that of his father's. Much debate is currently being waged in Jewish circles over the authenticity of the additional inscription, and only time will tell which side will win out.

Regarding the biblical account of the burial and resurrection of Jesus Christ in Matthew 27:16-28:15, archaeological evidence for Jesus' rising from the grave also comes from the famous "Nazareth Inscription" of Caesar. An edict engraved in marble by either Tiberius Caesar, who was Roman emperor at the time of Jesus' resurrection, or from Claudius Caesar who ruled shortly afterward, was found in Nazareth, the birthplace of Jesus, in 1878. Caesar's edict was a direct result of the Jews wanting the tomb of Jesus sealed shut and an armed guard stationed in front of the tomb so that his disciples would not come and steal Jesus' body and falsely proclaim that he had been raised from the dead. In spite of the sealed tomb and the armed guard maintaining a cordon around the burial site, Jesus' resurrection took place as he promised it would, and the frustrated and worried Jewish Sanhedrin bribed the Roman guards into swearing that thieves had stolen Jesus' body. Because of the false testimony of the Jewish Sanhedrin and Roman guards that Christ's disciples had stolen his body and then claimed he had risen from the dead, Caesar subsequently made it a capital crime to disturb a sealed sepulcher. Read the Nazareth Inscription of Caesar after Christ's resurrection:

> Proclamation of Caesar: It is my desire that graves and tombs remain sealed for the benefit of those who have made them and their children, family members and their religion. If, however, anyone accuses that another has either destroyed them, removed the buried, or has with ill intent, taken them to other places in order to wrong them, or removed the sealing on other stones, I

> order that person to be brought to trial, just as a man should respect the gods, so also with regards to men, for all should respect the buried. It is therefore forbidden for anyone to disturb them. Should this edict be violated, the offender is to be sentenced to capital punishment on the charge of violating a sepulcher.

Evidence that the tomb of Jesus Christ was empty and that the Sanhedrin had tried to cover it up by saying his body was stolen by Christ's disciples while the Roman guards were asleep[8] is corroborated by historical records outside of the Bible. Trypho, an early Jewish apologist and historian, in a dialogue with St. Justin Martyr, an early Christian saint (110-165 A.D.), speaks of "one Jesus, a Galilean deceiver, whom we crucified; but his disciples stole him by night from the tomb, where he was laid when unfastened from the cross, and now deceive men by asserting that he has arisen from the dead and ascended into heaven."[9]

Still more evidence for the resurrection of Jesus Christ exists on the walls of the Roman catacombs during the time of the great persecutions of Christians (64-310 A.D.), where early Christians frequently depicted the resurrection in paintings.

Biblical Archaeology: Old Testament and New

Biblical archaeology, both Old Testament era and New, perhaps does more to prove the truth of Judeo-Christianity than anything else. No other religion has more archaeological evidence to support its beliefs and historical records than Judeo-Christianity—not Islam, not Hinduism, not Buddhism, not any of them. For example, it was long believed by skeptics that the biblical accounts of King David, the ancestor of both Mary and Joseph through whom Jesus claimed royal descent, were fictitious, since no archaeological evidence existed with King David's name on it.

However, in 1993 an inscription of King David was found on a fragment of a monument known as the Tel Dan fragment, dating back to the ninth century B.C. in the ancient city of Dan, which mentioned King David and his dynasty that lasted approximately one hundred years after his death. Up until then, no archeological records existed outside of the Bible mentioning the name of King David, and many skeptics were convinced that the biblical story of David was just a myth. The inscription on the Tel Dan fragment refers to the "House Of David," the "King of Israel."

In addition to the Tel Dan fragment, a fragmented Moabite stone called the Mesha Stela, which was first discovered in 1868, was pieced together in1995, and the words "House of David" was discovered in line thirty-one, further proving that King David truly existed. Other archaeological evidence for the existence of David comes from the Pool of Gibeon mentioned in the biblical story of King David, which was discovered by a team of archaeologists in 1956.

Critics of the Bible have doubted for years the existence of the nation of Israel and the other Jewish kings mentioned in the Old Testament. But again, archaeological evidence keeps proving the Bible right and the skeptics wrong. The finding of a stone inscription on the Mernetptah Stela in 1896 from Egypt confirms that Israel was established both as a people and as a nation centuries before King David, just as the Bible records. The Mernetptah Stela is a 7½-foot stone inscription discovered in the temple of Pharaoh Mernetptah (1213-1203 B.C.), at Thebes in Egypt. The pharaoh Mernetptah[10] launched an invasion on the Jews inhabiting the west bank area of Canaan, and he recorded, probably inaccurately for propaganda purposes, the event on the stone with an inscription that read: "Israel is laid waste; his seed is not; Palestine has become a widow for Egypt."

In addition to the evidence for the existence of the nation Israel and for King David, biblical archaeology, which began in the nineteenth century, has so far confirmed the following kings of Israel and Judah mentioned in the Bible: Omri, Ahab, Jeroboam II, Jehu, Pekah, Hoshea, Ahaz, Ahaziah, Uzziah, Hezekiah, Jehoiachin, and Manasseh.

As the years have past, more and more evidence of the accuracy of the Bible has been discovered. It has proven the existence of hundreds of kings, peoples, cities, and nations recorded in the Bible that were once thought to be myths, including the Babylonian kings Hammurabi and Belshazzar, the lost kingdom of the Hittites, the ancient walls of Jericho, and the Sumerian and Babylonian accounts of the great flood of Noah. In the country of Jordan, in the ancient area known as Moab at an excavation site known as the Deir 'Alla, 119 broken plaster fragments were found inscribed with an ancient text written in Aramaic, the language of Jesus, with the words "Warnings given by Balaam, son of Beor. A seer of the Gods." Balaam was long thought to be a mythological character, but these fragments verified his existence and the accounts of Balaam in the Book of Numbers. In 1897, archaeologists found the pools of Bethesda and Siloam, also thought to be mythological, that were mentioned in the Gospel of John.

New Testament characters have also been verified by biblical archaeology. For instance, archaeologist working on a first-century road discovered the inscription "Erastus, Commisionar of Public Works."

Erastus was the friend and disciple of Simon Peter mentioned in Romans 16:23 and 2 Timothy 4:20 as director of Corinth's public works. As recently as 2004, what is widely believed to be the cave of John the Baptist was discovered near the Jordan River in Israel by a father and his daughter while they were on a hike, complete with stone baptismal font and pictures of the Baptist on the walls. It is the most spectacular archaeological find in years.

Of all the evidences of Judeo-Christianity, biblical archaeology is unique in that it doesn't require belief in what other people say; rather, it provides visible, tangible proof of reality. In fact, so many are the discoveries of archaeologists that verify the accounts of the Bible that it would take volumes of books just to document them all. There are many credible books and magazines out there on biblical archaeology; one only has to be willing enough to seek them out and read them. One excellent book is *Bible Believers Archeology: Historical Evidence that Proves the Bible* by John Argubright. And, of course, don't forget the highly respected scientific magazine *Biblical Archaeological Review*.

Fulfilled Bible Prophecies

In all the writings and stories of Buddha, Confucius, Lao-tse, the Dali Lama, Krishna, and Muhammad, you will not find a single example of a predicted prophecy come true. Even in the Koran, the book that Muslims revere as the inspired word of God, there are absolutely no examples of a prophecy come true, except Muhammad's promise to return to Mecca, which was self-fulfilling like General Douglas MacArthur's "I will return" promise to the Filipinos in WWII. One of the best ways to judge which religion is true and comes from God, who is all-knowing and all-seeing, is through fulfilled prophecies. The Bible has not just one or two fulfilled prophecies, which by chance could have come true, but thousands of specific prophecies that have all come true.

In the Old Testament alone, there are over two thousands specific prophecies come true, like the destruction of Tyre and Babylon and God's vow never to let them be rebuilt again. George W. Bush's recent war in Iraq, which ousted Saddam Hussein, who had promised to rebuild the ancient city of Babylon and actually had architectural models and plans for its rebuilding, prevented the nullification of God's vow. For those of you who criticize Bush for invading Iraq, he may just have well been doing the will of God to keep Babylon from being rebuilt and spreading her terror to Israel all over again.

In the Old Testament, there are literally hundreds of messianic prophecies that have all come true in the person of Jesus Christ, especially in the Book of Isaiah, which prophesies in detail the life, passion, and death of Jesus Christ. So accurate, in fact, is the Book of Isaiah that critics used to claim it must have been written by Christians long after Jesus died. But that objection was forever laid to rest by the providential discoveries of the Dead Sea Scrolls in the Qumran caves in 1949, which included the Book of Isaiah, intact and practically identical to what is in the Bible of today, proving it was written long before the birth of Jesus.

Many of the stories and events contained within the Old Testament can be said to foreshadow or prefigure the coming of God in the person of Jesus Christ. For instance, Abraham's sacrifice of his son Isaac on an altar is an obvious foreshadow of the sacrifice of the Son of God on the cross. Joshua leading the people of Moses into the Promised Land is likewise a prefiguration of salvation through Jesus (Jesus, by the way, is the Greek form of Joshua, which means "Savior"). The story of Joseph who, while in the court of Egypt, saves his people from famine and death is another foreshadow of the royalty and saving power of Jesus the Savior. Likewise, the role of Mary the Mother of God and Queen of Heaven is prefigured in the Old Testament stories of Esther, Judith, Jael, the Thebezite woman, and other courageous and heroic women.

Bible prophecies that have come true can also be found in the New Testament. Jesus himself made a famous prediction, which came true in Matthew 24 when he lamented over the coming destruction of Jerusalem and its temple for having rejected its messiah. On passing by the temple, Jesus prophesied to his disciples that no temple stone would be left unturned, a prediction that actually came true when the Roman emperor Titus sacked and burned Jerusalem on August 30, 70 A.D., leaving no stone left unturned in the Jewish temple, just as Jesus said would happen.[11] The Jewish temple has never been rebuilt and will never be rebuilt because Jesus is the true Temple of God, which was raised up or rebuilt on the third day, exactly as Christ predicted.

Fulfilled prophecies that prove Jesus Christ is God have not been limited to the Bible. Many other prophecies have come from apparitions of Christian saints, especially the Blessed Virgin Mary. In my book *The Woman and the Dragon: Apparitions of Mary,* I document numerous apparitions and fulfilled prophecies of Mary, Mother of God, especially her apparitions at Fatima, Portugal. Mary's prophecies in Fatima, which were given to three Portuguese children in 1917, correctly predicted the end of WWI, the rise of communism in Russia and its spread throughout the world, the coming of a much worse second world war, an ominous light in the sky heralding its coming (the 1938 aurora borealis), the annihilation of many nations, the martyrdom of many priests and religious, the assassination attempts on Pope John Paul II, the fall of the Soviet Union and the conversion of Russia to her Immaculate Heart, and a brief period of peace thereafter, the time we are now in. Another well-known apparition of Mary in La Salette, France, in 1846 gave a lengthy prophecy of the events leading up to the end times, including the rise of Hitler and the looming sex scandals and crisis within the Catholic Church.

Raised from the Dead

Jesus' resurrection from the dead is an astonishing miracle, to be sure, but is by no means the only documented case in history of someone being brought back to life after being dead. In the Old Testament, Elias and Elijah raised the dead (4 Kings 4:32-35, 13:21), and in the New Testament, Jesus raised the widow's son (Luke 7:11-15), the daughter of Jairus (Mark 5:23), and his close friend Lazarus (John 11:43-44). St. Peter raised Dorcas (Acts 9:40), and St. Paul raised Eutychus (Acts 20:7-12). Jesus very much wanted his disciples to raise the dead and commanded them to do so: "Cure the sick, raise the dead, heal the leprous, expel demons."[12]

The raising of the dead has been performed numerous times since the first century, and we have more than enough documented evidence to prove these miracles really happened. The Greek historian Appolonius wrote that St. John the Evangelist raised a dead man to life at Ephesus.[13] St. Irenaeus (130-220), Bishop of Lyon and disciple of Polycarp, who was a disciple of St. John the Evangelist, wrote in his *Against the Heresies* that "Some persons that were dead have been raised and have continued among us many years."[14] The secular historian Palladius in *The Lausiac History* around the year 420 wrote of a monk named Macarius the Egyptian who raised a man from the dead as proof to those who did not believe in the resurrection.[15] St. Augustine of Hippo (354-430) wrote in his *City of God* of the various people who were raised from the dead by the relics of St. Stephen, the first Christian martyr.[16] Among the many saints who raised the dead in the name of Jesus Christ are the Empress St. Helena (250-330), mother of Emperor Constantine the Great, St. Hilary of Poitiers (315-368), St. Ambrose (340-397), St. Martin of Tours (316-397), St. Patrick (381-461), St. Benedict (480-547), St. Bernard of Clairvaux (1090-1153), St. Stanislaus of Cracow (1030-1079), St. Malachy (1095-1148), St. Cyril (1191-1235), St. Hyacinth (1185-1257), St. Anthony of Padua (1195-1231), St.

Dominic (1170-1221), St. Elizabeth of Hungary (1207-1231), St. Zita (d. 1271), St. Catherine of Sweden (1331-1381), St. Colette of Corbi (1381-1447), St. Frances of Rome (1384-1440), St. Catherine of Siena (1347-1380), St. Vincent Ferrer (1350-1419), St. Bernadine of Siena (1380-1444), St. Juan Capistrano (1386-1456), St. Joan of Arc (1421-1431), St. Francis of Paola (1416-1507), St. Peter of Alcántra (1499-1562), St. Ignatius Loyola (1491-1556), St. Teresa of Avila (1515-1582), St. Francis Xavier (1506-1552), St. Martin de Porres (1579-1639), St. Louis Bertrand (1526-1581), St. Philip Neri (1515-1595), St. Rose of Lima (1586-1617), St. Francis Solanus (1549-1610), St. Marianne de Jesus of Quito (1618-1645), St. Joseph of Cupertino (1603-1663), St. Paul of the Cross (1694-1775), and St. John Bosco (1815-1888), just to name a few.

The ranks of the dead brought back to life by these saints include persons who were drowned, hanged, mutilated, and even those who were already decayed or reduced to skeletons, or dead for several years. They include adults, children, unbaptized infants, executed criminals, persons raised from the dead to testify in criminal trials or to testify to a religious truth, and of persons who would have been condemned to hell but were brought back to life for another chance. An excellent reference book on the historical evidence of the resurrection of the dead by Christian saints, and their testimony to the literal existence of heaven, hell, and purgatory is Father Albert Hebert's *Raised From the Dead.*

In addition to Church records, many excellent secular books, such as Raymond Moody's *Life After Life*, have been written on near-death experiences. These books document testimonies of people who were clinically dead and whose souls left their bodies and hovered over their bed, where they could hear every word and see every action of those trying to resuscitate them. The majority of these people who have had near-death experiences recounted going through a long tunnel towards a bright light and ending up in places of heavenly bliss or places of hellish punishment and torment before being brought back to life. So many are the cases of people being resurrected or coming back to life after being declared clinically dead that it is inconceivable how anyone could believe that death ends everything. Those who do are either woefully ignorant of the facts or haters of the truth who refuse to believe in the evidence.

The Incorruptibles

Evidence that Jesus' body was preserved from decay and rose again from the grave on the third day is demonstrated by the great number of

Christian saints who have likewise died but whose bodies haven't undergone the natural process of putrefaction in the grave. These saints are called the "incorruptibles" because God has miraculously preserved their bodies from corruption, so as to be visible witnesses of the resurrection from the grave. Bodies of the incorruptibles are different than those that have undergone embalming with preservatives, such as Egyptian mummies, or from preservation by the natural elements, such as glacial ice, in that they are soft, flexible, lifelike, and sometimes exude sweet-smelling fragrances and oils, or even fresh blood, whereas embalmed or naturally preserved bodies are stiff, dry, discolored, leathery, withered, skeletal, and more often than not, totally unrecognizable.

The ranks of the saints who were found to be incorruptible after being exhumed are many and include St. Cecilia (d. 117), St. Agatha (d. 251), St. Etheldreda (630-679), St. Cuthbert (d. 687), St. Werburgh (d. 699), St. Guthlac (667-714), St. Withburga (d. 743), St. Wunibald (702-761), St. Alphege of Canterbury (954-1012), St. Romuald (952-1027), St. Coloman (d.1012), St. Edward the Confessor (1004-1066), St. Waltheof (d. 1159), St. Ubald of Gubbio (1100-1160), St. Idesbald (d. 1167), St. Isadore the Farmer (1080-1172), St. Benezet (d. 1184), St. Hugh of Lincoln (1140-1200), St. Edmund Rich of Canterbury (1180-1240), St. Rose of Viterbo (1235-1252), St. Sperandia (1216-1276), St. Zita (1218-1278), St. Albert the Great (1206-1280), St. Margaret of Cortona (1247-1297), St. Nicholas of Tolentino (1245-1305), St. Clare of Montefalco (1268-1308), St. Agnes of Montepulciano (1268-1317), St. Peregrine Laziosi (1265-1345), St. Catherine of Siena (1347-1380), St. Frances of Rome (1384-1440), St. Bernardine of Siena (1380-1444), St. Herculanus of Piegaro (d. 1451), St. Rita of Cascia (1381-1457), St. Antoninus (1389-1459), St. Didacus of Alcala (1400-1463), St. Catherine of Bologna (1413-1463), St. Catherine of Genoa (1447-1510), St. Anthony Maria Zaccaria (1502-1539), St. Angela Merici (1474-1540), St. John of God (1495-1550), St. Frances Xavier (1506-1552), St. Stanislaus Kostka (1550-1568), St. Louis Bertrand (1526-1581), St. Teresa of Avila (1515-1582), St. Charles Borromeo (1538-1584), St. Benedict the Moor (1526-1589), St. Catherine dei Ricci (1552-1590), St. John of the Cross (1542-1591), St. Pascal Baylon (1540-1592), St. Philip Neri (1515-1595), St. Germaine Cousin (1579-1601), St. Mary Magdalen de'Pazzi (1566-1607), St. Camillus de Lellis (1550-1614), St. Rose of Lima (1586-1617), St. Frances de Sales (1567-1622), St. Josaphat (1580-1623), St. Jeanne de Lestonnac (1556-1640), St. Jane Frances de Chantal (1572-1641), St. John Southworth (1592-1654), St. Andrew Bobola (1590-1657), St. Vincent de Paul (1580-1660), St. Pacificio of San Severino (1653-1721), St. Veronica Giuliani (1660-1727), St. Lucy Filippini (1672-1732), St.

Teresa Margaret of the Sacred Heart (1774-1770), St. Julie Billiart (1751-1816), St. Vincent Pallotti (1795-1850), St. Jean Vianney (1786-1859), St. Madeleine Sophie Barat (1779-1865), St. Pierre Julien Eymard (1811-1868), St. Catherine Labouré (1806-1876), St. Bernadette Soubirous (1844-1879), and St. Charles Makhlouf (1828-1898). Many of the bodies of these incorruptibles are kept in crystal coffins and displayed in shrines and churches built in their honor, as public evidence of their sanctity. An excellent reference book on the subject of saintly incorruption is *The Incorruptibles* by Joan Carroll Cruz.

The Sacred Stigmata

The reality of Christ's crucifixion for the sins of mankind is also attested to by the numerous instances of chosen souls bearing the physical wounds of Christ's crucifixion, the Sacred Stigmata. Stigmata are wounds in one or more of the five areas of the body where Jesus was ripped and punctured during his passion and crucifixion. They include the torso and extremities where Jesus was scourged with an iron-studded whip, the forehead where Jesus was punctured from the crown of thorns, the hands (or wrists) and feet where nine inch nails were hammered into Jesus to secure him to the cross, and the side of the chest where Jesus was pierced by the spear of Longinus.

Stigmatists are a living representation of the crucified Christ to a disbelieving world and are a gift from God to help strengthen our faith. In the last century alone, the century with so much evil an unbelief in it, over two dozen stigmatists have appeared around the world and have been thoroughly investigated by scientists and medical doctors who are experts in anatomy, psychology, and psychiatry. The most famous of these stigmatists is St. Padre Pio (1887-1968), a Franciscan monk from San Giovanni Rotondo, Italy, who had the stigmata of Christ for fifty years (1918-1968), the longest know case of stigmata. Padre Pio, now called St. Pio (Pope John Paul II canonized him in June of 2002), was an exceptionally gifted saint who underwent tremendous physical and spiritual suffering and torments from the devil to win souls back to Christ. As an especially holy soul, Padre Pio had the gifts of discernment of spirits, miraculous healings, fragrant smells (the odor of sanctity), bilocation, reading of hearts and minds, knowledge of other's sins before confession, speaking in foreign languages not learned, conversations with his guardian angel, surviving on little or no food (inedia), and the stigmata of Jesus Christ in the hands, feet, and side.

The stigmata of Padre Pio have been thoroughly investigated by medical science and have been determined to have no known natural causes. Unbelieving scientists who know that there is no natural explanation for St. Pio's stigmata have claimed that his stigmata was "psychogenically" caused by his staring at the crucifix for hours at a time. Padre Pio had a very witty reply to this by saying that it would be about as likely as "growing horns by staring at a cow in the field." Hundreds of miracles have been attributed to the prayers and intercession of St. Padre Pio, and just as many books and videos have been published and produced on his extraordinary life, too many to list here. St. Pio is the most celebrated saint of the twentieth century and was a living witness to the crucifixion of Jesus Christ.

Over three hundred chosen souls have received the stigmata of Christ. They include St. Francis of Assisi (1182-1226), the first known stigmatist, St. Gertrude the Great (1256-1302), St. Catherine of Siena (1347-1380), St. Lydwine of Shiedam (1380-1433), St. Rita of Cascia (1381-1457), St. Teresa of Avila (1515-1582), St. Catherine dei Ricci (1522-1590), St. Mary Magdalen dei Pazzi (1566-1607), St. Rose of Lima (1586-1617), St. Margaret Mary Alacoque (1647-1690), St. Veronica Guiliani (1660-1727), St. Mary Frances of the Five Wounds (1734-1791), Anne Catherine Emmerich (1774-1824), Louise Lateau (1850-1883), St. Gemma Galgani (1873-1903), Sister Josefa Menendez (1890-1923), Mary Rose Ferron (1902-1936), St. Faustina Kowalska (1905-1938), Berthe Petit (1870-1943), Alexandrina da Costa (1904-1955), Therese Neumann (1898-1962), Adrienne von Speyr (1902-1962), and Marthe Robin (1902-1981). There are even stigmatists who are still living, such as Sister Agnes Sasagawa of Akita, Japan, Maria Esperanza of Betania, Venezuela, and Christine Gallagher of Ireland. A good reference book on the subject of stigmata is *They Bore the Wounds of Christ: The Mystery of the Sacred Stigmata,* by Michael Freze, S.F.O.

Miracles of the Bread and Wine

Miracles involving the Holy Eucharist and the consecrated wine testify to the Real Presence of Jesus Christ in the bread and wine of the Mass: body, blood, soul, and divinity. These miracles include the communion host turning into real flesh and/or bleeding, the wine turning into real blood, hosts that taste sweet and emit a heavenly fragrance, stolen hosts bleeding when stabbed by unbelievers, hosts levitating or flying through the air, surviving fire, defying the natural decay process, turning hard as flint in the mouths of sinners and unbelievers, miraculously appearing out of thin air, and passing through solid objects. In addition, visions of Jesus have appeared to Sunday congregations as the consecrated host was raised up or put on display in monstrances. Saints and mystics have received communion from Jesus, Mary, and the angels, and have levitated or had miraculous visions and locutions of Jesus and Mary while receiving the Eucharist.

Evidence for many of these miracles have survived intact to this very day. Hosts that have turned into real human flesh and/or bled, and wine that has turned into real human blood, have lasted for centuries without preservatives or being hermetically sealed in airtight containers, defying scientific explanation. Centuries old altar cloths stained with the blood of the bread and wine have miraculously survived unaltered without decomposing. Stolen hosts that were recovered after being profaned have not disintegrated after hundreds of years of being exposed to the elements.

And not all these miracles have occurred many years ago without any surviving eyewitnesses. Numerous Eucharistic miracles have taken place within the last decade or so, and have been thoroughly documented and investigated by science. For instance, on December 8, 1991, the feast of the Immaculate Conception of Mary, in front of hundreds of eyewitness in an outdoor Mass being celebrated in Betania, Venezuela, a large communion host, at the moment of consecration and breakage into four smaller pieces by Father Otty Ossa Aristiza'bal, began to bleed real human blood from what appeared to be a small puncture wound. Betania, by the way, is Spanish for "Bethany," which was the small town where Jesus raised Lazarus from the dead.

This Church-approved miraculous event was videotaped as it happened to allay any possibility of fraud or deception.[17] The Betania communion wafer was subsequently tested at a laboratory in Caracas, Venezuela, which revealed that the blood was of human origin. After thorough testing, the miraculous host was sent back to the convent of Augustinian nuns in Los Teques for safekeeping and public display. On November 13, 1998, photographer Daniel Sanford went to see the bleeding host of Betania and witnessed another miracle, which he caught on tape. The bleeding host was enveloped in flames and in the center was a pulsating, bleeding heart. This miraculous vision happened for about thirty seconds and was recorded on tape for the whole world to see. Free copies of the videotape can be obtained by writing to Daniel.[18]

Since September 14, 1995, in Rome, Eucharistic miracles have been occurring to an arthritic invalid named Marisa Rossi on a regular basis in front of hundreds of eyewitness. Marisa has had numerous visions from Jesus and Mary, who have given her communion hosts and public messages, including warnings of impending catastrophes for the world if it doesn't repent and accept Jesus Christ and obey his word. Besides having hosts appear from out of nowhere to Marisa and the congregation, these same communion hosts have repeatedly bled on more than one occasion, the first time that has happened in Church history. Despite the skepticism even within the Church by some of its unbelieving members, the local bishop, Claudio Gatti, with approval by Pope John Paul II, has approved these Eucharistic miracles as worthy of belief.

Other credible reports of Eucharistic miracles have come to us from Naju, Korea, where on September 22, 1995, the mystic Julia Kim received a consecrated host that turned into flesh and blood. On April 26, 1997, in Constantia, New York, the mystic Emma de Guzman at the Basilica of the Sacred Heart received communion, which likewise turned into real flesh and blood. On April 10, 1994, in Yardville, New York, at the moment of consecration by Father Robert Rooney, a consecrated host began to bleed. A similar event happened in Barbeau, Michigan, on February 4, 1996. On June 9 and July 14 of 1970 in Stich, Germany, blood fell from the chalice at the moment of consecration by the priest and stained the altar cloths. The stained cloths were taken to the Polytechnic Institute of the University of Zurich and examined by the director of Chemical Laboratory, the chief of the Blood Control Laboratory, and the chief of the Laboratory for the Analysis of Hemorrhage and Coagulation. Their findings all agreed that the stains were from human blood and could not be scientifically explained.

Of course, the most famous Eucharistic miracle of all time was the Miracle of Lanciano, Italy, which occurred in the eight century. A priest who was having severe doubts about the true presence of Jesus in the Eucharist was celebrating Mass one day in the Church of St. Frances. At the moment of consecration, the outer two-thirds edge of the host suddenly turned into human flesh and the communion wine into human blood. The Eucharistic flesh and the blood from the wine were subsequently placed in crystal reliquaries, which were not hermetically sealed. By a miracle of God, the flesh and the coagulated blood have remained intact for over twelve centuries, although the center one-third of the host has long since disintegrated. In 1970, Professor Odoardo Linoli, university professor at large in chemistry and microscopy and head physician of the United Hospitals of Arezzo, Italy, teamed up with Dr. Ruggero Bertelli, professor emeritus of human anatomy at the University of Siena, Italy, to study the contents of the Lanciano reliquaries. They determined that the donut-shaped flesh was striated myocardium (heart muscle) of human origin, which could not have been so precisely cut with a knife from a cadaver's heart and then placed around a host, and the coagulated blood was of human origin with the correct percentage of human serum, proteins, and clotting fractions—all without any trace of preservatives.

The numbers of documented cases of Eucharistic miracles during the last two millennia are substantial. There have been miracles in Braine, France (1153), Augsburg, Germany (1194), Alatri, Italy (1228), Santarem, Portugal (13th century), Florence, Italy (1230 and 1595), Daroca, Spain (1239), Olmütz, Czechoslovakia (1242), Regensburg, Germany (1257), Bolsena-Orvieto, Italy (1263), Paris, France (1270 and 1290), Slavonice, Czechoslovakia (1280), Offida, Italy (1280), Hassett, Belgium (1317), Sien, Italy (1330 and 1730), Blanot, France (1331), Amsterdam, Netherlands (1345), Macerata, Italy (1356), Brussels, Belgium (1370), Middleburg-Louvain, Belgium (1374), Seefeld, Austria (1384), Dijon, France (1433), Avignon, France (1433), Turin, Italy (1453), Morrovalle, Italy (1560), Alcalá de Henares, Spain (1597), Faverny, France (1608), Paterno, Italy (1772), Bordeaux, France (1822), and Dubra, Poland (1867). The purpose, of course, behind all these documented cases of Eucharistic miracles is to prove that Jesus is truly in the Eucharist: body, blood, soul, and divinity. It is Christ's most ardent wish that we partake in the celebration of his sacrifice by eating his body and drinking his blood, as is evident by the Last Supper:

> During the meal Jesus took bread, blessed it, broke it, and gave it to his disciples and said: 'Take this and eat it, this is my body.' Then he took a cup, gave thanks, and gave it to them. 'All of you must drink from it, for this is my blood, the blood of the new covenant, to be poured out in behalf of many for the forgiveness of sins.'[19]

Note that Jesus said that this "is" his body and "is" his blood, not "represents" or "symbolizes" his body and blood. This is a very important distinction. By instituting a new covenant with mankind at the Last Supper, Jesus became the new sacrificial lamb who died on the cross in atonement for our sins. Receiving Holy Communion, therefore, is not an option for Christians; it is a requirement. We *must* do it as Jesus commanded: "Verily I say to you, that unless you eat the Flesh of the Son of Man and drink his Blood, you have no life in you: Those who eat My flesh and drink My Blood have eternal life, and I will raise them up on the last day."[20]

So important are the celebration of the Mass and the receiving of Holy Communion that God has generously given us more than enough miracles to prove that he is truly in the bread and the wine we receive. An excellent reference book on the history of Eucharistic miracles is Joan Carroll Cruz's *Eucharistic Miracles.*

The Holy Shroud of Turin

Still more evidence of the divinity of Jesus Christ comes from existence of holy relics, the most recognized of which is the Holy Shroud of Turin, the burial cloth of Jesus Christ mentioned in Matthew 27:59, Mark 15:46, Luke 23:53, John 19:39, and John 20:5-6. Forensically, the Holy Shroud of Turin, now kept in Turin, Italy, by the Roman Catholic Church, shows a three-dimensional imprinted image, front and back, of a crucified man between thirty and thirty-three years of age, 5'11", and 175 pounds. The remarkable image can only be discerned as being that of a man when observed from a distance, as when one gets nearer it seems to fade into an unrecognizable pink mist on the time-yellowed linen. The yellow tinge, however, does not penetrate into the pinkish image.

For those of you that believe the Shroud of Turin is a fake, here is a postmortem forensic report on the shroud's image by medical doctors. Medical observation of the man's face on the shroud reveal tumefactions that seem to be ematomas, especially on the right cheek, which indicate that the bearded, long-haired man was severely beaten just hours before burial. The nasal septum of the man is deviated by a fracture, indicating a severe blow to the nose. On the forehead, on the nape of the neck, and on the hair imprint front and back are numerous small-diameter puncture wounds with sinuous blood flows, consistent with the forcible positioning of a crown of thorns. On the limbs, chest, and back areas are over a hundred excoriated lacerations with round indentations about the size of 00 buckshot, which indicate scourging with two cat-of-ninetails whips interlaced with bar-bell shaped lead weights, typical of a first-century Roman-style flagellum. On the scapular region of the back leading to the armpits and the inside areas of the upper arms are two quadrangular abrasions, indicative of the crucified man carrying the horizontal beam of a large wooden cross for quite some distance. Roman executioners often required the condemned to carry their own cross to the site of their crucifixion. On the right side of the thorax about an inch and a half below the mamillary papilla, between the fifth and sixth ribs, is a large ovoid puncture wound with blood stains surrounded by a watery halo, characteristic of the postmortem effusions that empty the thoracic cavity of its blood and water contents, which would have been caused by the thrusting of a Roman lance into the body of the deceased. The pulmonary edema caused by the large collection of water into the lungs is characteristic of the tremendous trauma to the body of the crucified victim and the subsequent failure of his heart at the time of death. On the posterior portion of the shroud around the kidney area of the back are two large stains characteristic of uremic poisons seeping through the skin from the kidneys, common in cases of shock and massive kidney failure.

In the anterior view of the shroud, the left hand is folded over the right hand and shows a puncture wound in the "space of Destot," the area of junction between the wrist and hand where nails were hammered into the cross so that the weight of the body upon collapsing would not tear away from the cross. On the posterior imprint of the shroud we can clearly see the puncture wound of a single nail being hammered into the feet, which were overlaid upon each other on the cross. The knee areas on the anterior imprint also show signs of abrasions, where the crucified man apparently collapsed under the weight of the cross.

Dr. Pierre Barbet, a well-respected French physician, wrote a detailed study of the wounds on the shroud in his book *A Doctor at Calvary*. He found that all the wounds on the man were anatomically correct, that signs of rigor mortis had begun to set in, but that no signs of decay were evident. The man's abdomen was distended, indicating asphyxiation and paralysis of the chest muscles as he compensated by trying to breath through the diaphragm and stomach muscles, characteristic of someone's arms being stretched out of the sockets while being nailed to a cross. Two different scourgers apparently made the 90 to 120 flagellation marks. The scourger on the right was taller and whipped harder and, therefore, made deeper, higher lacerations.

The Holy Shroud of Turin has been a contentious object of debate and fascination by scientists for centuries, for it seems to be the burial cloth of Jesus Christ. In 1978, twenty scientists from the U.S. arrived in Turin with seventy-two crates of instruments to study the shroud. Among the team of scientists were pathologists, a medical school professor, image-enhancement specialists, physicists, chemists, spectroscopy experts, archaeologists, thermography and microanalyst experts, and x-ray specialists. Tests were conducted over a five-day period but it took the team three years to finalize their observations and results. After all their tests, the scientists could not explain how the image was formed, but made several important conclusions.

The image was made from a real corpse and not a statue or a human model. The shroud was woven as a single piece except for a 3½-inch strip on the left side. The image was a reddish monochrome tint, pale and subtle and was only on the surface and did not extend throughout the cloth to the other side. The image of the man is three dimensional and in the form of a photographic negative. The cloth was folded by an unknown means that did not distort the image. The yellowish tinge that appeared on the linen did not penetrate into the image. The body was not moved, unwrapped, or rewrapped, which would have disfigured the blood clots and stains on the shroud. Coins were placed on the eyes of the dead man, having twenty-four "coincidences" similar to a coin issued by Pontius Pilate in 29-32 A.D. known as the "lepton," a common custom during the time of Jesus. The image was not formed by vapors, sweat, blood, spices, ammonia, or urea. The blood marks contained iron oxide, protein, and porphyrin characteristic of human blood. No paints, pigments, stains, oils, powders, dyes, or inks were used to create the image. The image was not affected by dousing with water in 1532 when the corners of the folded shroud caught on fire. Finally, the image cannot be changed by chemical agents, dissolved, or bleached.

Historical records of the Holy Shroud of Turin date only as far back as the fourteenth century, but its existence before then is connected to the Holy Edessan Image" also known as the "Holy Mandylion," which was venerated as the burial shroud of Jesus, the same one mentioned in the Gospels of Luke and John. The burial shroud of Jesus was believed to have been taken by Thaddeus, "one of the seventy,"[21] to Edessa (now Urfa in eastern Turkey) around 32-33 A.D. Thaddeus supposedly delivered the shroud to King Agbar V, ruler of Edessa, for safekeeping, since Edessa was not under control of the Roman Empire. A twelfth-century Latin Codex in the Vatican (Codex 5696, vol. 35) records that Jesus gave a message to King Agbar that he was sending him the Holy Shroud for guardianship: "I send you a cloth on which the image not only of my face but of my whole body has been divinely transformed."

The Holy Edessan Image was also called the Holy Mandylion, which means "little handkerchief," because it was folded eight times and showed only the face of the crucified Christ. Byzantine icons and coins of later years bear a remarkable resemble to the face on the shroud, suggesting that Christ's image on the Mandylion was well known and highly revered by the Eastern Church.

Around 57 A.D., antichristian persecution led by Ma'nu VI scattered across Turkey. The Holy Shroud was sealed in a container and hidden away in a niche above Edessa's west gate, where it was lost for generations. In 325, Bishop Eusebius visited Urfa, and in an effort to translate Syrian archives into Greek, discovered the documented correspondence between King Agbar and Jesus about the safekeeping of the shroud. Hidden away in the wall, the Holy Shroud was lost to history until 525, when a flood destroyed Edessa. Workmen repairing the wall of Edessa discovered the shroud hidden in a small compartment above the west gate. Ancient Byzantine and Russian art have depicted the finding of the Holy Mandylion in the Edessa wall. In 639, the Muslim Arabs captured Edessa, and it is believed that the shroud was moved to

Jerusalem for safekeeping. The historian Archulphus was a pilgrim to Jerusalem in 640, where he saw and kissed the "winding-sheet of the Lord which was placed over his head in the sepulcher." And in 800, St. John Damascene mentions the holy shroud as one of the venerated relics of the early Christians of that time.

The shroud shows up again in Edessa around 942, when the Christian Byzantine general Curcuas captured Edessa, and the Archbishop Abramius of Somastasa arranged for the town to hand over the shroud in order to avoid destruction. The shroud was then removed from Edessa and was placed in the Byzantine emperor's huge collection of relics in the Pharos Chapel in Constantinople. In 942, the shroud arrived at the Church of Our Lady of Blachernae, and was displayed in the throne of Blachernae Palace until it was moved to Pharos Chapel. Paintings of that era begin to show Christ reclining as on the shroud, instead of the customary images of him wrapped up mummy-style. In 1171, the Byzantine emperor Manuel Commenus showed the relics of Jesus' passion and crucifixion to King Amaury of Jerusalem, which included the nails, the crown of thorns, the sponge, and the shroud. In 1204, Robert de Clari, a knight from Picardy who joined the Roman Catholic Crusades against Muslims, took part in the capture of Constantinople on the way to liberate the Holy Land. He reported on the relics and riches he saw in Constantinople, which included two pieces of the true cross, two of the nails of the crucifixion, the head of Longinus's spear, a tunic, a vial of blood, a crown, and a linen cloth with Jesus' face (the Veil of Veronica).

It is generally believed that the Holy Shroud ended up in the hands of Roman Catholics sometime before or after the fall of Constantinople in 1261, during the time of the Crusades. The shroud was put on display in the Cathedral of Saint-Etienne during the fourteenth century. In 1349, a fire broke out in the cathedral and the shroud disappeared once again, and was probably stolen. In 1357, the shroud wound up in the hands of Geoffrey de Charney, a member of the Knights Templar, given to him by King Philip VI, who had probably gotten the shroud from the one who stole it from the cathedral. When Geoffrey died, his widow put the shroud on exhibition in the small French town of Lirey for financial reasons. It was about this time that reproductions of the Holy Shroud began to be painted on linen cloths about the size of the true shroud at St. Gommaire, Belgium, and at Lisbon, Portugal. The amateurish, one-dimensional artistry of that era made it very easy to distinguish the real shroud from the copies.

In 1452, Marguerite de Charney, the wife of Geoffrey de Charney II, bequeathed the shroud to Anne de Lusignan, the wife of the Duke of

Savoy. The Duke of Savoy built a chapel at the Chambery to house the shroud, and several exhibitions took place. When not on display, the shroud was kept in a silver box in the chapel. In 1532, a fire broke out at the Chambery, and the silver box that encased the shroud overheated and dripped molten silver on its corners, miraculously missing the image. Two Franciscan priests saved it from the fire by dousing it with water. The Poor Clare Nuns, who sewed fourteen large triangular patches and eight smaller ones over the burn holes, repaired the shroud.

In 1578, the shroud was taken by the House of Savoy to Turin, Italy, and deposited in the chapel adjoining the Cathedral of St. John. Finally, in 1694, the shroud was moved to the Royal Chapel of Turin Cathedral, where it remains to this day. The House of Savoy owned the shroud until 1983, when Italy's last king, the exiled Umberto II of the House of Savoy, bequeathed it to the Vatican.

Secundo Pia photographed the Holy Shroud for the first time in 1898, and was amazed to find that the image of the crucified man was a photographic negative, an impossible to make image during the fourteenth century because there was no knowledge of positive and negative photography back then. This is perhaps the greatest proof that the image on the Shroud of Turin is not a forgery. With the announcement of the presence of the photographic negative image, physicians and scientists around the world began to do careful studies of the physiognomy of the person pictured on the shroud. The physicians Dr. Pierre Barbet in France, Dr. David Willis in England, Dr. Hermann Moedder in Germany, Dr. Judica-Cordiglia in Italy, Dr. Robert Bucklin in the United States, and others, concluded that the man on the shroud had indeed been crucified.

In 1988, carbon-14 dating was simultaneously done on the Shroud of Turin at the three different locations: the University of Oxford, England, the University of Zurich, Switzerland, and the University of Arizona at Tucson. Preliminary tests results dated it between 1260 and 1390 A.D., and skeptics and unbelievers rejoiced in this announcement as proof that the Shroud of Turin was a medieval forgery and that the crucified body of Jesus Christ was not on the image.

An extremely important scientific discovery, however, was made five years later in 1993 by Dr. Leonicio A. Garza-Valdes, a pediatrician in San Antonio, Texas, an adjunct professor of microbiology at the Health Sciences Center of the University of Texas, and an authority on Mayan archaeology and artifacts. His discovery provided scientific proof that the Shroud of Turin is much older than what the 1988 carbon-14 dating tests showed. Under high-power scanning electron microscopy, Dr.

Garza-Valdes discovered the presence of a shiny, yet invisible to the naked eye, "bioplastic coating" of polyhydroxyalkanoate and mineral crystals on the surface of the shroud fibers, which had built up over time, and which caused the carbon-14 dating to date the shroud newer than it actually is. Bioplastic coatings are the result of the accumulation of billions of microbiological organisms that congregate over long periods of time, like coral reefs in the ocean. They are a completely natural process and can be found, as Dr. Garza-Valdes has proven, on all ancient artifacts and relics. When the scientists in 1988 did their carbon-14 dating, they did not take into account that they were dating both the old fibers in the shroud and the newer bioplastic coatings on the fibers, which threw their results off by centuries. The Shroud of Turin, therefore, is actually much older than what the 1988 Carbon dating has shown.

Further examination of the shroud by Dr. Garza-Valdes uncovered the presence of tiny splinters of oak on the occipital area near the back of the head on the shroud image, the presence of vinegar-forming bacteria, and the presence of blood that contains human male DNA. These additional findings suggest that they come from the splinters of the cross, the vinegar that was given to Jesus to drink, and the blood of Jesus Christ.[22]

Other findings bolster the conclusion that the image on the shroud is that of Jesus Christ. In the 1970s, Dr. Max Frei, a Swiss criminologist, discovered and identified pollen grains in the dust on the shroud, which were unique to plants of the Holy Land and Turkey, providing evidence that the shroud had some time been exposed to the air of these places. In 1978 the American Shroud of Turin Research Project (STURP), comprised of about thirty scientists from the Los Alamos National Scientific Laboratory, the Air Force Weapons Laboratory, the Jet Propulsion Laboratory, and other highly respected institutions, conducted a five-day examination of the Shroud of Turin, and concluded that no paint had been used on the shroud image, and that the blood stains were indeed made of human blood. The only logical conclusion is that the Shroud of Turin is indeed the burial cloth of Jesus.

Apparitions of Mary, Mother of God

Like the Holy Shroud of Turin, a miraculous image of the Blessed Virgin Mary has been given to us by God to strengthen our faith in Jesus. In December of 1531, a poor Aztec peasant in central Mexico named Juan Diego had an apparition from the Blessed Virgin Mary on

Tepeyac Hill, just outside of modern-day Mexico City, requesting him to go to the local bishop and petition him to build a shrine in her honor. Of course, the bishop did not believe Juan Diego, but politely requested that Juan go back and ask the lady to give him a sign so that he too may believe in her. Juan Diego went back to Tepeyac Hill and pleaded with the Virgin Mary for a sign, and she told him to take his *tilma* or cactus fiber cloak and gather the beautiful Castilian roses that were growing on the hill, and take them to the bishop. It is impossible for roses to grow in wintertime in the rocky, high terrain of cactus and mesquite, so Juan eagerly gathered up the roses in his cloak, thinking that they were the sign the bishop requested, and took them back to him. As Juan unfolded his cloak, the roses fell to the floor and a miraculous image of the Blessed Virgin appeared on Juan's *tilma.* The bishop, the governor, and another dignitary who was present all fell on their knees and praised God when they saw Mary's miraculous image, which was the true sign from God that Bishop Ramirez y Fuenleal had requested. A church was subsequently built to Our Lady of Guadalupe and millions of Indians converted to Christianity as a result of Mary's miraculous image on Juan's *tilma.*

Juan Diego's *tilma* has been thoroughly investigated by science and proven to be authentic. Despite exposure to the harsh Mexican climate, the handling and touching by thousands of admirers and investigators, the smoke of incense and countless votive candles, a disastrous flood in 1629, nitric acid spilled on it in 1791, and a bomb explosion on November 14, 1921, by anti-government terrorists attempting to destroy it, Juan Diego's *tilma* has miraculously held together, without sizing or varnish, for over 450 years.[23] The *tilma* is truly a miracle because Maguey cactus fibers have an approximate life span of only twenty years before decomposing.[24]

A Painters' Commission under the Viceroy, the Marquis of Mancera, commissioned in 1663 to study the sacred image of Our Lady of Guadalupe, testified under oath that "it is impossible, humanly, for an artificer to paint or produce a thing so excellent on a cloth as coarse as is the *tilma* or *ayate* on which appears the divine picture." The Commission added, "The imprinting of the said Picture of Our Lady of Guadalupe on the *ayate* or *tilma* of the said Juan Diego, was, and must be understood and be declared to have been, a supernatural work and a secret reserved to the Divine Majesty."[25]

A special committee of three professors from the Royal University examining the image also declared that "The continuance through so many years of the Holy Picture's freshness of form and color, in the presence of such opposing elements, cannot have a natural cause. Its sole principle is He who alone is able to produce miraculous effects above all the forces of nature."[26]

In 1756, celebrated painter Miguel Cabrera and other artists declared that the colors, style, and detail of the sacred image would be impossible to reproduce. "The plan of this Holy Picture," Cabrera said, "is so singular, so perfectly accomplished, and so manifestly marvelous, that we hold it for certain that anyone who has any knowledge whatever of our art must, on seeing it, at once declare it to be a miraculous portrait."[27]

Modern scientific examination of the image of Mary on the cloak of Juan Diego has reaffirmed its miraculous origin. First of all, the coarse fibers of the cloak make it an impossible surface to paint on, much like painting on shredded wheat cereal. Second, the mysterious changing colors of the sublime picture on the *tilma* have made it impossible to reproduce by painting or photographing. In fact, every investigation to date, whether by microscope, infrared radiation, or computer-enhanced photography, has pointed to its supernatural origin.

The skeptics' hypothesis that the image on the cloak was a painting was disproved in 1936 by a German professor living in Mexico by the name of Fritz Hahn. Hahn took a couple of the fibers from the image and delivered them to the German Nobel Prize winner in chemistry, Richard Kuhn, director of the Kaiser Wilhelm Institute in Heidelberg. Kuhn thoroughly tested the fibers and announced that the colors on the fibers were unknown to science, being neither animal, vegetable, nor mineral dyes.[28] The use of synthetic dyes, of course, was ruled out because that technology had not yet been invented in 1531. In 1946, powerful microscopic examination of the image revealed that there were no brush strokes. Again in 1954 and 1966, Mexican professor Francisco Camps Ribera undertook an exhaustive microscopic study and concluded the same thing.[29]

On May 29, 1951, a draftsman by the name of J. Carlos Salinas Chavez was examining a large photograph of the face of the sacred image under a powerful magnifying glass and noticed that the bust of a bearded man appeared in the cornea of the right eye of Mary. The discovery prompted the archbishop of Mexico City, Luis Marie Martinez, to set up a special committee to investigate the claim. Chavez's discovery was confirmed and announced to the public on December 11, 1955, in an announcement that the figure of the man was none other than Juan Diego, positively identified by a contemporary painting of Juan. The following July, two eye doctors, Javier Torroello Buene and Rafael Torifa Lavoignet, verified the human figure in both eyes, and noted that the reflected figures followed the curvature of the cornea exactly as seen in a real human eye. Furthermore, when shining an ophthalmoscope through the pupils of Mary's eyes, Dr. Lavoignet noted on July 23,1956, that they reflected light exactly as real eyes do. Performing the same experiment on the eyes of numerous other paintings and photographs did not replicate the same results.

In 1962, another amazing discovery was made on the *tilma.* While examining a photograph of the eyes of Mary magnified twenty-five times, Dr. C. Wahlig O.D., and his wife Isabelle, an optician, found the figures of two other people behind Juan Diego, assumed to be looking at the bishop at the moment his cloak was unfolded, which was captured in the reflection of Mary's eyes, a scientifically inexplicable phenomenon. Among those who took part in this experiment with Dr. Wahlig were Dr. Francis T. Avignone, a practicing optometrist and formerly a lecturer in practical optics and optometry at Columbia University; Edward Gebhardt, a television engineer with the National Broadcasting Company, who took the photographs; Dr. Michael Wahlig, his son; Alexander Wahlig, M.D., his brother and an eye surgeon; H. G. Noyes, M.D., an ophthalmologist and a former lecturer in optical sciences at Columbia University; Dr. Glen Fry, in charge of optical developments for the U.S. Government during WWII; and Dr. Italo Manelli, professor of physics and head of the University of Pisa, Italy.[30]

Special significance was given to the findings of Dr. Wahlig of the three persons in the eyes of Mary because two years earlier a 1533 painting of the scene of Juan Diego unfolding the *tilma* in front of Bishop Ramirez y Fuenleal, discovered hidden behind the altar of the Church of Our Lady of Guadalupe when it was restored in 1620, shows an uncanny resemblance to the reflected figures in both facial features and standing positions.

In 1979, Professors Philip Callahan of the University of Florida and Jody Smith of Pensacola, took infrared photographs of Juan's *tilma* to determine if there were any underlying sketches or drawings beneath the image of Mary. They concluded that there were no under-drawings, no varnish, and no sizing on the image; without which the picture would have surely disintegrated years ago. Their conclusion was similar to Dr. Wahlig's: the picture is a miraculous image inexplicable by science.

The *tilma* of Juan Diego, like the Holy Shroud of Turin, has been thoroughly investigated by science. It has been concluded that science cannot explain the images on either one. The only acceptable conclusion is that both the Holy Shroud of Turin and the *tilma* of Juan Diego are miracles from God.

Notes

Introduction

1. *Discoveries and Opinions of Galileo,* Translated with an Introduction and Notes by Stillman Drake, Anchor Books, 1957.

2. *Cosmos* the television series starring Carl Sagan as the main narrator traveling through time on a space ship was awarded Emmy and Peabody awards for excellence in broadcasting.

3. Carl Sagan, *Cosmos,* Random House, New York, 1980, 4.

4. Sagan was a member of CODESH (Coalition for the Defense of Secular Humanism) and *Free Inquiry* magazine. Sagan and his second wife Ann Druyan were also avid pot smokers. Ann Druyan was a one-time director of NORML (National Organization for the Appeal of Marijuana Laws).

5. Carl Sagan, *The Demon-Haunted World: Science as a Candle in the Dark,* 1995, Random House, New York, p.143.

6. Ibid., 28.

7. Ibid., 31.

8. Ibid., 32-33.

9. Ibid., 26.

10. Revelation 13:11-18.

Chapter One

1. The historical revisionist movie *Gladiator* was based on the cruelty of the Roman Empire during the reign of Emperor Commodus, son of Marcus Aurelius. In true life, Marcus Aurelius persecuted Christians much more than his son Commodus, and died of the plague rather than at the hands of his son.

2. Fanatic followers of John Duns were thought to be "stupid"; the modern pejorative of "dunce" comes from them.

3. Ockham died from the plague in 1347 while seeking reconciliation with Pope Clement VI for his excommunication.

4. Ficino was ordained as a priest in 1473 and became a canon of the Cathedral of Florence. His *Theologica Platonica* (1482), a study of the immortality of the soul, demonstrates Ficino's knowledge of St. Thomas Aquinas; it also takes account the Plotinian cosmology and of the influence of the stars on human lives. His commentary on Plato's Symposium introduced the notion of platonic love. This concept of a special friendship based on love of God was seminal in the literature of the late Renaissance.

5. Asclepius is also the name of the Greek god of medicine who was killed by Zeus for raising a man from the dead.

6. The French printer, Etienne Dolet, was hanged and burned at the stake in Paris on August 3, 1546, for publishing the works of humanist reformers such as Erasmus and Melanchthon.

7. "If you would be a real seeker after truth, it is necessary that at least once in your life you doubt, as far as possible, all things." Rene Descartes, *Principles of Philosophy.*

8. This famous quote of Descartes originated when he supposed that a "malicious demon of the utmost power and cunning bent on deceiving me in every possible." Perhaps "the sky, the earth, colors, shapes, sounds and all external things" are merely "the delusions of dreams which he has devised to ensnare my judgments." His answer to his own supposition was "Let the demon deceive me as much as he may. . . I am, I exist is certain, so long as it is put forward by me or conceived in my mind."

9. The word "science" wasn't coined until 1831. Scientists that time were known as "natural philosophers."

10. Paine was an Anglo-American philosopher and revolutionary whose pamphlet, *Common Sense,* greatly influenced public opinion during the American Revolutionary War. During the Revolution, Paine wrote a series of pamphlets, entitled *The American Crisis,* that George Washington ordered read to his troops to improve their morale; the first of these opened with the famous words "These are the times that try men's souls." Paine returned to Great Britain in 1787, and in 1791-92 he published *The Rights of Man,* in two parts, which advocated the overthrow of the British monarchy. The British government indicted Paine for treason, and he fled to France. In France, Paine was elected a deputy to the National Convention, which overthrew the French Monarchy in 1789. By favoring the exile, rather than the execution, of King Louis XVI, however, he offended Maximilien de Robespierre, the leader of the radical faction, and he was imprisoned for 11 months until Robespierre's downfall in 1794. Paine's book *The Age of Reason,* which was published in 1794, was seen as too antichristian, which gained him

a bad reputation and caused him to lose many friendships. Paine returned to America in 1802 and lived his last years in ostracism and poverty.

11. Joseph Priestley (1733-1804), the man with the religious-sounding name, was a British chemist and dissenting preacher who forsook orthodox Christianity for unitarianism.

12. Students of political history will remember Montesquieu for his theory that a relationship exists between an area's climate, geography, and general circumstances and the form of government that evolves.

13. Diderot's voluminous writings include the novels *La religieuse (The Nun,* 1796), an attack on convent life; *Le neveu de Rameau (Rameau's Nephew,* 1761, or later), a social satire; and *Jacques le fataliste (Jacques the Fatalist,* 1796), which explored the psychology of free will and determinism. *Lettre sur les aveugles* (1749), about the way the blind learn, and the dramatic philosophical dialogue *Le reve d'Alembert* (1830) contain his atheistic beliefs. A pioneer in aesthetic criticism, he founded Les Salons in1759, a journal for which he wrote criticisms of the annual Paris art exhibitions.

14. French mathematician, philosopher, and encyclopedist. Born in Paris, d'Alembert was the illegitimate son of the French writer Claudine Guerin de Tencin (1681?-1749) and was left as an infant on the steps of the Chapel of Saint Jean le Rond, from which he received his name. He was educated at the College Mazarin, where he excelled in mathematics, physics, and astronomy. In 1751, he became associated with the French encyclopedist Denis Diderot in editing the great French Encyclopedie. Although he withdrew from the editorship in 1758 because of government interference with the publication, d'Alembert continued to contribute articles on science and philosophy.

15. French physician and philosopher. Forced to flee from France and then from Leiden because of his materialistic teachings in his *Histoire naturelle de l'ame* (1745) and *L' Homme-machine* (1747). Founded asylum with Frederick the Great in Berlin. Believed that psychical phenomena are due to organic changes in brain and nervous system, that the only pleasures are those of the senses, that life should be spent in enjoyment of pleasures, that the soul ceases to exist with the death of the body. Also wrote *Discours sur le bonheur* (1748), *Les Animaux plus que machines* (1750), and *Le Petit Hommealongue queue* (1751).

16. Hume, David, "Principle Writings on Religion Including Dialogues Concerning Natural Religion and the Natural History of Religion," Oxford University Press, 1998, 192 ed., 217.

17. Samuel Clark (1675-1729), English philosopher. Chaplain to bishop of Norwich (1698); to Queen Anne (1706); disciple of Isaac Newton.

18. Ibid., 597.

19. Ibid., 234.

20. Immaterialism is the belief that the essential nature of reality lies only in consciousness or reason, that only mental states or entities are knowable, and that only what is perceptible is real,

21. Kant claimed his ancestors came from Scotland and their family name was Cant.

22. Letter to Lavater, April 28, 1775.

23. Kant, Immanuel, *Religion Within the Boundaries of Pure Reason,* (1793).

24. New Agers are noted for saying "I am god."

25. Matthew 24:24

26. Kant, *Fundamental Principles of the Metaphysic of Morals.*

27. The Young Hegelians, however, thought that Hegel betrayed his own philosophy of freedom of consciousness in his work *The Philosophy of the Right,* which described a rational and free community not unlike the authoritarian Prussian monarchy.

28. Bauer, Bruno. 1809-1882. German theologian and historical writer. Lecturer on theology (1834-42); deprived of license (1842) for radical literary criticism of the Bible; later an avowed atheist and socialist. Author of *Kritik der evangelischen Geschichte des Johannes* (1840), *Kritik der evangelischen Geschichte der Synoptiker* (1840), *Christus und die Caesaren* (1877), etc.

29. Schmidt, Johann Kaspar (1806-1856). Pseudonym Max Stirner. German philosopher. Made living as a translator of Adam Smith's *Wealth of Nations*; author of *Der Einzige und sein Eigentum* (1845) and other works that inspired many nineteenth- and twentieth-century anarchists and revolutionaries.

30. Hegel disagreed with Kant's transcendent and unknowable "things in themselves," believing that ultimate reality, which he called the Mind or Absolute Spirit, was indeed knowable. "All we know is all there is," Kant often said.

31. Hegel, did not like to use the terms thesis, antithesis, and synthesis.

32. Hegel thought that Christianity was the epitome of world religions.

33. Hegel was a supporter of the Enlightenment and the French Revolution.

34. de Lubac, Henri, *The Drama of Atheistic Humanism,* Ignatius Press, San Francisco, 1995, 33.

35. Before it was called communism it was known as "scientific socialism."

36. Marx is known as the "Red Doctor."

37. Charles Fourier (1772-1837) French philosopher, salesman, and socialist, born in Besancon, and educated at the university there. His socialist ideas were based on an evolutionary process. History, he said, is a process of social evolution with eight distinct periods. These periods are ascending in superiority to a peak stage: the Societary Economy, which he called "the reign of harmony." The four principles of Fourierism are: nurture over nature; everybody works; social ownership of production; no private property.

38. The duc de Saint-Simone (1760-1825), who called himself "Citizen Bonhomme" or "Good Man," had a revelation during the French Revolution where Charlemagne supposedly appeared to him and said he would be the most famous philosopher and intellectual of history. The Duke's four-point plan was this: all privileges will be abolished; work will be permitted for everyone; rewards will be determined by merit not birth; the society of the future will be run by scientists and economists, not politicians.

39. Robert Owen, a British self-made capitalist and cotton-mill owner, tried to set an example by transforming the squalid Scottish mill town New Lanark into a model industrial community between 1815 and 1828.

40. de Lubac, Henri, *The Drama of Atheistic Humanism,* Ignatius Press, San Francisco, 1995, 39.

41. Ibid., 41.

42. Bentham, Jeremy, *An Introduction to the Principles of Morals and Legislation,* ed. H. L. Hart and J. H. Burns, Oxford University Press, 1996.

43. As a member of the British Parliament, John Stuart Mill was considered a political radical for his time, advocating such liberal ideas as public ownership of natural resources, equality for women, compulsory education, and birth control. His advocacy of women's rights in the debates on the Reform Bill of 1867 and his work *On the Subjection of Women* (1869) led to the formation of the women's suffragette movement in Great Britain.

44. The term eugenics, which means "good birth," was first coined by a cousin of Charles Darwin named Sir Francis Galton, a British statistician who believed that all human traits and personalities were inherited, and who measured the size of people's head to determine their IQ and "feeble-mindedness."

45. Adler, Eric, *The Kansas City Star,* April 27, 2002.

46. *The Gay Science,* [125], 1882.

47. *The Gay Science,* Book V, [342], 1887.

48. *The Antichrist,* [62], 1888.

49. *The Antichrist,* [2], 1888.
50. *The Antichrist,* [7], 1888.
51. *The Antichrist,* [5], 1888.
52. *The Antichrist,* [3], 1888.
53. Letter to Peter Gast, October 27, 1887.
54. *Beyond Good and Evil,* [212], 1886.
55. Ibid.
56. *Ecce Homo,* 1888.
57. "Why I am a Destiny," [1].
58. *The Antichrist,* [51], 1888.
59. Ibid.
60. Matthew 12:36-37.
61. 1 Timothy 4:1-5.
62. Mother Theresa of Romania chose to establish her Sisters of Charity in Calcutta, India, so that she could work amongst the poorest of the poor, those whom no one wanted.
63. *The Antichrist,* [20], 1888.
64. *The Antichrist,* [42], 1888.
65. *The Antichrist,* [2], 1888.
66. Letter to Peter Gast, October 27, 1887.
67. Leni Riefenstahl's epic documentary of the 1934 Nazi rally in Nuremberg, *Triumph of the Will,* glorified Hitler's determined will to resurrect Germany to greatness again in the Third Reich.
68. Known as one of the greatest logicians and thinkers of his time, Peirce's works include *Photometric Researches* (1878) and *Studies in Logic* (1883). His scientific essays appear in *Chance, Love, and Logic,* published posthumously in 1923.
69. Although not one to believe in metaphysics, Peirce was deeply involved in semiotics, a general philosophical theory of the importance and meaning of signs and symbols.
70. William James, *A Pluralist Universe* (1909).
71. 1933 May/June issue of *New Humanist* (Vol. VI: 3, 1-5). The following is how the Humanist Manifesto appeared:

> The time has come for widespread recognition of the radical changes in religious beliefs throughout the modern world. The time is past for mere revision of traditional attitudes. Science and economic change have disrupted the old beliefs. Religions the world over are under the necessity of coming to terms with new conditions created by a vastly increased knowledge and experience. In every field of human activity, the vital movement is now in the direction of candid and explicit humanism. In order that religious

humanism may be better understood we, the undersigned, desire to make certain affirmations which we believe the facts of our contemporary life demonstrate. There is a great danger of a final, and we believe fatal, identification of the word *religion* with doctrines and methods which have lost their significance and which are powerless to solve the problems of living in the Twentieth Century. Religions have always been means for highest values in life. Their end has been accomplished through the interpretation of the total environing situation (theology or world view), the sense of values resulting therefrom (goal or ideal), and the technique (cult), established for realizing the satisfactory life. A change in any of these factors results in alteration of the outward forms of religion. This fact explains the changefulness of religions throughout the centuries. But through all changes religion itself remains constant in its quest for abiding values, an inseparable feature of human life. Today man's larger understanding of the universe, his scientific achievements, and his deeper appreciation of brotherhood have created a situation which requires a new statement of the means and purposes of religion. Such a vital, fearless, and frank religion capable of furnishing adequate social goals and personal satisfactions may appear to many people as a complete break with the past. While this age does owe a vast debt to the traditional religions, it is nonetheless obvious that any religion that can hope to be a synthesizing and dynamic force for today must be shaped for the needs of this age. To establish such a religion is a major necessity of the present. It is a responsibility which rests upon this generation. We therefore affirm the following:

First: Religious humanists regard the universe as self-existing and not created.

Second: Humanism believes that man is a part of nature and that he has emerged as the result of a continuous process.

Third: Holding an organic view of life, humanists find that the traditional dualism of mind and body must be rejected.

Fourth: Humanism recognizes that man's religious culture and civilization, as clearly depicted by anthropology and history, are the product of a gradual development due to his interaction with his natural environment and with his social heritage. The individual born into a particular culture is largely molded by that culture.

Fifth: Humanism asserts that the nature of the universe depicted by modern science makes unacceptable any supernatural or cosmic guarantees of human values. Obviously humanism does not deny the possibility of realities as yet undiscovered, but it does insist that the way to determine the existence and value of any and all realities is by means of intelligent inquiry and by the assessment of their relation to human needs. Religion must formulate its hopes and plans in the light of the scientific spirit and method.

Sixth: We are convinced that the time has passed for theism, deism, modernism, and the several varieties of "new thought."

Seventh: Religion consists of those actions, purposes, and experiences which are humanly significant. Nothing human is alien to the religious. It includes labor, art, science, philosophy, love, friendship, recreation—all that is in its degree expressive of intelligently satisfying human living. The distinction between the sacred and the secular can no longer be maintained.

Eighth: Religious humanism considers the complete realization of human personality to be the end of man's life and seeks its development and fulfillment in the here and now. This is the explanation of the humanist's social passion.

Ninth: In place of the old attitudes involved in worship and prayer the humanist finds his religious emotions expressed in a heightened sense of personal life and in a cooperative effort to promote social well-being.

Tenth: It follows that there will be no uniquely religious emotions and attitudes of the kind hitherto associated with belief in the supernatural.

Eleventh: Man will learn to face the crises of life in terms of his knowledge of their naturalness and probability. Reasonable and manly attitudes will be fostered by education and supported by custom. We assume that humanism will take the path of social and mental hygiene and discourage sentimental and unreal hopes and wishful thinking.

Twelfth: Believing that religion must work increasingly for joy in living, religious humanists aim to foster the creative in man and to encourage achievements that add to the satisfactions of life.

Thirteenth: Religious humanism maintains that all associations and institutions exist for the fulfillment of human life. The intelligent evaluation, transformation, control, and direction of such associations and institutions with a view to the enhancement of human life is the purpose and program of humanism. Certainly religious institutions, their ritualistic forms, ecclesiastical methods, and communal activities must be reconstituted as rapidly as experience allows, in order to function effectively in the modern world.

Fourteenth: The humanists are firmly convinced that existing acquisitive and profit-motivated society has shown itself to be inadequate and that a radical change in methods, controls, and motives must be instituted. A socialized and cooperative economic order must be established to the end that the equitable distribution of the means of life be possible. The goal of humanism is a free and universal society in which people voluntarily and intelligently cooperate for the common good. Humanists demand a shared life in a shared world.

> *Fifteenth and last:* We assert that humanism will: (a) affirm life rather than deny it; (b) seek to elicit the possibilities of life, not flee from it; and (c) endeavor to establish the conditions of a satisfactory life for all, not merely for the few. By this positive morale and intention humanism will be guided, and from this perspective and alignment the techniques and efforts of humanism will flow.
>
> So stand the theses of religious humanism. Though we consider the religious forms and ideas of our fathers no longer adequate, the quest for the good life is still the central task for mankind. Man is at last becoming aware that he alone is responsible for the realization of the world of his dreams, that he has within himself the power for its achievement. He must set intelligence and will to the task.

72. Dewey publicly defended the Russian communist Leon Trotsky against his rival Joseph Stalin, and supported the British socialist philosopher Bertrand Russell.

73. Dewey, John, *Human Nature and Conduct,* Henry Holt, 1922.

74. In 1896, Dewey wrote "The Reflex Arc Concept in Psychology," which attacked passive stimulus-response learning for active student participation learning.

75. Dewey, *Democracy and Education,* 1916; MW 9:369.

76. Russell, Bertrand, *Why I am Not a Christian: And Other Essays on Religion and Related Subjects,* edited by Paul Edwards, Simon & Schuster, 1957, p. 17. It was taken from Russell's lecture on March 6, 1927, at Battersa Town Hall, which was put on by the south London Branch of the National Secular Society.

77. Ibid., 21.

78. Ibid., v.

79. Ibid., 65.

80. Sartre, Jean-Paul, *No Exit*

81. Sartre, Jean-Paul, *Being and Nothingness,* 23.

82. Sartre, *Nausea.*

83. Sartre's other major existential works include the unfinished series *Les chemins de la liberte (The Roads of Liberty),* comprising *The Age of Reason* (1945), *The Reprieve* (1945), and *Troubled Sleep* (1949), *The Respectful Prostitute* (1946), *The Condemned of Altona* (1959), and his autobiography, *The Words* (1964).

Chapter Two

1. Cuvier did not believe in evolution. To prove his point he pointed to a four-thousand-year-old mummy of an Egyptian ibis, which had not changed at all.

2. In 1871, Darwin wrote *Descent of Man* and *Selection in Relation to Sex.*

In his *Origin* Darwin does not mention the evolution of man from apes but hints at it. Wallace goaded Darwin into writing it but disagrees with Darwin saying that natural selection cannot account for modern man. Wallace believed that the physical body of man came from evolution but not the spiritual being of man which came from God. Darwin believed that everything could be explained by natural selection, acknowledging that even monkeys laugh and smile.

3. Bertram Boltwood studied with Rutherford in Canada, and discovered that radioactive materials decay at a constant rate, so measure the radioactive half-life of radioactive particles to determine their age. Half-life of Uranium^{238} to Pb^{206} dates the earth to 4.5 billion years old. By the 1920s, the age of the earth was agreed to be 4.5 billion years old.

4. Gregor Mendel (1822-1884) was a native of Morovia, Austria, who joined a monastery near Brno in the Czech area. Mendel loved science but failed to become a degreed scientist because of the stress of studying and taking examinations, which caused him several nervous breakdowns. At his monastery in Brno, Mendel was keenly interested in botany and the mode of inheritance in hybrid plants, so he carefully designed experiments crossing thousands of pea plants with different characteristics over a period of ten years to observe the results. It was Mendel's work with pea plants that enabled him to discover the true laws of genetics (Mendelian genetics), in that genes are inherited in pairs, one from the mother and one from the father, and that some genes are dominant and some are recessive. Mendel's journal on genetics was published in 1866, but at that time no one thought it important because he was working with hybrid species. Mendel tried one more time, but again he failed to get noticed by the scientific community. (As far as we know, Darwin never heard of Gregor Mendel.) Mendel would go to his grave as an insignificant nobody, and it wouldn't be until 1900 when Mendel's work was rediscovered by three independent researchers who validated his results the hard way by redoing them all over, that Mendelian genetics was finally accepted by the scientific community; which, in turn, provided a helpful answer on how Darwin's slight modifications in traits could be passed on to succeeding generations without being totally lost in the gene pool. Mendelian genetics, therefore, was another important ally to the Darwinian evolutionists.

5. The term "gene" was coined by Wilhelm Johanssen in 1909.

6. The book, *Voices for Evolution,* iv, published by the NCSE (National Center for Science Education in Berkeley) contains the lists of about seventy scientific, educational, civil liberty, and liberal religious organizations that demand the teaching of evolution in public schools as a virtual scientific fact, and adamantly oppose to any teaching of creationism as an alternative theory. The prestigious National Academy of Sciences

adopted an official resolution on April 27, 1976, which affirmed "freedom of inquiry and expression," and that the search to uncover the mysteries of the universe "should be conducted under conditions of intellectual freedom, without religious, political, or ideological restrictions." But in their 1984 booklet "Science and Creationism: A view from the National Academy of Science," they state that "creationism has no place in a scientific curriculum at any level." John Ankerberg and John Weldon, *Darwin's Leap of Faith: Exposing the False Religion of Evolution* (Eugene, Ore.: Harvest House Publishers, 1998), 105.

7. Not until the 1950s did we know much about molecular biochemistry. In 1955, the first chemical structure of a protein, insulin, was reported. In 1957, the first 3-D structure of a protein, Sperm whale myoglobin, was likewise reported.

8. Isaac Manly, M.D., *God Made: A Medical Doctor Looks at the Reality of Creation,* (Joplin, Mo.: College Press, 1994), 13, 116.

9. The theory that the earth has undergone a series of sudden geological catastrophes, resulting in the extinction of numerous species like the dinosaur, is currently the most widely accepted theory amongst modern geologists and paleontologists.

10. Environmentalists and birth control advocates who recommend population control to save the earth are almost always Malthusian alarmists. The recent advances in biotechnology and genetic engineering, however, have greatly increased our ability to produce food, and there is no danger of a food shortage. Famines today are the result of repressive dictatorial regimes that purposely starve their people to keep themselves in power, not the result of a shortage of food in the world. As far as the earth being overcrowded, the entire population of the world, if stood back to back, could be placed in Jacksonville, Florida.

11. Charles Darwin, *Autobiography of Charles Darwin: 1809-1882,* ed. Nora Barlow (New York: W. W. Norton & Company, 1996) 120.

12. The term "mutation" originated from the Dutch Botanist Hugo DeVries.

13. Hybrids and mixes that don't normally form on their own in real life can be artificially produced by breeders in both plants and animal, such as the breeding of horses with donkeys to get mules.

14. Darwin delayed twenty years the publication of his book mainly because of his wife's belief in God, his own self-doubts, his dislike of controversy, the fight it would cause with the clergy, and his acknowledgment of the many deficiencies in his theory of evolution.

15. Critics of Darwin say his ideas of evolution by means of natural selection were not new and that other minor evolutionists, such as William Wells, Patrick Matthew, Edward Blyth, and Geoffrey St. Hillaire, preceded him.

16. The amniotic egg of a reptile with its shell, amnion, allantois, yolk sac, is vastly different than the amphibian egg with its jelly and yolk mass. The amniotic egg has a tough impervious shell, an egg white (albumen) and the formation of a special acid to yield its water, the excretion of nitrogenous wastes in the form of water insoluble uric acids, an amniotic cavity in which the embryo floats, the formation of the allantois from the future floor of the hindgut as a receptacle for waste products and later to serve the function of a respiratory organ, the development of a caruncle or tooth for the hatchling to break out of its shell, a sufficient quantity of yolk to feed the embryo before hatching, changes in the urogenital system of the adult female to allow fertilization of the egg before hardening of the shell.

17. Hereditary units (genes) were first discovered in 1866 by Gregor Mendel.

18. I wonder how many people would believe in evolution if Darwinists called it a primordial "sewer" rather than a "soup"?

19. Evolutionists often refer to the "primordial soup" in which life first arose, where organic molecules formed amino acids, which combined to form proteins, which combined to form the first one-celled organism. This primordial soup was supposedly a thick, condensed mixture of organic compounds. But no such evidence exists in the oldest-known sedimentary rocks, the dawn rocks of Greenland that are about 3.9 billion years old, that there was ever any primordial soup. Geological evidence shows that the early earth was covered with water, not soup, just as the Book of Genesis says.

20. Darwin, C., from Darwin, F. (1888) ed, The Life and Letters of Charles Darwin, 3 vols., John Murray, London, op cit, vol. 3, 18.

21. For amino acids to have formed by themselves in a reducing environment, as in the Urey-Miller experiment, the earth had to be totally devoid of oxygen. But all the scientific evidence suggests that oxygen was present when the earth came into existence.

22. Richard Dawkins, *The Selfish Gene* (London: Oxford University Press, 1989), 20.

23. Ibid., 192-93.

24. Evolutionists like Richard Dawkins know this dilemma and speculate that zinc was the first enzyme to replicate DNA and RNA, which is pure speculation without any scientific proof.

25. Richard Dawkins, *The Blind Watchmaker: Why the Evidence of Evolution Reveals a Universe Without Design* (New York: W. W. Norton & Company, 1996), 139.

26. Ibid., 159-60.

27. Harold Morowitz, a Yale University physicist, calculated the odds of

one single bacterium arising spontaneously from the basic building blocks of life to be 1 chance in $10^{100,000,000,000}$. (From Ankerberg, John, & Weldon, John, Darwin's Leap of Faith: Exposing the False Religion of Evolution, 1998, Harvest House Publishers, Eugene Oregon.)

28. David J. Rodabaugh, Ph.D., associate professor of mathematics at the University of Missouri, calculated that the probability of life arising by itself anywhere in the universe is $10^{2,999,942}$. David Rodabaugh, Ph.D., "The Queen of Science Examines the King of Fools," *Creation Research Society* Quarterly, June 1975, 14.

29. Dawkins, *The Blind Watchmaker.*

30. Ibid., 49.

31. Ibid., 50.

32. Ibid., 94.

33. Charles Darwin, *The Variation of Plants and Animals Under Domestication,* 1868.

34. Ibid.

35. Haeckel called his missing link *Pithecanthropus,* which means "ape-man."

36. Darwin, *Origin,* 492.

37. British embryologist Sir Gavin de Beer discussed the problem of Darwinists with homologous structures in *Homology, an Unresolved Problem.*

38. J. Buettner-Janusch, *An Introduction to the Primates in Evolutionary and Genetic Biology of Primates,* 2 vols. (1963). See Ed J. Buettner-Janusch, Academic Press, New York, vol.1, 3.

39. Discovered in Germany in 1861, this is claimed by the evolutionists to have been the first bird, which it is not.

40. Fossils of Ichthyostega have been discovered in Greenland and show it to have been an animal of nearly three feet in length, with a broad, flat head like an alligator, and legs, and a fish-like tail.

41. This is another fossil species found in Greenland much like the Ichthyostega except it is shorter in length, has gills, and possesses eight digits on the front limbs and seven on the back limbs, which is not in keeping with land animals who have five digits on all feet.

42. Darwin, *Origin,* 307.

43. Richard Goldschmidt, *The Material Basis of Evolution* (Patterson, NJ: Pageant Books 1940).

44. Wallace disagreed with Darwin's *Descent,* saying that natural selection alone cannot account for the evolution of modern man. Wallace believed that the physical body of man came from evolution but not the spiritual being of man, which he believed came from God.

45. The scientific name given to Java Man is *Homo erectus.*

46. Samantha Weinberg, *A Fish Caught in Time* (New York: Harper Collins, 2000), 140.

47. Ibid., 137.

48. Jonathon Wells, *Icons of Evolution: Science or Myth* (Washington, D.C.: Regnery Publishing, 2000), 130.

49. Ibid., 145-46.

50. Ibid., 146.

51. A. C. Chandler, *Introduction to Parasitology* 1961, 10th ed., J. Wiley and Sons, New York, 16.

52. G. Cuvier, cited in W. Coleman, *George Cuvier: Zoologist* (Cambridge: Harvard University Press, 1964), 171-72.

53. Darwin. C. (1881) in Darwin, F. (1888) ed, *The Life and Letters of Charles Darwin,* 3 vols., John Murray, London, op cit, vol. 3, 248.

54. Darwin, *Origin,* 181, 192.

55. Ibid., 182.

56. Trilobites are primitive arthropods, which are related to crabs and other shelled creatures.

57. Leslie Barinerd Arey, Ph.D., Sc.D., LL.D., *Developmental Anatomy: A Textbook and Laboratory Manual of Embryology,* (rev 7th ed. Philadelphia and London, W. B. Saunders Company, 1974), 529.

Chapter Three

1. Ironically, the teaching of Islam, which is currently the in vogue subject with liberal educators in America, is mandatory in some schools.

2. In fact, Jesus talked about the real existence of Satan and hell more than any other subject.

3. Deuteronomy 28:28.

4. 1 Samuel 16:14.

5. Daniel 4:13-14.

6. Daniel 4:31-32.

7. Homer, *The Iliad and the Odyssey* (Barnes and Noble Books, 1999).

8. A young David soothed the tormented mind of King Saul by playing a harp for him.

9. Hippocrates, *On the Sacred Disease.* Trans. Francis Adams, rendered into HTML by Steve Thomas (Adelaide: University of Adelaide Library Electronic Texts Collection, 1988), http://etext.library.adelaide.edu.au/h;hippocrates/h7w/disease.html.

10. Morton Hunt, *The Story of Psychology* (Doubleday, 1993).

11. Alfred M. Freedman, Harold L. Kaplan, and Benjamin J. Sadock, *Modern Synopsis of Comprehensive Textbook of Psychiatry/II,* 2nd ed. (Baltimore: The Williams & Wilkins Co., 1976), 14.

12. Ibid., 15-18.

13. John Cottingham, *The Cambridge Companion to Descartes* (Cambridge: University of Cambridge Press, 1992).

14. The name Maximilian is derived from the Latin word maximus, which means greatest or maximum.

15. Hunt, *The Story of Psychology*, 97-102.

16. Ibid., 101.

17. Ibid., 102-7.

18. Ibid., 122-26.

19. Ibid., 127-43.

20. Ibid., 144-65.

21. Ibid.

22. Ibid., 208-23.

23. Ibid., 224-41.

24. Kansas City Star, 12-30-93.

25. Hunt, *The Story of Psychology*, 232.

26. Social Welfare History Archives of the University of Minnesota.

27. Hunt, *The Story of Psychology*, 233.

28. Ibid., 248-53.

29. Ibid., 253-62.

30. Ibid., 261.

31. Hunt, *The Story of Psychology*, 268-75. B. F. Skinner, *About Behaviorism* (New York: Vintage Books, 1948).

32. One of the largest treatment centers for clerical sex abuse of minors and children is the Servants of the Paraclete treatment center in New Mexico.

33. Sigmund Freud, *The Interpretation of Dreams* (MacMillan, 1913). Freud, *On Dreams* (Rebman, 1914). Freud, *Dream Psychology: Psychoanalysis for Beginners* (McCann, 1920). Freud, *Dreams in Folklore* (International University Press, 1958).

34. Elizabeth Loftus and Katherine Ketcham, *The Myth of Repressed Memory: False Memories and Allegations of Sexual Abuse* (St. Martin's Press, 1996).

35. *Newsweek* magazine, March 14, 1994; *Time* magazine, April 17, 1995.

36. Dee Brown, *Satanic Ritual Abuse: A Therapist's Handbook* (Blue Moon Press, 1994). Colin Ross and Elizabeth Loftus *Satanic Ritual Abuse: Principles of Treatment* (Toronto: University of Toronto Press, 1995).

37. Freud was an ardent believer in Darwinism.

38. E. Michael Jones, *Degenerate Moderns: Modernity as Rationalized Sexual Behavior* (Ignatius Press, 1993), 196.

39. Peter Gay, *Freud: A Life for Our Times* (W. W. Norton & Company, 1998), 44.

40. Modern Synopsis of Comprehensive Textbook of Psychiatry/II, second edition, 230.

41. Gay, *Freud: A Life for Our Times,* 56.

42. Ibid., 56.

43. Ibid., 61.

44. Ibid., 57.

45. Ibid., 149.

46. Ibid., 62.

47. Ibid., 63.

48. Ibid., 65.

49. Jones, *Degenerate Moderns,* 215.

50. Ibid., 156.

51. Ibid., 180.

52. Ibid., 159.

53. Peter Gay, *The Freud Reader* (W. W. Norton & Company, 1995), 429.

54. Ibid., 435.

55. Ibid., 504.

56. Ibid., 508-9.

57. Ibid., 697.

58. Ibid., 703-4.

59. Ibid., 704-5.

60. Ibid., 786-87.

61. Ibid., 790.

62. A Freudian could really have fun psychoanalyzing that bad oral habit.

63. R. M. Jurevich, Ph.D., *The Hoax of Freudianism: A Study of Brainwashing the American Professionals and Laymen* (Philadelphia: Dorrance & Co., 1974).

64. The ancient practice of alchemy in the Middle Ages was linked to the search for the elixir of life and the quest to turn base metals into gold. It has been linked to superstition, mythology, mysticism, and the occult in the past, but gave rise to modern chemistry during the Scientific Revolution of the 1600s.

65. Murray Stein, *Jung's Treatment of Christianity: The Psychotherapy of a Religious Tradition* (Chiron Publications, 1985), 89.

66. C. G. Jung, *Memories, Dreams, Reflections* (New York: Vintage Books, 1989), 150.

67. Jung, *Treatment of Christianity,* 10-11.

68. C. G. Jung, *Psychology and Western Religion,* translated by R. F. C. Hull, Bollingen Series (Princeton University Press 1984): 88.

69. Ibid., 25.

70. Ibid., 11-12.

71. Ibid., 9.

72. Ibid., 13-24.

73. Pre-Christian Greece saw the advent of Gnostic cultists who were distinguished by their pretension to mysticism, esoteric insights, their emphasis on secret knowledge of spiritual truth (gnosis) rather than faith, and their conviction that matter was intrinsically evil. Gnosticism is the heretical belief that the Unknown God didn't create the world; that the world is a mistake caused by a fall and split within the deity; and that spiritual man is an alien to the natural world and is "related to the deity and becomes conscious of his deepest Self when he hears the word of revelation. Not sin or guilt, but unconsciousness, is the cause of evil."

74. Jung, *Psychology and Western Religion,* 30.

75. Ibid., 50.

76. Ibid., 250.

77. C. G. Jung, *Psychology and Western Religion,* 60-96.

78. Ibid., 64.

79. Ibid., 74.

80. Ibid., 92.

81. Ibid., 84.

82. Jung, *Psychology and Western Religion,* 118.

83. Ibid., 120-21.

84. Ibid., 121-42.

85. Jung, *Treatment of Christianity,* 136.

86. C. G. Jung, *The Psychology & Alchemy* (Princeton University Press), 34.

87. Jung, *Memories, Dreams, Reflections,* 11-12.

88. Ibid., 13.

89. Ibid., 17.

90. Ibid., 18-19.

91. Ibid., 36.

92. Ibid., 37-39.

93. Ibid., 39.

94. Ibid., 40.

95. Ibid., 93.

96. Ibid., 55.

97. Ibid., 44-45.

98. Ibid., 51.

99. Ibid., 35-36.

100. Ibid., 105.

101. Ibid.

102. Jung's maternal grandfather, three uncles, a great-uncle, and an aunt all experienced visions and paranormal phenomena.

103. C. G. Jung, *Psychology and the Occult,* translated by R. F. C. Hull (Bollingen Series, Princeton University Press, 1977), 6-91.

104. Ibid.

105. Ibid.

106. Ibid., 116, 108-25.

107. Ibid., 97.

108. Jung, *Memories, Dreams, Reflections,* 155.

109. Ibid., 177.

110. Ibid., 181.

111. Ibid., 183.

112. Ibid., 184-85.

113. David Michael Lindsey, *The Woman and the Dragon: Apparitions of Mary* Gretna, La: Pelican Publishing Company, 2001), 88-89.

114. Robert Hopcke, *Jung, Jungians, and Homosexuality* (Boston and London, Shambhala, 1991). Besides the feminine spirit of woman inside men, Jung also had a masculine spirit for women he called the *animus.*

115. Jung, *Memories, Dreams, Reflections,* 187.

116. Ibid., 190.

117. Ibid., 190-91.

118. Ibid., 192.

119. Jung, *Psychology and the Occult,* 146-52.

120. Jung, *Memories, Dreams, Reflections,* 229-30.

121. Ibid., 313.

122. Ibid., 198-99.

123. Ibid., 280-82.

124. Ibid., 213-14.

125. Ibid., 220.

126. Ibid., 289-93.

127. Ibid., 356.

128. For the complete Secret of La Salette, refer to my book, *The Woman and the Dragon: Apparitions of Mary,* 88-91.

129. Leviticus 18:22.

130. Matthew 18:6, Mark 9:42, Luke 17:2.

131. Romans 1:25-27.

132. 2 Timothy 3:1-3.

133. Mark 8:34-38.

134. Genesis 3:4-5.

135. Revelation 13:18.

136. In her 1846 Secret of La Salette, Mary forewarned us that evil books would be abundant on earth, so we should be very careful of what we read. The former address of Bantam Doubleday Dell and Warner Books on 666 Fifth Avenue in New York City is a great symbolic reminder of this fact.

137. Stephen Larsen and Robin Larsen, *A Fire in the Mind* (Doubleday, 1991).

138. Ibid., 15.

139. Ibid., 16-19.

140. Ibid., 23-24.

141. Ibid., 32.

142. Helen Blavatsky, Henry Olcott, William Judge, and others founded the Theosophical Society in New York in 1875 for the promotion of gnosticism, freemasonry, Rosicrucianism, the Kabbala, Eastern mysticism, spiritism, occultism, and antichristianity. Theosophy's links with satanism and the occult were evident when Blavatsky founded and edited the magazine *Lucifer* before she died in 1891.

143. Larsen, *A Fire in the Mind,* 41-50.

144. Ibid.

145. Ibid., 65-69.

146. Ibid., 70-78.

147. Ibid., 79-86.

148. Ibid., 86-94.

149. Ibid., 95-102.

150. Matthew 24:22-24.

151. Thomas Mann's book *The Magic Mountain* was a favorite of Campbell. Mann's writings on the ancient Gnostic manuscript known as the "Gospel of Thomas," which was found at Nag Hammadi in Egypt in 1945 and has been touted by critics of the Catholic Church as the true secret teachings of Jesus Christ. Two of the Gnostic quotations in the Gospel of Thomas that were favorites of Campbell were "Split a piece of wood; I am there. Lift up the stone, and you will find me there," and "The Kingdom of God will not come by expectation."

152. Larsen, *A Fire in the Mind,* 103-20.

153. Ibid., 138-51.

154. Ibid., 231-35.

155. Joseph Campbell, *The Hero's Journey: Joseph Campbell on His Life and Work* (New World Library, 2003), 58.

156. Campbell did not believe in the reincarnation of the soul or life after death.

157. John M. Mahers and Dennie Briggs, ed., *An Open Life: Joseph Campbell in Conversation with Michael Toms* (New York: Harper Perennial, 1989), 78.

158. Larsen, *A Fire in the Mind,* 317-19.

159. Ibid., 320-26.

160. Diane K. Osbon, ed., *Reflections on the Art of Living: A Joseph Campbell Companion* (New York: Harper Perennial, 1991), 59-60.

161. Joseph Campbell, *The Hero with a Thousand Faces* (Princeton University Press, 1990), 25.

162. Ibid., 4.
163. Ibid., 256.
164. Ibid., 249.
165. Osbon, *Reflections,* 146.
166. Campbell, *Hero with a Thousand Faces,* 71.
167. Osbon, *Reflections,* 123.
168. Carl Jung's writings on the psychological interpretations of religion as originating in the primitive archetypes of man's unconscious mind, perhaps more than any other famous author or teacher, convinced Campbell that all religions, including Christianity, were just myths.
169. Larsen, *A Fire in the Mind,* 362-64.
170. Ibid., 273-74.
171. Ibid., 392-422.
174. Osbon, *Reflections,* 414.
175. Joseph Campbell, *The Masks of God: Occidental Mythology,* vol. 3, (Penguin Books, 1972), 7.
176. Ibid., 362.
177. Joseph Campbell, *The Masks of God: Creative Mythology,* vol. 4 (Penguin Books, 1976), 629-30.
178. Joseph Campbell, *Myths to Live By,* (Penguin Books, 1993).
179. Joseph Campbell, *The Mythic Image,* Princeton University Press, 1981.
180. Larsen, *A Fire in the Mind,* 480.
181. Campbell often said sarcastically that Christians think of myths as "other people's religion."
182. Larsen, *Hero's Journey,* 172.
183. Joseph Campbell, *The Power of Myth, with Bill Moyers* (Doubleday, 1988), 22.
184. Ibid., 39.
185. Ibid., 57.
186. Ibid., 148.
187. Ibid., 161.
188. Mahers, *An Open Life* , 94.
189. Ibid., 29.
190. Ibid., 28-29.
191. Osbon, *Reflections,* 156.
192. Mahers, *An Open Life,* 28.
193. Campbell, *The Power of Myth,* 229.
194. Ibid., 148.
195. The last name of Charles Manson, for instance, who was released from prison in 1966 and immediately formed a band of a dozen or so murdering, thieving, doped-up hippie apostles, is "Man's Son" or "Son of Man," the name Jesus referred to himself more often than any other

name. Madonna, on the other hand, appeared in the world about the same time as the real Madonna was appearing to visionaries in Bosnia and Rwanda before their civil wars started in order to warn the world about future evils and punishments. Madonna the singer's message was one of materiality, self-indulgence, unbridled sex, and blasphemous immorality, rather than one of repentance and conversion.

196. Larsen, *A Fire in the Mind,* 488.

197. Ibid., 487-88.

198. Maya Deren, who was called the "mother of underground film," was a Russian-born photographer, actress, dancer, lecturer, film maker, film critic, and author who became the first American to receive the Cannes Grand Prix International Award for Avant-Garde Film. Like her name, which means "illusion," her 1940s films were characterized by very odd, surrealistic, and dream-like fantasies of imagination.

199. Osbon, *Reflections,* 282.

200. Varying forms of the name of "Christ" also had an uncanny way of popping up in Campbell's life. Besides having the same initials as Jesus Christ, an unusual number of Joseph Campbell's close female associates had various feminized versions of Christ's name. The wife of his single most important teacher, Heinrich Zimmer, was named Christiane. The name of the wife of comparative religionist Mircea Eliade, Campbell's good friend and colleague, was Christinel. A student of his at Sarah Lawrence who suffered from hallucinations and later married a good friend of his, Stan Grof, was named Christina Valier.

201. Larsen, *A Fire in the Mind,* 561-62.

202. Ibid., 483-486.

Chapter Four

1. St. Matthew was a tax collector and was well trained in keeping accurate accounts and documentation, which was useful in recording the activities and sayings of Jesus Christ.

2. Flavius Josephus, *Jewish Antiquities,* Book 18, chapter 3:3, ed. H. St. J. Thackeray (Cambridge: Harvard University Press, 1998).

3. Josephus, *Jewish Antiquities,* Book 20, chapter 9.

4. Josephus, *Jewish Antiquities,* Book 18, chapter 5.

5. *The Ante-Nicene Fathers,* vol. 8 (Grand Rapids, MI: Wm. B. Eerdmans, 1951), 460-461.

6. Josephus, *Jewish Antiquities,* Book 18, chapter 2.

7. Jesus' brother Joseph was actually Jesus' cousin on his father's side, as cousins were called brothers in Jesus' time because there was no such word as cousin in Aramaic, the language Jesus spoke. Jesus' paternal uncle, Alphaeus, the brother of St. Joseph who was the husband of the Blessed Virgin Mary, had four boys named James the Lesser, Joseph,

Judas Thaddeus, and Simon. James the Lesser and Judas Thaddeus were two of the original twelve apostles of Christ mentioned in the Bible. The other two cousins, Joseph and Simon of Alphaeus, were skeptical of Jesus at first, but later become two of his most devoted disciples. The Apostle Judas Thaddeus, who wrote the Epistle of Jude, is not to be confused with the other apostle named Judas Iscariot, who betrayed Christ.

8. The excuse that the Roman guards were asleep is highly unlikely because the punishment for a guard sleeping on the job was death.

9. D. James Kennedy, *Why I Believe* (Nashville: Word Publishing, 1980), 117.

10. Mernetptah was the son of Ramses II, and many scholars think that he was the pharaoh under whom the Jewish Exodus took place. Scholars also believe that the "Victory Stela" of Mernetptah was just boastful propaganda and a misrepresentation of historical fact, but it does prove that the Israelite people definitely existed.

11. Jesus' prediction of the total destruction of the Jewish Temple actually occurred on August 30, 70 A.D., on the exact same date as the destruction of Solomon's Temple by the Babylonians.

12. Matthew 10:8.

13. Father Albert J.Hebert, S. M., *Raised From the Dead* (TAN publishers, 1986), 33.

14. Ibid., 33.

15. Ibid., 36.

16. Ibid., 43.

17. Video *Betania, Land of Grace: Apparitions of the Blessed Mother in Venezuela,* Marian Communications, P.O. Box 300, Lincoln University, PA 19352.

18. To request a copy, write to Daniel Sanford, 329 Cypress Ave., Woodlynne, NJ 08107.

19. Matthew 26:26-28.

20. John 6:53-54.

21. Luke 10:1.

22. Dr. Leonicio A. Garza-Valdes, *The DNA of God* (New York: Doubleday, 1999) 3, 83-85, 190-91.

23. Ibid., 137-40.

24. Ibid., 72.

25. Ibid., 78.

26. Ibid.

27. Ibid., 84-85.

28. Francis Johnston, *The Wonder of Guadalupe* (Rockford, Ill: Tan Books and Publishers, Inc., 1981), 121.

29. Ibid.

30. Ibid., 128-29.

Bibliography

Allen, Thomas B. *Possessed: The True Story of an Exorcism.* New York: Doubleday, 1993.

Amorth, Gabriele. *An Exorcist Tells His Story.* Trans. Nicoletta V. MacKenzie. San Francisco: Ignatius Press, 1999.

Ankerberg, John, and John Weldon. *Darwin's Leap of Faith: Exposing the False Religion of Evolution.* Eugene, Ore.: Harvest House Publishers, 1998.

Aquinas, St. Thomas. *Summa Contra Gentiles.* 4 vols. Notre Dame: University of Notre Dame Press, 1976.

Aquinas, St. Thomas. *Summa Theologica.* 5 vols. Chicago: Encyclopedia Britannica, 1990.

Arey, Leslie Brainerd, Ph.D., Sc.D., LL.D. *Developmental Anatomy: A Textbook and Laboratory Manual of Embryology.* 7th ed. Philadelphia: W.B. Saunders Co., 1974.

Argubright, John. *Bible Believers Archeology: Historical Evidence that Proves the Bible.* Vol. 2. Longwood, Fla.: Xulon Press, Inc., 2003.

Aristophanes. *Four Plays by Aristophenes: The Clouds, The Birds, Lysistrata, The Frogs.* Ed. William Arrowsmith. New York: Meridan, 1994.

Aristotle. *The Complete Works of Aristotle.* Vols. 1 and 2. Ed. Jonathan Barnes. Princeton, N.J.: Princeton University Press, 1997.

Arnaldez, Roger, and David Streight. *Averroes: A Rationalist in Islam.* Notre Dame: University of Notre Dame Press, 2000.

Augustine. *The Confessions of St. Augustine.* Ed. Albert Cook Outler. Mineola, NY: Dover Publications, 2002.

——. *City of God.* Ed. David Knowles. Trans. Henry Bettenson. New York: Penguin Classics, 1972.

Avincenna, Ibn S. *Canon of Medicine.* Chicago: Kazi Publications, 1999.

Aziz, Robert. *C. G. Jung's Psychology of Religion and Synchronicity.* New York: State University of New York Press, 1990.

Barbet, Dr. Pierre. *Doctor at Calvary: The Passion of Our Lord Jesus Christ as Described by a Surgeon.* Harrison, NY: Roman Catholic Books, 1993.

Beckett, Samuel. *Waiting for Godot.* New York: Grove Press, 1954.

Behe, Michael J. *Darwin's Black Box: The Biochemical Challenge to Evolution.* New York: Simon & Schuster Inc., the Free Press, 1996.

Bentham, Jeremy. *An Introduction to the Principles of Morals and Legislation.* Ed. H. L. Hart and J. H. Burns. London: Oxford University Press, 1996.

Bentham, Jeremy, and John Stuart Mills. *The Classical Utilitarians.* Indianapolis: Hackett Publishing, 2003.

Bobgan, Martin, and Deidre Bobgan. *Psycho Heresy: The Psychological Seduction of Christianity.* Santa Barbara: EastGate Publishers, 1987.

Boethius. *Tractates & The Consolation of Philosophy.* Trans. S.J. Tester, E.K. Rand, and H.F. Stewart. Cambridge: Harvard University Press, 1988.

Broad, William J., and Nicholas Wade. *Betrayer of the Truth: Fraud and Deceit in the Halls of Science.* New York: Simon & Schuster Adult Publishing Group, 1983.

Brown, Dee. *Satanic Ritual Abuse: A Therapist's Handbook.* Huntingdon, Penn.: Blue Moon Press, 1994.

Buettner-Janusch, J. *An Introduction to the Primates in Evolutionary and Genetic Biology of Primates.* 2 vols. New York: Academic Press 1964.

Burenhult, Gören. *The First Humans: Human Origins and History to 10,000 B.C.* San Francisco: Harper, 1994.

Caird, Rod, and Dr. Robert Foley. *Ape Man: The Story of Human Evolution.* New York: MacMillan, 1994.

Campbell, Joseph. *A Skeleton Key to Finnegan's Wake.* San Diego: Harcourt, 1944.

——. *Baksheesh and Brahman: Indian Journal.* Ed. Robin, Stephen Larsen, and Anthony Van Couvering. New York: Harper, 1995.

——. *Historical Atlas of World Mythology.* Vol. 1. New York: Van der Marck, 1988.

——. *Mythic Worlds, Modern Words: On the Art of James Joyce.* Ed. Edmund L. Epstein. New York: Harper, 1993.

——. *Myths to Live By.* Reprinted by Penguin Books, New York, 1993. Originally published by Viking Press, New York, 1972.

——. *Tarot Revelations.* San Francisco: Alchemy Books, 1980.

——. *The Flight of the Wild Gander.* New York: Viking, 1969.

——. *The Hero's Journey: Joseph Campbell on His Life and Work.* Novato, Cal.: New World Library, 2003.

——. *The Hero with a Thousand Faces.* Princeton, NJ: Princeton University Press, 1990.

——. *The Inner Reaches of Outer Space: Metaphor as Myth and Religion.* reprinted by New World Library, 2002.

——. *The Masks of God.* 4 vols. New York: Penguin Books 1972-76.

——. *The Mythic Dimension: Selected Essays.* Ed. Antony Van Couvering. San Francisco: Harper, 1997.

——. *The Mythic Image.* Princeton: Princeton University Press, 1981.

——. *The Mythic Imagination.* New York: Harper, 1998.

——. *The Power of Myth with Bill Moyers.* New York: Doubleday, 1988.

——. *Renewal Myths and Rites of the Primitive Hunters and Planters.* Woodstock: Spring Publications, 1989.

——. *Transformations of Myths Through Time.* San Francisco: Perennial Library, 1990.

——. *The Universal Myths: Heroes, Gods, Tricksters, and Others.* Ed. Alexander Eliot and Mircea Eliade. New York: New American Library, 1990.

——. *Thou Art That: Transforming Religious Metaphors.* Ed. Eugene Kennedy. New York: New World Library, 2001.

——. *Where the Two Came to Their Father: A Navajo War Ceremonial.* New York: Pantheon Books, 1943.

——. *The Way of Myth: Talking with Joseph Campbell, with Fraser Boa.* Boston: Shambhala, 1994.

Carlson, Elof. *The Unfit: History of a Bad Idea.* Woodbury, NY: Cold Spring Harbor, 2002.

Cassier, Ernst. *Kant's Life and Thought.* New Haven: Yale University Press, 1981.

Chandler, A. C. *Introduction to Parasitology.* 10th ed. New York: J. Wiley and Sons, 1961.

Cohen, I. L. *Darwin Was Wrong: A Study in Probabilities.* Greenvale, NY: New Research Publications, 1984.

Coleman, Christopher B., trans. *The Treatise of Lorenzo Valla on the Donation of Constantine.* Toronto: University of Toronto Press, 1993.

Comte, Auguste. *Auguste Comte and Positivism.* Ed. Gertrude Lenzer. Somerset, NY: Transactions Publishers, 1997.

Corrington, Robert S. *An Introduction to C. S. Peirce: Philosopher, Semiotician, and Ecstatic Naturalist.* Lanham, Md.: Rowman and Littlefield Publishers, 1997.

Cruz, Joan Carrol. *Eucharistic Miracles.* Rockford, Ill.: TAN Publishers, 1987.

——. *The Incorruptibles.* Rockford, Ill.: TAN Publishers, 1977.

Darwin, Charles. *Autobiography of Charles Darwin: 1809-1882.* Ed. Nora Barlow. New York: W.W. Norton & Company, 1996.

——. *Descent of Man and Selection in Relation to Sex.* Amherst, NY: Prometheus Books, 1997.

——. *The Life and Letters of Charles Darwin.* Ed. Francis Darwin. Honolulu: University Press of the Pacific, 2001.

——. *On the Origin of the Species by Means of Natural Selection or the Preservation of Favored Races in the Struggle for Life.* New York: Oxford University Press, 1996.

——. *The Variations in Plants and Animals Under Domestication.* Stark, Kans.: De Young Press, 1997.

Davies, Brian, and Gill Evans, ed. *Anselm of Canterbury: The Major Works of.* London: Oxford University Press, 1998.

Dawkins, Richard. *River Out of Eden: A Darwinian View of Life.* Boulder, Colo.: Basic Books, 1995.

——. *The Blind Watchmaker: Why the Evidence of Evolution Reveals a Universe Without Design.* New York: W. W. Norton & Company, 1996.

——. *The Selfish Gene.* New York: Oxford University Press, 1989.

Debus, Allen G. *Man and Nature in the Renaissance.* New York: Cambridge University Press, 1978.

Dewey, John. *The Philosophy of John Dewey: Two Volumes in One.* Ed. John J. McDermott. Chicago: University of Chicago Press, 1989.

de Lubac, Henri. *The Drama of Atheistic Humanism.* San Francisco: Ignatius Press, 1995.

della Mirandola, Pico. *On the Dignity of Man.* Trans. Douglas Carmichael, Paul J. Miller, and Charles G. Wallis. Indianapolis: Hackett Publishing, 1998.

Denton, Michael. *Evolution: A Theory in Crisis.* Bethesda, Md.: Adler & Adler, 1985.

——, ed. *The Cambridge Companion to Descartes.* New York: Cambridge University Press, 1992.

Descartes, Rene. *Discourse on Method and Meditations on First Philosophy.* Trans. Donald Cress. Indianapolis: Hackett Publishing, 1999.

——. *The Philosophical Writings of Rene Descartes.* Vols. 1 and 2. Trans. Dugald Murdoch and John Cottingham. New York: Cambridge University Press, 1985.

Drake, Stillman, ed. and trans. *Discoveries and Opinions of Galileo.* New York: Anchor Books, 1957.

Edinger, Edward F. *The Christian Archetype: A Jungian Commentary on the Life of Christ.* Toronto: Inner City Books, 1987.

Empiricus, Sextus. *Selections From the Major Writings.* Ed. Phillip P. Halle. Trans. Sanford G. Etheridge. Indianapolis: Hackett Publishing, 1998.

Engels, Friedrich. *Feuerbach: The Roots of the Socialist Philosophy.* Trans. Austin Lewis. Honolulu: University Press of the Pacific, 2001.

Epicurus. *The Epicurus Reader: Selected Writings and Testimonia.* Ed. Lloyd Gerson. Trans. Brad Inwood. Indianapolis: Hackett Publishing 1994.

Erigena, John Scotus. *Treatise On Divine Predestination.* Vol. 5 of *Notre Dame Texts in Medieval Culture Series.* Trans. Mary Brennan. Notre Dame: University of Notre Dame Press, 2003.

——. *Periphyseon: Division of Nature.* Trans. I. P. Sheldon-Williams and John J. O'Meara. Baltimore, Md.: Dumbarton Oaks, 1987.

Faust, Ernest, Paul Russell, and Rodney Jung. *Craig and Faust's Clinical Parasitology.* 8th ed. Philadelphia: Lea & Febiger, 1977.

Feuerbach, Ludwig. *The Essence of Christianity.* Trans. George Eliot. Amherst, NY: Prometheus Books, 1989.

France, Peter. *Diderot.* 11th ed. New York: Oxford University Press, 1983.

Freedman, Alfred M., Harold L. Kaplan, and Benjamin J. Sadock. *Modern Synopsis of Comprehensive Textbook of Psychiatry/II.* 2nd ed. Baltimore: The Williams & Wilkins Co., 1976.

Freze, Michael, S. F. O. *They Bore the Wounds of Christ: The Mystery of the Sacred Stigmata.* Huntington, Ind.: Our Sunday Visitor Publishing, 1989.

Freud, E. L. *The Cocaine Papers.* New York: Stonehill, 1975.

——. *The Letters of Sigmund Freud.* Boulder, Colo.: Basic Books, 1960.

Freud, Sigmund and Josef Breuer. *Studies on Hysteria.* Boulder, Colo.: Basic Books, 1957.

Freud, Sigmund. *An Autobiographical Study.* New York: W. W. Norton, 1952.

——. *Beyond the Pleasure Principle.* New York: Liveright, 1961.

——. *Civilization and Its Discontents.* Mineola, NY: Dover, 1994.

——. *Group Psychology and the Analysis of the Ego.* New York: W. W. Norton, 1975.

——. *Inhibition, Symptoms and Anxiety.* New York: W. W. Norton, 1959.

——. *Leonardo Da Vinci and a Memory of His Childhood.* New York: W. W. Norton, 1966.

——. *Moses and Monotheism.* Knopf, 1939.

——. *On Dreams.* New York W. W. Norton, 1980.

——. *The Ego and the Id.* New York: W. W. Norton, 1962.

——. *The Future of an Illusion.* New York: W. W. Norton, 1975.

——. *The Interpretation of Dreams.* New York: Gramercy, 1996.

——. *The Origin and Development of Psychoanalysis.* Washington, D.C.: Regnery Gateway, 1987.

——. *The Origins of Psycho-Analysis: Letters to Wilhelm Fliess, Drafts and Notes.* Boulder, Colo.: Basic Books, 1977.

——. *The Problem of Anxiety.* New York: W. W. Norton, 1936.

——. *The Psychopathology of Everyday Life.* New York: W. W. Norton, 1965.

——. *Three Essays on the Theory of Sexuality.* Boulder, Colo: Basic Books, 1962.

——. *Totem and Taboo: Some Points of Agreement Between Savages and Neurotics.* New York: W. W. Norton, 1952.

Garza-Valdes, Dr. Leonicio A. *The DNA of God.* New York: Doubleday, 1999.

Gay, Peter. *A Godless Jew: Freud, Atheism and the Making of Psychoanalysis.* New Haven, Conn.: Yale University Press, 1987.

——. *Freud: A Life for Our Times.* New York: W. W. Norton, 1998.

——. *Deism: An Anthology.* Magnolia, Mass.: Peter Smith, 1985.

——. *The Freud Reader.* New York: W. W. Norton & Company, 1995.

Goldschmidt, Richard. *The Material Basis of Evolution.* revised ed., Yale University Press, 1993. Originally published by Pageant Books in Paterson, NJ, 1960.

Goodman, L. Evan. *Avicenna.* New York: Routledge, 1992.

Gosselin, Mia. *Nominalism and Contemporary Nominalism.* Boston: Kluwer Academic Publishers, 1990.

Graham, Elsie. *Goethe.* Berlin: Walter de Gruyter, Inc., 1977.

Harvey, Van A. *Feuerbach and the Interpretation of Religion.* Vol. 1. New York: Cambridge University Press, 1997.

Hebert, Father Albert J., S. M. *Raised From the Dead.* Rockford, Ill.: TAN Publishers, 1986.

Hippocrates. *Prognostic: Regimen in Acute Diseases, Sacred Diseases, Art, Breaths, Law, Decorum, Physician, Dentition.* Cambridge: Loeb Classical Library, Harvard University Press, 1992.

Hollander, Samuel. *The Economics of Thomas Robert Malthus.* Toronto: University of Toronto Press, 1997.

Homer. *The Iliad and the Odyssey.* Ed. Samuel Butler. New York: Barnes and Noble Books, 1999.

Hondrich, Ted, ed. *The Oxford Companion to Philosophy.* Oxford University Press, 1995.

Hopcke, Robert H. *Jung, Jungians, and Homosexuality.* Boston: Shambhala Publications, Inc., 1991.

Huizinga, Johan. *Erasmus and the Age of Reformation.* London: Phoenix Press, 2002.

Hull, R. F. C. and William McGuire, ed. *C. G. Jung Speaking: Interviews and Encounters.* Princeton University Press, 1977.

Hume, David. *Principle Writings on Religion Including Dialogues Concerning Natural Religion and the Natural History of Religion.* New York: Oxford University Press, 1998.

Hunt, Morton. *The Story of Psychology.* New York: Doubleday, 1993.

Johanson, Donald and Blake Edgar. *From Lucy to Language.* New York: Simon & Schuster, 1996.

Johnson, Philip. *Darwin on Trial.* Washington, D.C.: Regnery Gateway, 1991.

Johnston, Francis. *The Wonder of Guadalupe.* Rockford, Ill.: Tan Books and Publishers, Inc., 1981.

Jones, E. Michael. *Degenerate Moderns: Modernity as Rationalized Sexual Behavior.* San Francisco: Ignatius Press, 1993.

Jung, C. G. *Answer to Job.* Vol. 283, ed. 50. Princeton: Princeton University Press, 1972.

——. *Aspects of the Feminine.* Princeton: Princeton University Press, 1982.

——. *Four Archetypes: Mother/Rebirth, Spirit/Trickster.* Princeton: Princeton University Press, 1972.

——. *Man and His Symbols.* New York: Dell Publishing, 1970.

——. *Memories, Dreams, Reflections.* New York: Vintage Books, 1989.

——. *On the Nature of the Psyche.* Princeton: Princeton University Press, 1970.

——. *Psychology and the Occult.* Trans. R. F. C. Hull, Bollingen Series, Princeton University Press, 1977.

——. *Psychology and Western Religion.* Trans. R. F. C. Hull. Princeton: Princeton University Press, 1984.

——. *The Psychology of the Transference.* Princeton: Princeton University Press, 1976.

——. *The Collected Works of C. G. Jung.* 19 vols. Princeton: Princeton University Press, 1976.

——. *General Index to the Collected Works.* Vol. 13 of *The Collected Works of C. G. Jung.* Princeton: Princeton University Press, 1979.

James, William. *Pragmatism.* Ed. Bruce Kuklick. Indianapolis: Hackett Publishing, 1997.

Jordanova, L. J. *Lamarck.* New York: Oxford University Press, 1984.

Jurevich, R. M., Ph.D. *The Hoax of Freudianism: A Study of Brainwashing the American Professionals and Laymen.* Philadelphia: Dorrance & Co., 1974.

Kant, Immanuel. *Cambridge Companion to Kant.* Ed. Paul Guyer. New York: Cambridge University Press, 1992.

Kaufmann, Walter. *The Portable Nietzsche.* New York: Viking Penguin Books, 1982.

Kennedy, D. James. *Why I Believe.* Nashville: Word Publishing, 1980.

Klare, Roger. *Gregor Mendel: Father of Genetics.* Berkeley Heights, N.J.: Enslow Publishers, 1997.

Kohn, Alexander. *False Prophets: Fraud and Error in Science and Medicine.* New York: Barnes and Noble Books, 1997.

Kramnick, Isaac, ed. *The Portable Enlightenment Reader.* New York: Penguin Books, 1995.

Larsen, Stephen, and Robin Larsen. *A Fire in the Mind.* New York: Doubleday, 1991.

Leaman, Oliver. *Moses Maimonides.* Richmond, England: Curzon Press 1997.

Leibniz, Baron Gottfried Wilhelm von. *Leibniz.* Ed. Catherine Wilson. Burlington, Vt.: Ashgate Publishing, Ltd., 2001.

Levy, Bernard Henri. *Sartre: The Philosopher of the Twentieth Century.* Trans. Andrew Brown. Molden, Mass.: Polity Press, 2003.

Lindsey, David Michael. *The Woman and the Dragon: Apparitions of Mary.* Gretna, La.: Pelican Publishing Company, 2001.

Loftus, Elizabeth, and Katherine Ketcham. *The Myth of Repressed Memory: False Memories and Allegations of Sexual Abuse.* New York: St. Martin's Press, 1996.

Lurie, Edward. *Louis Agassiz: A Life in Science.* Baltimore, Md.: John Hopkins University, 1998.

Lyell, Sir Charles. *Life, Letters, and Journals of Sir Charles Lyell.* Fanborough, England: Gregg International, 1968.

Lyons, Sherrie L. *Thomas Henry Huxley: The Evolution of a Scientist.* Amherst, NY: Prometheus Books, 1999.

Magee, Bryan *The Philosophy of Schopenhauer.* New York: Oxford University Press, 1997.

Mahers, John M. and Dennie Briggs, ed. *An Open Life: Joseph Campbell in Conversation with Michael Toms.* New York: Harper Perennial, 1989.

Manly, Isaac, M.D. *God Made: A Medical Doctor Looks at the Reality of Creation.* Joplin, Mo.: College Press, 1994.

Martin, Malachi. *Hostage to the Devil: The Possession and Exorcism of Five Contemporary Americans.* New York: HarperCollins, 1992.

Martinich, Aloysius P., and Jeremy Black. *Thomas Hobbes.* New York: St. Martin's Press, 1997.

Marx, Karl, and Friedrich Engels. *Das Kapital.* Abridged. Chicago: Regnery, 1996.

——. *The Collected Works, 1882-89.* New York: International Publishers Company, Inc., 1991.

——. *The Communist Manifesto.* New York: Signet Classics, 1998.

——. *The German Ideology: Including Thesis on Feuerbach and Introduction to the Critique of Osbon.* New York: Harper Perennial, 1991.

——. *Political Economy.* Amherst, NJ: Prometheus Books, 1998.

——. *The Marx-Engels Reader.* Ed. Robert C. Tucker and David P. McLellan. New York: W. W. Norton & Company, 1978.

——. *The Portable Karl Marx.* Ed. Eugene Kamenka. New York: Penguin, 1983.

Mason, Jeffrey Moussaieff. *The Complete Letters of Sigmund Freud to Wilhelm Fliess.* Cambridge: Harvard University Press, 1985.

Matsumura, Molleen, ed. *Voices for Evolution.* Berkeley: National Center for Science Education, 1995.

Menzies, *The Ante-Nicene Fathers.* Vol. 8. Grand Rapids, Mich.: William Eerdmans, 1951.

McGuire, William, and Wolfgang Sauerlaender. *The Freud/Jung Letters: the Correspondence Between Sigmund Freud and C. G. Jung.* Princeton: Princeton University Press, 1974.

Mitha, Farouk. *Al-Ghazali and the Ismailis: A Debate on Reason and Authority in Medieval Islam.* London: I. B. Tauris & Co., 2001.

Monticone, Diane Kollar. *Montesquieu and His Reader: A Study of the Esprit des Lois.* Lanham, Md.: University Press of America, 1990.

Moody, Raymond. *Life After Life.* New York: Harper, 2001.

Nietzsche, Friedrich. *The Portable Nietzsche.* Ed. and trans. Walter Kaufman. New York: Viking Penguin, 1982.

Osborne, Diane K., ed. *A Joseph Campbell Companion: Reflections on the Art of Living.* New York: Harper, 1991.

Paine, Thomas. *Age of Reason.* Secaucus, NJ: Carol Publishing Group, 1997.

Plato. *Complete Works of Plato.* Ed. John M. Cooper. Indianapolis: Hackett Publishing, 1997.

——. *Great Dialogues of Plato: Complete Texts of the Republic, Apology, Crito, Phaido, Ion and Meno.* Vol.1. Trans. W. H. Rouse. New York: Signet Classics, 1999.

Raby, Peter. *Alfred Russell Wallace: A Life.* Princeton: Princeton University Press, 2002.

Richardson, William J. *Heidegger: Through Phenomenology to Thought.* New York: Fordham University Press, 2002.

Ridley, Mark. *The Darwin Reader.* 2nd ed. New York: W. W. Norton & Company, 1996.

Robinson, Daniel N. *An Intellectual History of Psychology.* Madison: University of Wisconsin Press, 1986.

Rosemann, Phillip. *Peter Lombard.* New York: Oxford University Press, 2004.

Ross, Colin, and Elizabeth Loftus. *Satanic Ritual Abuse: Principles of Treatment.* Toronto: University of Toronto Press, 1995.

Rudwick, Martin. *George Cuvier: Fossil Bones and Geological Catastrophes.* Chicago: University of Chicago Press, 1997.

Rupke, Nicolaas A. *Richard Owen: Victorian Naturalist.* New Haven, Conn.: Yale University Press, 1994.

Russell, Bertrand. *A History of Western Philosophy.* New York: Simon & Schuster, 1972.

——. *Basic Writings of Bertrand Russell.* Ed. Lester E. Dennon. New York: Routledge, 2001.

——. *Why I am Not a Christian: And Other Essays On Religion and Related Subjects.* New York: Simon & Schuster, 1957.

Sagan, Carl, and Ann Druyan. *Comet.* New York: Random House, 1985.

——. *Shadows of Forgotten Ancestors.* New York: Random House, 1992.

Sagan, Carl. *Billions and Billions: Thoughts on Life and Death at the Brink of the Millennium.* New York: Random House, 1997.

——. *Broca's Brain: Reflections on the Romance of Science.* New York: Random House, 1979.

——. *Cosmos.* New York: Random House, 1980.

——. *The Dragons of Eden: Speculations on the Evolution of Human Intelligence.* New York: Random House, 1977.

——. *The Demon-Haunted World: Science as a Candle in the Dark.* New York: Random House, 1995.

——. *Pale Blue Dot: A Vision of the Human Future in Space.* New York: Random House, 1994.

Salaman, Clement, and Jean-Pierre Mahe, trans. *The Way of Hermes: New Translations of the Corpus Hermeticum and the Definitions of Hermes Trismegistus to Asclepius.* Rochester, Vt.: Inner Traditions International, Ltd., 2000.

Sedgwick, Sally, ed. *The Reception of Kant's Critical Philosophy: Fichte, Schelling, and Hegel.* New York: Cambridge University Press, 2000.

Seldon, Steve. *Inheriting Shame: The Story of Eugenics and Racism in America.* Vol. 23. New York: Teachers College Press, Columbia University, 1999.

Sellers, Wilfrid. *Empiricism and the Philosophy of the Mind.* Cambridge: Harvard University Press, 1996.

Skinner, B. F. *Walden Two.* New York: MacMillan Co., 1948.

Smith, Martin F. *Lucretius: On the Nature of Things.* Trans. W. H. D. Rouse Boston: Harvard University Press, 1988.

Spellman, W. M. *John Locke.* New York: St. Martin's Press, 1997.

Spiegelman, J. M., Ph.D. *Catholicism and Jungian Psychology.* Phoenix: Falcon Press, 1988.

Spinoza, Baruch. *The Cambridge Companion to Spinoza.* Ed. Don Garrett. New York: Cambridge University Press, 1995.

Stein, Dr. Gordon, ed. *An Anthology of Atheism and Rationalism.* New York: Prometheus Books, 1980.

Stein, Murray. *Jung's Treatment of Christianity: The Psychotherapy of a Religious Tradition.* Wilmetle, Ill.: Chiron Publications, 1985.

Stroud, Barry. *Hume.* London: Routledge, 1993.

Taylor, Charles. *Hegel.* New York: Cambridge University Press, 1990.

Voltaire, Francois Marie Arouet de. *Candide.* New York: Penguin Books, 1947.

——. *The Portable Voltaire.* New York: Penguin, 1976.

Wallis, Rich T. and Lloyd P. Gerson. *Neoplatonism.* Indianapolis: Hackett Publishing, 1995.

Wegener, Alfred. *The Origins of Continents and Drift.* Trans. John Biram. Mineola, NY: Dover Publications, 1966.

Wehr, Demaris S. *Jung and Feminism: Liberating Archetypes.* Boston: Beacon Press, 1987.

Weinberg, Samantha. *A Fish Caught in Time.* New York: Harper Collins, 2000.

Wells, Jonathan. *Icons of Evolution: Science or Myth.* Washington, D.C.: Regnery Publishing, 2000.

Westfall, Richard S. *The Construction of Modern Science: Mechanics and Mechanisms.* New York: Cambridge University Press, 1977.

Whiston, William. *The Works of Josephus, Complete and Unabridged.* Peabody, Mass.: Hendrickson Publications, 1988.

White, Michael. *The Pope and the Heretic: The True Story of Giordano Bruno, the Man Who Dared to Defy the Roman Inquisition.* New York: HarperCollins Publisher, 2003.

Wilder-Smith, A. E. *The Natural Sciences Know Nothing of Evolution.* San Diego: Master Books, 1981.

Winchester, Simon. *The Map that Changed the World: William Smith and the Birth of Modern Geology.* New York: HarperCollins, 2002.

Wittgenstein, Ludwig. *The Cambridge Companion to Wittgenstein.* Ed. Hans Sluga and David G. Stern. New York: Cambridge University Press, 1996.

Wolker, Robert. *Rousseau.* New York: Oxford University Press, 2001.

Zagorin, Perez. *Francis Bacon.* Princeton, NJ: Princeton University Press, 1999.

Zimmer, Carl. *Evolution: The Triumph of an Idea.* New York: HarperCollins, 2001.

Index